Biogeochemistry of Chelating Agents

ACS SYMPOSIUM SERIES **910**

Biogeochemistry of Chelating Agents

Bernd Nowack, Editor
Swiss Federal Institute of Technology

Jeanne M. VanBriesen, Editor
Carnegie Mellon University

Sponsored by the
ACS Divisions of Environmental Chemistry, Inc. and
Geochemistry, Inc.

American Chemical Society, Washington, DC

Library of Congress Cataloging-in-Publication Data

Biogeochemistry of chelating agents / Bernd Nowack, editor, Jeanne M. VanBriesen, editor.

p. cm.—(ACS symposium series ; 910)

Sponsored by the ACS Divisions of Environmental Chemistry, Inc. and Geochemistry, Inc. at the 226th National Meeting of the American Chemical Society, New York, N.Y., Sept. 7–11, 2003.

Includes bibliographical references and index.

ISBN 0–8412–3897–9 (alk. paper)

1. Chelates—Congresses. 2. Biogeochemistry—Congresses. 3. Environmental chemistry—Congresses.

I. Nowack, Bernd II. VanBriesen, Jeanne M. III. American Chemical Society. Meeting. (226th : 2003 : New York, N.Y.) IV. American Chemical Society. Division of Environmental Chemistry, Inc. V. American Chemical Society. Division of Geochemistry, Inc. VI. Series.

QD474.B54 2005
577′.14—dc22 2005041113

The paper used in this publication meets the minimum requirements of American National Standard for Information Sciences—Permanence of Paper for Printed Library Materials, ANSI Z39.48–1984.

PRINTED IN THE UNITED STATES OF AMERICA

Foreword

The ACS Symposium Series was first published in 1974 to provide a mechanism for publishing symposia quickly in book form. The purpose of the series is to publish timely, comprehensive books developed from ACS sponsored symposia based on current scientific research. Occasionally, books are developed from symposia sponsored by other organizations when the topic is of keen interest to the chemistry audience.

Before agreeing to publish a book, the proposed table of contents is reviewed for appropriate and comprehensive coverage and for interest to the audience. Some papers may be excluded to better focus the book; others may be added to provide comprehensiveness. When appropriate, overview or introductory chapters are added. Drafts of chapters are peer-reviewed prior to final acceptance or rejection, and manuscripts are prepared in camera-ready format.

As a rule, only original research papers and original review papers are included in the volumes. Verbatim reproductions of previously published papers are not accepted.

ACS Books Department

Contents

Biological Reactions

Transport and Fate

Remediation Applications

Indexes

Preface

This book is derived from a symposium sponsored by the American Chemical Society (ACS) Divisions of Environmental Chemistry, Inc. and Geochemistry, Inc., *Biogeochemistry of Chelating Agents,* which was organized for the 226th ACS National Meeting in New York, September 7–11, 2003. Papers were solicited with a call for papers, as well as with direct contact with researchers. The symposium consisted of 32 oral presentations in four half-day sessions, and an evening poster session with eight additional papers.

Chelating agents are used in numerous technical and industrial applications. Due to their strong interaction with metals, chelating agents have received considerable attention related to possible adverse effects on ecosystems and biological systems. The symposium focused on the complex chemical and biological reactions of anthropogenic chelating agents in natural systems. The goal of the symposium was to bring together researchers who were working with chelating agents from a variety of disciplines—chemistry, environmental chemistry, microbiology, agrochemistry, phytoremediation, bioremediation, wastewater treatment, industry (pulp and paper, textile, and oil production)—to discuss the unique chemical properties of these compounds. Topics for papers included structure–function relationships, influence of speciation, carboxylic and phosphonate chelating agents, metal mobilization in the environment, enhanced phytoremediation via chelants, and biodegradation of chelates.

Chapters for the book were solicited from the symposium's oral and poster presentations as well as from additional researchers in the field to provide a balanced presentation. The book is organized into five sections: speciation, analytical methods, biological reactions, transport and fate, and remediation applications. Chapter 1 provides a comprehensive overview of the compounds, their occurrence in the environment, and their research history. The mix of review chapters, original research, and reports of new work presented in the following five sections best reflect the current state of the subject. Chapters 2 and 3 are devoted to the importance and calculation of the speciation of chelating agents, the basis for understanding these compounds in the environment. Chapters 4–7 comprise the analytical methods section whereas Chapters 8–12 discuss biological reactions (biodegradation and bioavailability). The core of the book is Chapters 13–21 dealing with transport and fate of chelating agents in water, groundwater, and soils. A final section, Chapters 22–25, is devoted to the use of chelating agents for the remediation of polluted soils.

This book is the first comprehensive compilation of the environmental chemistry of chelating agents, their interactions with organisms, and their use in remediation applications. The contents of the book should appeal to scientists and practitioners in the field of environmental science and engineering.

Acknowledgments

We gratefully acknowledge the ACS Division of Environmental Chemistry, Inc. and BASF, Ludwigshafen, Germany, for financial support for speakers from overseas countries. We also acknowledge the Donors of the ACS Petroleum Research Fund for support of the symposium. Gratitude is also expressed to the 47 peer reviewers whose comments contributed significantly to improved final versions of the chapters.

Bernd Nowack

Institute of Terrestrial Ecology
Swiss Federal Institute of Technology (ETH)
CH–8952 Schlieren
Switzerland
nowack@env.ethz.ch (email)

Jeanne M. VanBriesen

Department of Civil and Environmental Engineering
Carnegie Mellon University
Pittsburgh, PA 15213–3890
jeanne@cmu.edu (email)

The complete set of extended abstracts for the symposium are part of the Preprints of Extended Abstracts for this ACS Division (Volume 42, Number 3) and are available through the Division's Business Office, ACS Division of Environmental Chemistry, Inc., 1810 Georgia Street, Cape Girardeau, MO 63701–3816.

Chapter 1

Chelating Agents in the Environment

Bernd Nowack[1] and Jeanne M. VanBriesen[2]

[1]Institute of Terrestrial Ecology, Swiss Federal Institute of Technology (ETH), CH--8952 Schlieren, Switzerland
[2]Department of Civil and Environmental Engineering, Department of Biomedical Engineering, Carnegie Mellon University, Pittsburgh, PA 15213

Anthropogenic chelating agents of the types aminopoly-carboxylates and phosphonates are ubiquitous in the environment. In this chapter we describe these compounds and present the most important representatives and their uses. Typical concentrations in natural systems (e.g., surface and ground waters, wastewaters, and drinking water) are described along with a brief summary of their potential effects in the environment. The increasing interest in and research on these compounds is evaluated in terms of publication volume and topic area.

Introduction

The environmental fate of chelating agents has received considerable attention. EDTA (ethylenediaminetetraacetic acid) for example occurs at a higher concentration in European surface water than any other identified anthropogenic organic compound (*1*). Chelating agents have the potential to perturb the natural speciation of metals and to influence metal bioavailability (*2, 3*) and their presence at high concentrations may lead to the remobilization of metals from sediments and aquifers, consequently posing a risk to groundwater and drinking water (*4*). The largest concern, however, is that many chelating agents (e.g. EDTA or phosphonates) are not or only slowly biodegradable and are therefore rather persistent in the environment (*5, 6*).

Strong chelating agents occur in natural waters predominantly in the form of metal complexes. A discussion of the fate of a chelating agent *always* has to address the presence of metals and how they interact with the chelates. There are many studies emphasizing the importance of chelation on *metal* bioavailability, plant uptake, toxicity, transport, adsorption, distribution and fate (*7*). Conversely, *chelating agents* are also affected by the presence of metals resulting in different reactivities of metal-chelates.

Environmental concentrations, usage, biodegradation and toxicology of aminocarboxylates have been thoroughly reviewed (*2, 8-19*). Much less is known about the fate and behavior of the corresponding phosphonates in the environment. The most cited review about phosphonates in the environment is several years old and therefore does not cover the newest literature (*6*). Recent reviews discuss the environmental chemistry of phosphonates (*20, 21*) or their toxicology and risk assessment (*22*).

In this chapter, we introduce some terminology and discuss the many different types of chelates and their uses. Further, we introduce information on chelate distribution in the environment as well as examine the increasing interest in these compounds over the past 50 years.

The compounds: Description and Some Definitions

The interaction between (metal)-cations with molecules or anions containing free electron pairs is called complex formation. The coordinating compounds are called ligands (ligare Latin = to bind). They can occupy one, two, three or more positions in the inner coordination sphere of the central ion and are referred to as monodentate, bidentate and so on. Complex formation with multidentate ligands, the chelating agents (or chelants or chelators), is called chelation, and the complexes that are formed are chelates. The term chelate was first applied in 1920 by Morgan and Drew (*23*) who stated: "The adjective chelate, derived from the great claw or chela ("chely") of the lobster or other crustaceans, is suggested for these caliperlike groups which function as two associating units and fasten on to the central metallic atom so as to produce heterocyclic rings." A closer look at the linguistics of the word "chele" reveals that the original Greek meaning is that of a horse's hoof and by extension, anything resembling the hoof of a horse, e.g. a crab's or scorpion's claw (*24*). Table 1 gives a short overview of the different terms that are used in this respect to chelating agents. In a series of excellent reviews on crystallography of chelants (*25*) it was demonstrated that their interaction with cations leads to the formation of stable five-membered chelate rings, as shown on Figure 1. The structural data on the phosphonic analogs are very poor, but indirect spectroscopic data reveal tentatively a similar structure.

Table 1: Definitions

ligand	coordinating compound
chelating agent	multidentate ligand
chelate	the metal-chelating agent complex
chelant	synonym for chelating agent
chelator	synonym for chelating agent
chelon	synonym for chelating agent
APC	abbreviation for aminopolycarboxylates
complexon	synonym for APCs

The topic of this book is anthropogenic chelating agents of the groups of the aminopolycarboxylates and the phosphonates. The aminocarboxylates are characterized by one or more tertiary or secondary amines and two or more carboxylic acid groups. The names, abbreviations and structures of the most important compounds are shown in Table 2. EDTA and NTA are the best known representatives of this group. EDDS, IDSA and MGDA have received recently some attention as possible replacements for EDTA and DTPA. ED3A is an important breakdown product of EDTA and DTPA (*26*). It is not used commercially, although LinearAlkyl-ED3A (L-ED3A) is a recent addition to the commercial list. Since the pioneering work of Schwarzenbach the APCs are also known as "Complexones" (*27*). Table 3 shows some important breakdown products of APCs.

Table 2: Aminopolycarboxylate chelating agents (APCs)

Acronym	Name	Structure
ADA	β-alanine diacetic acid	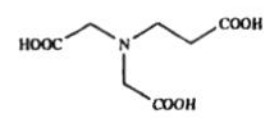
DTPA	diethylenetriaminepentaacetic acid	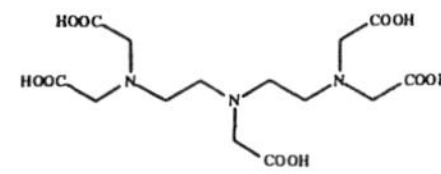
EDDHA	ethylenediaminedi(o-hydroxyphenylacetic) acid	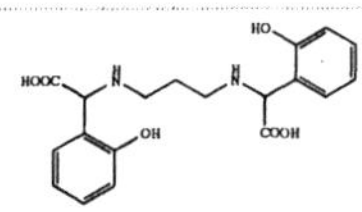 EDDHA
EDDS	ethylenediaminedisuccinic acid	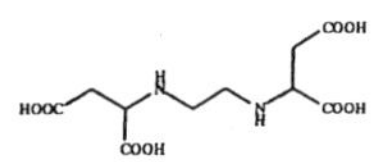
EDTA	ethylenediaminetetraacetic acid	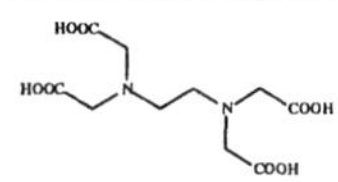
HEDTA	N-(hydroxyethyl)-ethylenediaminetriacetic acid	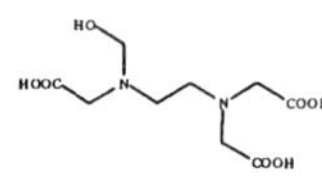
IDSA	iminodisuccinic acid	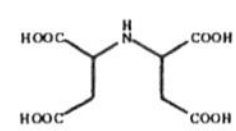
L-GLDA	glutamate-N,N-diacetic acid	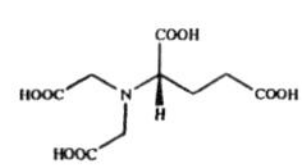
MGDA	methylglycinediacetic acid	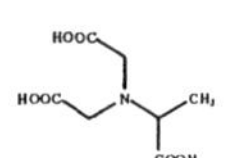
NTA	nitrilotriacetic acid	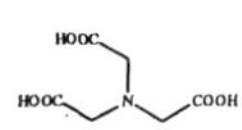
PDTA	1,2-diaminopropanetetraacetic acid	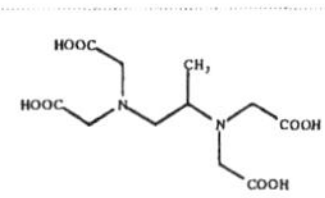

Table 3: Important breakdown products of APCs

Acronym	Name	Structure
DTTA	diethylenetriamine-N,N,N',N''-tetraacetic acid	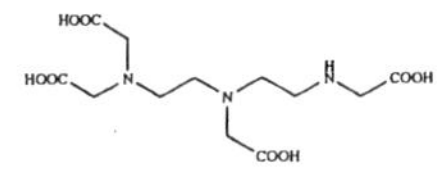
ED3A	ethylenediaminetriacetic acid	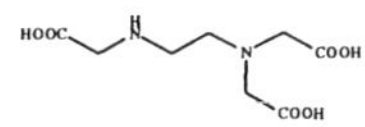
EDDA	ethylenediaminediacetic acid	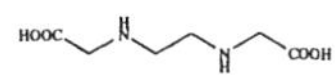
EDMA	ethylenediaminemonoacetic acid	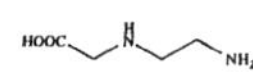
IDA	iminodiacetic acid	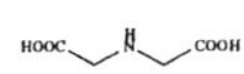
3KP	3-ketopiperazinediacetic acid	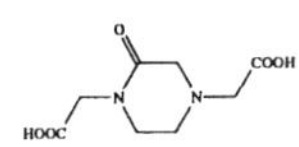

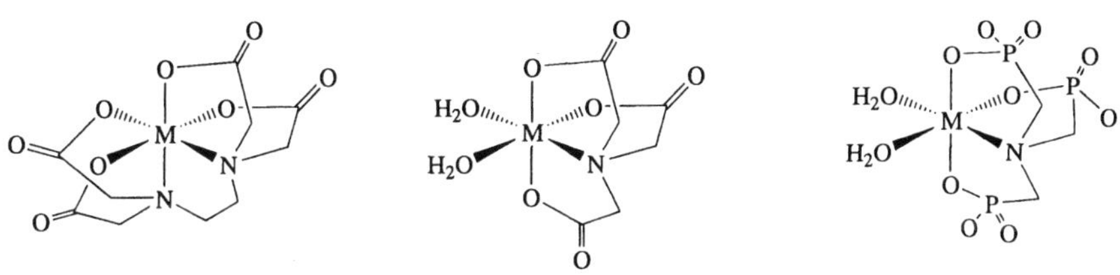

metal-EDTA metal-NTA metal-NTMP

Figure 1. Ideal octahedral structure of metal-EDTA, metal-NTA, and metal-NTMP complexes.

The phosphonates, compounds containing the phosphonic acid moiety -C-$PO(OH)_2$, are often structure-analogues of the aminopolycarboxylates, for example NTA and NTMP, EDTA and EDTMP or DTPA and DTPMP. In addition mixed carboxylates-phosphonates (e.g. PBTC) are also used in industrial applications. The mixed carboxylate-phosphonate N-

Table 4: Phosphonate chelating agents

Acronym	other Acronyms	Name	Structure
DTPMP	DETPMP, DTPPH, DETPMPA, DETPMPO, ETPP, Dequest 2060	diethylenetriamine pentakis (methylene-phosphonic acid)	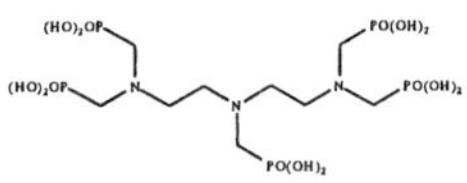
EDTMP	EDTP, EDTPH, ENTMP, EDTMPA, TePMEDA, EDTPO, EDITEMPA, Dequest 2041	1,2-diaminoethane tetrakis (methylene-phosphonic acid)	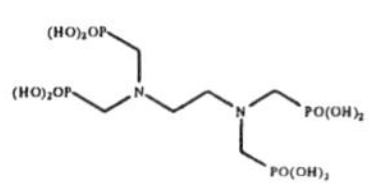
HEDP	HEDPA, HEDPO, HEBP, EHDP, Dequest 2010	1-hydroxy ethane(1,1-diylbis-phosphonic acid)	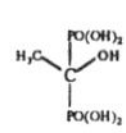
NTMP	ATMP, NTP, NTPH, NTPO, TPMA, NTTA, Dequest 2000	nitrilotris (methylene-phosphonic acid)	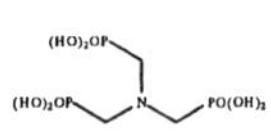
PBTC	PBTCA, Bayhibit AM	phosphono butane-tricarboxylic acid	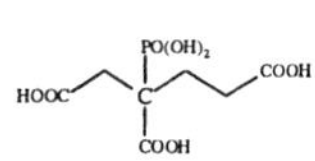
PMG	Glyphosate, Roundup	N-phosphonomethyl-glycine	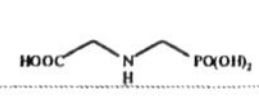

phosphonomethylglycine (Glyphosate) is the most widely used herbicide in the world (*28*). Table 4 shows the names, abbreviations, and structures of the most important phosphonates. These compounds are known under many different abbreviations that vary between the disciplines and countries and have changed with time.

Production and Use

The total worldwide use of synthetic aminopolycarboxylic acids (including EDTA and the related compounds DTPA and NTA) was 200,000 tons in 2000 (*29*). Table 5 shows the latest available data on the consumption of chelating agents.

Table 5: Consumption of chelating agents (in tons/year)

	Europe	Year	ref	US	year	ref
EDTA	34,550	1999	(*17*)	50,000	1987	(*9*)
NTA	19,890	1999	(*17*)			
DTPA	14,350	1999	(*17*)			
HEDTA	2,000	1981	(*30*)	18,000	1981	(*30*)
phosphonates	16,000	1999	(*17*)	30,000	1998	(*31*)

Figure 2 shows the uses of three chelating agents. The different compounds have very different uses that also vary between countries or continents. A summary of the uses of APCs and phosphonates can be seen in Table 6.

Phosphonates are used as chelating agents in many applications, e.g. in pulp, paper and textile industry to complex heavy metals in chlorine-free bleaching solutions that could inactivate the peroxide (*6*). In medicine phosphonates are used to chelate radionuclides for bone cancer treatments (*32*) and to treat various bone and calcium metabolism diseases (*33*). Phosphonates are not only chelating agents but also very potent inhibitors of mineral precipitation and growth (*34*). This effect is effective at concentrations well below the amount needed to chelate all metals (*35*). An important industrial use of phosphonates is in cooling waters, desalination systems and in oil fields to inhibit scale formation, e.g. barium sulfate or calcium carbonate precipitation (*36*). In detergents phosphonates are used as a combination of chelating agent, scale inhibitor and bleach stabilizer (*37*).

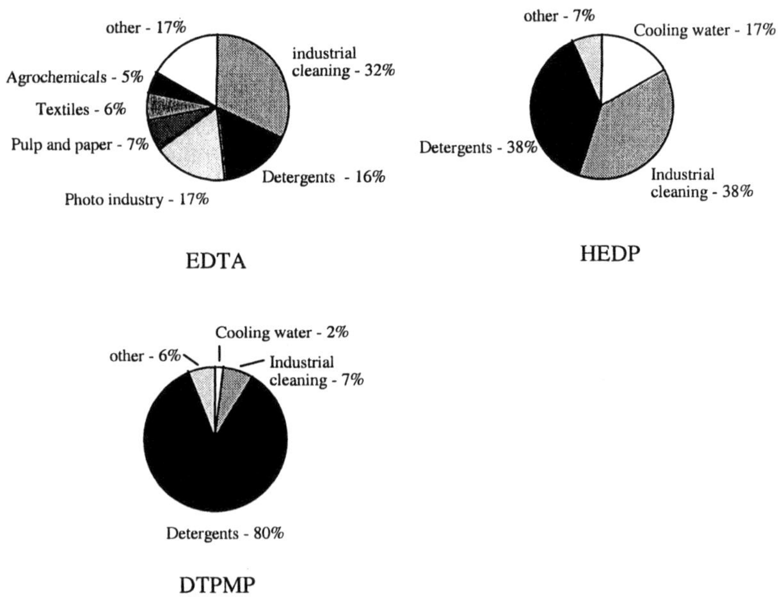

Figure 2: Uses of EDTA, HEDP and DTPMP, after (8) and (6)

EDTA, the most widely used chelating agent in the world, is used to control water hardness in boiler and chemical process water and to control aqueous metals that (a) interfere with bleaching in fabric and paper processing, (b) interfere with photographic developing solutions, and (c) compete in metal finishing and plating operations (*38, 39*). In consumer applications, EDTA is used to sequester trace metals to prevent catalytic reactions leading to rancidity, loss of flavor, and discoloration in food products and vitamins, and to control metals that destabilize cosmetics and pharmaceuticals (*38, 40*). EDTA is used medically to remove toxic metals (e.g., lead in children) and in dental procedures (*41*). Recently, medical uses have expanded to include treatment for heart disease and high blood pressure. Chelating agents have also been used in agriculture to increase the availability and transport of nutrient metals including Fe, Zn, and Cu (*42, 43*), for metal extraction from contaminated soils (*44-49*), to mobilize contaminant metals to enhance phytoremediation efforts (*50-52*), and for remediation at metal-contaminated sites (*53-57*). EDTA is also used in nuclear reactor decontamination and waste processing (*58, 59*). EDTA has recently been suggested for use as a groundwater tracer (*60*).

Table 6: Uses of chelating agents

Area	Compound	Use
detergents	APCs	complexation of Ca and Mg
		complexation of heavy metals
	phosphonates	inhibition of scale formation, bleach stabilizer
photo industry	EDTA, PDTA	oxidizing agent
pulp and paper	APCs, phosphonates	complexation of heavy metals
		bleach stabilizer
textile industry	phosphonates	complexation of heavy metals
	APCs	bleach stabilizer
metal plating	EDTA	electroless metal plating
food	EDTA	Stabilizer, antioxidant, Fe-fortification
personal care	EDTA, phosphonates	stabilizer, antioxidant
power plants	phosphonates	inhibition of scale formation
agriculture	EDTA, EDDHA	Fe, Cu and Zn fertilizers
	glyphosate	herbicide
oil production	phosphonates	inhibition of mineral precipitation
	EDTA	
dairy and beverage industry	APCs	cleaning of bottles and equipment
medicine	phosphonates	carrier of radionuclides
		treatment of bone and calcium metabolism diseases
	APCs	treatment of metal poisoning
		chelation therapy
		MRI contrasting agents
soil remediation	APCs	extraction of heavy metals, chelant enhanced phytoremediation
construction	phosphonates	cement modifier
flue gas cleaning	EDTA	removal of NO_x

Environmental Distribution

Widespread use of EDTA and its slow removal under many environmental conditions has led to its standing as the anthropogenic compound at the highest concentration in many European surface waters (*12, 29, 61-63*). River concentrations in Europe are reported as 10-100 μg/L and lake concentrations are in the 1-10 μg/L range (*8, 13, 64*). EDTA in surface waters was measured in the Midwestern United States as part of a study of endocrine disruptors from wastewater outfalls. Concentrations were reported for the Illinois River (51 μg/L), the Des Plains River (76 μg/L), and the Minnesota River (31 μg/L) (*65*). All of these rivers receive significant input of wastewater treatment effluents, and treated effluents that were released upstream of the river measurement points had EDTA concentrations between 170 - 436 μg/L (*65*). Evaluation of the Mississippi River and its tributaries in 1987-1992 found EDTA to be ubiquitous, with values ranging from 1-32 μg/L (*66*).

EDTA was detected in drinking water from an alluvial aquifer (at 8 μg/L), indicating migration of EDTA from the nearby surface waters (158 μg/L) through the subsurface (*67*). EDTA concentrations in U.S. groundwater receiving wastewater effluent recharge was reported at 1-72 μg/L, and EDTA was found to be a conservative tracer, with higher concentrations of EDTA corresponding to a greater percentage of reclaimed water in drinking water production wells (*65*).

No data about the environmental concentrations of phosphonates are available and only measurements for wastewaters have been reported (*68-71*). Recently, less well known chelating agents β-ADA, PDTA and HEDTA were found in the Rhein, Neckar and Elbe at concentrations of more than 1 μg/L (*72*).

Historical Perspective

The first chelating agent to be synthesized was NTA in 1862 (*86*). The industrial production of NTA started in 1936 at I.G. Farbenindustrie in Germany. The industrial production of EDTA followed in 1939, also in Germany. The complex-forming ability of EDTA was first described in the literature in 1942 (*87*). The unique properties of APCs arise from the "chelate effect", first described in the classical work of Schwarzenbach (*88*). The first phosphonate to be synthesized was a bisphosphonate in 1897 (*89*). Since the work of Schwarzenbach in 1949 (*90*), phosphonic acids have also been known as effective complexing agents.

Table 6: Concentrations of EDTA in the environment

compartment	concentration (nM)	max.	reference
WWTP[1)], effluent	30-2700	17 μM	*(73)*
WWTP, effluent	150-500		*(74)*
WWTP effluent, US	600-1500		*(66)*
Swiss river	5-150	0.7 μM	*(75, 76)*
River Rhein, Germany	1-70		*(61, 76)*
German rivers	2-200		*(29, 62, 64)*
US rivers	3-260		*(66)*
lake water, Germany	4-14		*(77)*
groundwaters	1-100		*(78, 79)*
drinking water	90		*(80)*
waste site	340		*(81)*

1) WWTP: wastewater treatment plant

Table 7: Concentrations of other chelating agents in the environment

compound	compartment	concentration (nM)	reference
DTPMP	WWTP[1)] effluent	120	*(68)*
DTPMP	WWTP influent	50-2000	*(68)*
NTMP, EDTMP	WWTP influent	50-800	*(68)*
NTA	WWTP influent	400-60'000	*(82)*
NTA	German rivers	3-84	*(29, 63, 64)*
NTA	drinking water	1-4	*(80, 83)*
DTPA	lakewater	20-45	*(84)*
DTPA	German rivers	3-170	*(29, 72, 85)*
PDTA	German rivers	4-30	*(72)*
HEDTA	German rivers	130	*(72)*

1) WWTP: wastewater treatment plant

The research on anthropogenic chelating agents in the environment started in 1950 with a publication on the use of EDTA in algal growth media (*91*) and a discussion of the bioavailability of metal-complexes. The research activity remained quite low until the beginning of the 1970s when 10-20 publications per year were published for the next 20 years that discussed chelating agents in the environment (see Figure 3). A major increase occurred in the middle of the 90s with a doubling of the amount of published chelating agent research.

In 1951 the first description of the use of Fe(III)EDTA as iron fertilizer for chlorotic plants was published (*92*). This observation has sparked significant research on the effect of chelating agents on metal solubility in soils, plant uptake, degradation, and mobility. Figure 4 shows that until the beginning of the 1970s most research about chelating agents in the environment dealt with soil related topics. Research on natural waters, lakes and rivers, started in the 1960s but only became important in the 1970s. By the end of the 1970s this topic had surpassed the soil related publications and stayed the most important research area until today. The soil related research remained on a steady rate but increased dramatically in the middle of the 1990s. This increase was due to the proposed use of chelating agents for soil remediation, both for extraction of metals and for chelant-enhanced phytoremediation. The 1970s also saw the increase in research devoted to biodegradation, analytics, and wastewater. The increase in the number of publications in the 1990s is also caused by a large number of published research devoted to chelating agents in groundwater, mainly EDTA, and the surface reactions that occur with natural iron and aluminum oxides.

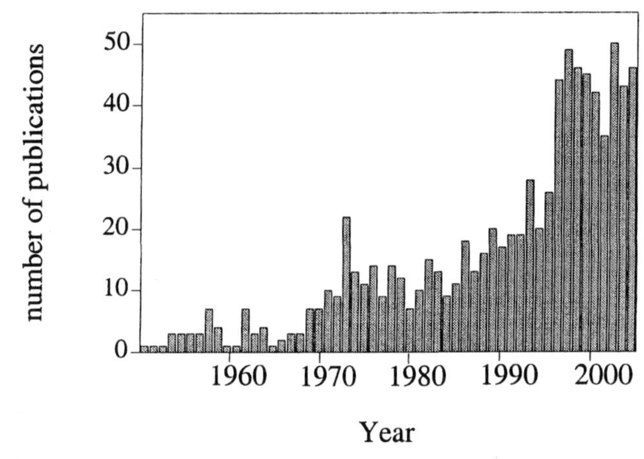

Figure 3: Number of publications about the environmental behavior of chelating agents (APCs and phosphonates) from 1950 to 2003.

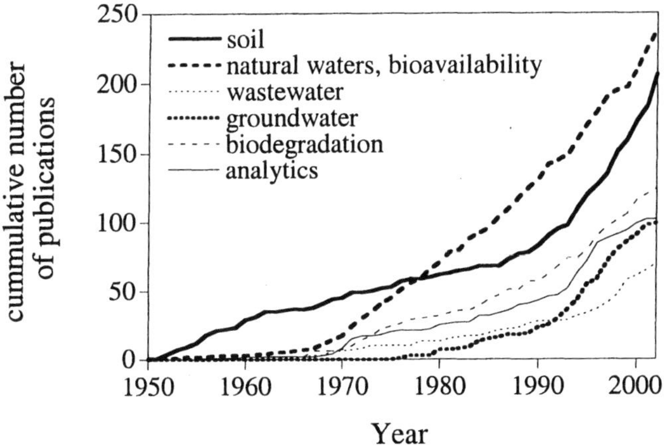

Figure 4: Cumulative number of publications about the environmental behavior of chelating agents (APCs and phosphonates). Natural water: including photochemistry, remobilization from sediments and bioavailability of complexes; groundwater: including surface reactions.

Effects in the Environment

Recalcitrant chelating agents are an environmental concern predominately because of their persistence (*5*) and strong metal chelation. The presence of chelating agents in high concentrations in wastewaters and surface waters has the potential to remobilize heavy metals out of river sediments and treated sludges (*74, 93-100*), although low and environmentally relevant concentrations seem to have only a very minor influence on metal solubility (*15, 98*). Elevated concentrations of chelating agents enhance the transport of metals (e.g. Zn, Cd, Ni, Cr, Cu, Pb, and Fe) in soils (*101-107*) and enhance the undesired transport of radioactive metals away from disposal sites (*10, 81, 108-110*). Low concentrations of chelating agents may either stimulate planktic algae growth or decrease it while high concentrations inhibit activity (*111-113*). Chelating agents are nontoxic (measured as acute exposure) to many forms of life (*114*); long-term low-level exposure effects are unknown (*13*). EDTA at elevated concentrations is toxic to bacteria due to chelation of metals in the outer membrane (*115*). EDTA ingestion at high concentrations by mammals changes excretion of metals (*116*) and can affect cell membrane permeability (*117*).

Strong chelating agents occur in natural waters predominantly in the form of metal complexes. A discussion of the fate of a chelating agent always has to address the presence of metals and how they interact with the chelates. There are many studies emphasizing the importance of chelation on metal bioavailability, plant uptake, toxicity, transport, adsorption, distribution and fate. Conversely, chelating agents are also affected by the presence of metals resulting in different reactivities of metal-complexes. Knowing the speciation of chelating agents in natural waters is therefore crucial for predicting their environmental fate (*15*).

Phosphonates have properties that differentiate them from other chelating agents and that greatly affect their environmental behavior (*20*). Phosphonates have a very strong interaction with mineral surfaces, which results in a significant removal in technical and natural systems. Due to this strong adsorption, little or no remobilization of metals is expected and only low concentrations are expected in the environment.

References

(1) Frimmel, F. H. *gwf Wasser Abwasser* **1989**, *130*, 106-112.

(2) Anderson, W. L.; Bishop, W. E.; Campbell, R. L. *CRC Crit. Rev. Toxicol.* **1985**, *15*, 1-102.

(3) Schowanek, D.; McAvoy, D.; Versteeg, D.; Hanstveit, A. *Aquat. Toxicol.* **1996**, *36*, 253-275.

(4) Müller, G.; Förstner, U. *Z. f. Wasser Abwasser Forsch.* **1976**, *9*, 150-152.

(5) Bucheli-Witschel, M.; Egli, T. *FEMS Microbiol. Rev.* **2001**, *25*, 69-106.

(6) Gledhill, W. E.; Feijtel, T. C. J. In *The Handbook of Environmental Chemistry*; Hutzinger, O., Ed.; Springer Verlag: Berlin, Heidelberg, 1992; Vol. Volume 3, Part F, pp 261-285.

(7) Stumm, W.; Morgan, J. J. *Aquatic Chemistry*; John Wiley and Sons, Inc. New York, 1996.

(8) Frimmel, F. In *Detergents in the environment*; Schwuger, M. J., Ed.; Marcel Dekker, Inc.: New York, 1997; pp 289-312.

(9) Wolf, K.; Gilbert, P. A. In *The Handbook of Environmental Chemistry*; Hutzinger, O., Ed., 1992; Vol. Volume 3, Part F, pp 243-259.

(10) Means, J. L.; Alexander, C. A. *Nucl. Chem. Waste Management* **1981**, *2*, 183-196.

(11) Kiessling, D.; Kaluza, U. In *Detergents in the environment*; Schwuger, M. J., Ed.; Marcel Dekker, Inc.: New York, 1997; pp 265-288.

(12) Nörtemann, B. *Appl. Microbiol. Biotechnol.* **1999**, *51*, 751-759.

(13) Sillanpää, M. *Rev. Environ. Contam. Toxicol.* **1997**, *152*, 85-111.

(14) Sillanpää, M.; Pirkanniemi, K. *Environ. Technol.* **2001**, 22, 791-801.

(15) Nowack, B. *Environ. Sci. Technol.* **2002**, *36*, 4009-4016.
(16) Oviedo, C.; Rodriguez, J. *Quim. Nova* **2003**, *26*, 901-905.
(17) Knepper, T. P.; Weil, H. *Vom Wasser* **2001**, *97*, 193-232.
(18) Knepper, T. P. *Trends Anal. Chem.* **2003**, *22*, 708-724.
(19) Williams, D. *Chemistry in Britain* **1998**, 48-50.
(20) Nowack, B. *Water Res.* **2003**, *37*, 2533-2546.
(21) Nowack, B. In *Phosphorous in Environmental Technologies: principles and applications*; Valsami-Jones, E., Ed.; IWA Publishing, London, 2004; pp 147-173.
(22) Jaworska, J.; van Genderen-Takken, H.; Hanstveit, A.; van de Plasche, E.; Feijtel, T. *Chemosphere* **2002**, *47*, 655-665.
(23) Morgan, G. T.; Drew, H. D. K. *J. Chem. Soc.* **1920**, *117*, 1456-1465.
(24) Haworth, D. T. *J. Chem. Educ.* **1998**, *75*, 47.
(25) Porai-Koshits, M. A.; Polynova, T. N. *Koordinats. Khim.* **1984**, *10*, 725-772.
(26) Ternes, T. A.; Stumpf, M.; Steinbrecher, T.; Brenner-Weiss, G.; Haberer, K. *Vom Wasser* **1996**, *87*, 275-290.
(27) Schwarzenbach, G.; Flaschka, M. *Die Komplexometrische Titration*; Enke, Stuttgart, 1965.
(28) Baylis, A. D. *Pest. Manag. Sci.* **2000**, *56*, 299-308.
(29) Schmidt, C. K.; Fleig, M.; Sacher, F.; Brauch, H. J. *Environ. Pollut.* **2004**, *131*, 107-124.
(30) Egli, T. *Microbiol. Sci.* **1988**, *5*, 36-41.
(31) Davenport, B.; DeBoo, A.; Dubois, F.; Kishi, A. "CEH Report: Chelating agents," SRI Consulting, 2000.
(32) de Klerk, J. M. H.; van Dijk, A.; van het Schip, A. D.; Zonnenberg, B. A.; van Rijk, P. P. *J. Nucl. Med.* **1992**, *33*, 646-651.
(33) Fleisch, H. *Bisphosphonates in bone desease. From the laboratory to the patient*; Academic Press, 2000.
(34) Gal, J. Y.; Bollinger, J. C.; Tolosa, H.; Gache, N. *Talanta* **1996**, *43*, 1497-1509.
(35) Ralston, P. H. *J. Petrol. Tech.* **1969**, *21*, 1029-1036.
(36) Pairat, R.; Sumeath, C.; Browning, F. H.; Fogler, H. S. *Langmuir* **1997**, *13*, 1791-1798.
(37) May, H. B.; Nijs, H.; Godecharles, V. *Household and Personal Product Industry* **1986**, *23*, 50-54.
(38) Othmer, K. In *Encyclopedia of Chemical Technology, 4th ed.*; Krosihwitz, J. I., Ed.; John Wiley and Sons: New York, 1993; pp 764-795.
(39) Considine, D. M. *Chemical and Process Technology Encyclopedia*; McGraw Hill Book Company: New York, 1974.
(40) Whittaker, P.; Vanderveen, J. E.; Dinovi, M. J.; Kuznesof, P. M.; Dunkel, V. C. *Regulatory Toxicol. Pharmacol* **1993**, *18*, 419-427.

(41) Mayfield, L.; Soderholm, G.; Norderyd, O.; Attstrom, R. *J. Clinical Periodontology* **1998**, *25*, 707-714.
(42) Abdulla, I.; Smith, M. S. *J. Sci. Fd. Agric.* **1963**, *14*, 98-109.
(43) Stewart, I. *Ann. Rev. Plant Physiol.* **1963**, *14*, 295-310.
(44) Peters, R. W. *J. Hazard. Mater.* **1999**, *66*, 151-210.
(45) Tandy, S.; Bossart, K.; Mueller, R.; Ritschel, J.; Hauser, L.; Schulin, R.; Nowack, B. *Environ. Sci. Technol.* **2004**, *38*, 937-944.
(46) Hong, P. K. A.; Li, C.; Banerji, S. K.; Regmi, T. *J. Soil Contam.* **1999**, *8*, 81-103.
(47) Elliott, H. A.; Brown, G. A. *Water Air Soil Pollut.* **1989**, *45*, 361-369.
(48) Van Benschoten, J. E.; Matsumoto, M. R.; Young, W. H. *J. Environ. Eng.* **1997**, *127*, 217-224.
(49) Heil, D. M.; Samani, Z.; Hanson, A. T.; Rudd, B. *Water Air Soil Pollut.* **1999**, *113*, 77-95.
(50) Huang, J. W.; Chen, J.; Berti, W. R.; Cunningham, S. D. *Environ. Sci. Technol.* **1997**, *31*, 800-805.
(51) Vassil, A. D.; Kapulnik, Y.; Raskin, I.; Salt, D. E. *Plant Physiol.* **1998**, *117*, 447-453.
(52) Epstein, A. L.; Gussman, C. D.; Blaylock, M. J.; Yermiyahu, U.; Huang, J. W.; Kapulnik, Y.; Orser, C. S. *Plant Soil* **1999**, *208*, 87-94.
(53) Reed, B. E.; Moore, R. E.; Cline, S. R. *J. Soil Contam.* **1995**, *4*, 243-267.
(54) Peters, R. W.; Shem, L. In *ACS Symposium Series*; Vadegrift, G., Reed, D., Tasker, I., Eds., 1992; Vol. 509, pp 70-84.
(55) Allen, H. E.; Chen, P. H. *Environ. Prog.* **1993**, *12*, 284-293.
(56) Yu, J.; Klarup, D. *Water Air Soil Pollut.* **1994**, *75*, 205-225.
(57) Ebbs, S. D.; Kochian, L. V. *Environ. Sci. Technol.* **1998**, *32*, 802-806.
(58) Ayres, J. A. *Decontamination of Nuclear Reactors and Equipment*; Ronald Press: New York, 1970.
(59) Piciulo, P. L.; Adams, J. W.; David, M. S.; Milian, L. W.; Anderson, C. I. *Release of Organic Chelating Agents From Solidified Decontamination Wastes*; National Technical Information Service: Springfield, VA, 1986.
(60) Byegard, J.; Skarnemark, G.; Skalberg, M. *J. Radioanal. Nucl. Chem.* **1999**, *241*, 281-290.
(61) Haberer, K. *gwf Wasser Abwasser* **1991**, *132*, 60-64.
(62) Pietsch, J.; Schmidt, W.; Sacher, F.; Fichtner, S.; Brauch, H. J. *Fresenius J. Anal. Chem.* **1995**, *353*, 75-82.
(63) Trapp, S.; Brüggemann, R.; Kalbfus, W.; Frey, S. *gwf Wasser Abwasser* **1992**, *133*, 495-504.
(64) Frimmel, F. H.; Grenz, R.; Kordik, E.; Dietz, F. *Vom Wasser* **1989**, *72*, 175-184.
(65) Barber, L. B.; Brown, G. K.; Zaugg, S. D. In *Analysis of Environmental Endocrine Disruptors,*; Keith, L., Jones-Lepp, T., Needham, L., Eds.; American Chemical Society, Washington DC, 1999; Vol. ACS Symposium Series 747.

(66) Barber, L. B.; Leenheer, J. A.; Pereira, W. E.; Noyes, T. I.; Brown, G. K.; Tabor, C. F.; Writer, J. H. In *Contaminants in the Mississippi River 1987-92*; Meade, R. H., Ed.; US Geological Survey, Reston, VA, 1995; Vol. US Geological Survey Circular 1133.
(67) Loyaux-Lawniczak, S.; Douch, J.; Behra, P. *Fresenius J. Anal. Chem.* **1999**, *364*, 727-731.
(68) Nowack, B. *Water Res.* **1998**, *32*, 1271-1279.
(69) Nowack, B. *Water Res.* **2002**, *36*, 4636-4642.
(70) Nowack, B. *J. Chromatogr. A* **2002**, *942*, 185-190.
(71) Nowack, B. *J. Chromatogr. A* **1997**, *773*, 139-146.
(72) Sacher, F.; Lochow, E.; Brauch, H. J. *Vom Wasser* **1998**, *90*, 31-41.
(73) Kari, F. G.; Giger, W. *Water Res.* **1995**, *30*, 122-134.
(74) Alder, A. C.; Siegrist, H.; Gujer, W.; Giger, W. *Water Res.* **1990**, *24*, 733-742.
(75) Kari, F. G.; Giger, W. *Environ. Sci. Technol.* **1995**, *29*, 2814-2827.
(76) Giger, W.; Schaffner, C.; Kari, F. G.; Ponusz, H.; Reichert, P.; Wanner, O. *EAWAG-News* **1991**, *32*, 27-31.
(77) "Die Entwicklung der NTA- und EDTA-Konzentrationen im Bodensee und einigen Bodensee-Zuflüssen von 1985 bis 1990," Internationale Gewässerschutzkommission für den Bodensee, 1991.
(78) Bergers, P. J. M.; de Groot, A. C. *Water Res.* **1994**, *28*, 639-642.
(79) Kuhn, E.; van Loosdrecht, M.; Giger, W.; Schwarzenbach, R. P. *Water Res.* **1987**, *21*, 1237-1248.
(80) Dietz, F. *gwf Wasser Abwasser* **1987**, *128*, 286-288.
(81) Means, J. L.; Crerar, D. A.; Duguid, J. O. *Science* **1978**, *200*, 1477-1481.
(82) Woodiwiss, C. R.; Walker, R. D. *Water Res.* **1979**, *13*, 599-612.
(83) Hansen, P. D. *Vom Wasser* **1986**, *66*, 167-176.
(84) Sillanpää, M.; Aimo, O. *Toxicol. Environ. Chem.* **1996**, *57*, 79-91.
(85) Wanke, T.; Eberle, S. H. *Acta Hydrochim. Hydrobiol.* **1992**, *20*, 192-196.
(86) Heintz, W. *Ann. Chem. Pharm.* **1862**, *122*, 257-294.
(87) Pfeiffer, P.; Offermann, W. *Ber. dtsch. chem. Ges.* **1942**, *75*, 1.
(88) Schwarzenbach, G. *Helv. Chim. Acta* **1952**, *35*, 2344-2363.
(89) von Baeyer, H.; Hofmann, K. A. *Ber. dtsch. chem. Ges.* **1897**, *30*, 1973-1978.
(90) Schwarzenbach, G.; Ackermann, H.; Ruckstuhl, P. *Helv. Chim. Acta.* **1949**, *32*, 1175-1186.
(91) Hutner, S. H.; Provasoli, L.; Schatz, A.; Haskins, C. P. *Proc. Am. Phil. Soc.* **1950**, *94*, 152-170.
(92) Jacobson, L. *Plant Physiol.* **1951**, *26*, 411-413.
(93) Barica, J.; Stainton, M. P.; Hamilton, A. L. *Water Res.* **1973**, *7*, 1791-1804.
(94) Samanidou, V.; Fytianos, K. *Water Air Soil Pollut.* **1990**, *52*, 217-225.
(95) Fischer, K. *Chemosphere* **1992**, *24*, 51-62.
(96) Nowack, B.; Kari, F. G.; Krüger, H. G. *Water Air Soil Pollut.* **2001**, *125*, 243-257.

(97) Kuhlmann, B.; Schöttler, U. *Chemosphere* **1992**, *24*, 1217-1224.
(98) Nowack, B.; Xue, H. B.; Sigg, L. *Environ. Sci. Technol.* **1997**, *31*, 866-872.
(99) Siegrist, H.; Alder, A.; Gujer, W.; Giger, W. *Wat. Sci. Technol.* **1989**, *21*, 315-324.
(100) Bordas, F.; Bourg, A. C. M. *Aquat. Geochem.* **1998**, *4*, 201-214.
(101) Baik, M. H.; Lee, K. J. *Ann. Nucl. Energy* **1994**, *21*, 81-96.
(102) Zachara, J. M.; Smith, S. C.; Kuzel, L. S. *Geochim. Cosmochim. acta* **1995**, *59*, 4825-4844.
(103) Jardine, P. M.; Mehlhorn, T. L.; Larsen, I. L.; Bailey, W. B.; Brooks, S. C.; Roh, Y.; Gwo, J. P. *J. Contam. Hydrol.* **2002**, *55*, 137-159.
(104) Mayes, M. A.; Jardine, P. M.; Larsen, I. L.; Brooks, S. C.; Fendorf, S. E. *J. Contam. Hydrol.* **2000**, *45*, 243-265.
(105) Li, Z.; Shuman, L. M. *Sci. Total Environ.* **1996**, *191*, 95-107.
(106) Wenzel, W. W.; Unterbrunner, R.; Sommer, P.; Sacco, P. *Plant Soil* **2003**, *249*, 83-96.
(107) Römkens, P.; Bouwman, L.; Japenga, J.; Draaisma, C. *Environ. Pollut.* **2002**, *116*, 109-121.
(108) Cleveland, J. M.; Rees, T. F. *Science* **1981**, *212*, 1506-1509.
(109) Killey, R. W. D.; McHugh, J. O.; Champ, D. R.; Cooper, E. L.; Young, J. L. *Environ. Sci. Technol.* **1984**, *18*, 148-157.
(110) Keiling, C.; Marx, G. *Radiochim. Acta* **1991**, *52/53*, 287-290.
(111) Sunda, W. G.; Engel, D. W.; Thuotte, R. M. *Environ. Sci. Technol* **1978**, *12*, 409-413.
(112) Manahan, S. E.; Smith, M. J. *Environ. Sci. Technol* **1973**, *7*, 829-833.
(113) Sunda, W. G.; Guillard, R. R. L. *J. Mar. Res.* **1976**, *34*, 511-529.
(114) Lanigan, R. S.; Yamarik, T. A.; Andersen, F. A. *Int. J. Toxicol.* **2002**, *21*, 95-142.
(115) Temple, G. S.; Ayling, P. D.; Wilkinson, S. D. *Microbios* **1992**, *72*, 7-16.
(116) Braide, V. B. *Gen. Pharmac.* **1984**, *15*, 37-41.
(117) Hancock, R. E. W. *Ann. Rev. Microbiol.* **1984**, *38*, 237-264.

Speciation

Chapter 2

Speciation of Chelating Agents and Principles for Global Environmental Management

David R. Williams

School of Chemistry, Cardiff University, P.O. Box 912, Cardiff, Wales CF10 3TB, United Kingdom

The chemical speciation prevailing between chelating agents and environmental components is pivotal to environmental impacts and their management. Such speciation may be analytically measured or simulated using thermodynamics. This leads to sustainability through biodegradation of organics and entrapment of inorganics within the geosphere. Through correct speciation management environmental risks to human and biosphere health can be reduced to acceptable levels.

Over the last half century the identifications of the causes of diseases and of environmental damage, and also the use of inorganic approaches to rectify these, have been considerably illuminated through emerging capabilities to research the pivotal chemical species. Furthermore, the 'duty of care' placed upon scientists to protect our environment from manufactured chemicals and to limit their effects upon humans therefrom is best realised from a speciation knowledge database (*1-4*). When scientists accept this challenge they must appreciate the mega mega dimensional scale of up to 10^{16} in magnitude between chemicals and the planet that they could affect as shown in Figure 1 (*5*).

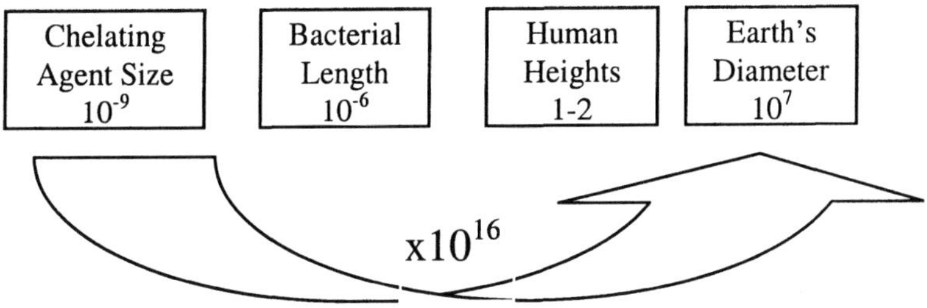

Figure 1. Relativity scale of chelates and their influence upon bacteria, humans, and our planet. All units are in metres.

Metals which, under ambient conditions, can complex, redox or emit radiation are potentially hazardous. Further, the labile nature of metal – ligand interactions makes them vulnerable to chelate ligand influence. However, the vast majority of such ligands (Latin '*ligare*' = to bind) introduced as used by industry are composed of simpler "organic" elements such as C, H, O, N, S, P, *etc* and, once biodegraded to CO_2, CH_4, H_2O, NH_3, *etc*, are environment acceptable and are recycled as part of the geosphere steady state titration originally described by Sillén in the 1960s (*6*).

Chemical speciation simulation of an industrial process and of the related environmental risks can provide mechanistic details, optimise reactant quantities and doses used in applications, minimise waste disposal, and replace non-biodegradable chelating (Greek 'chele' = claw of crab) agents with environmental-friendly substitutes (which increasingly are blends of two or more agents) (*7,8*).

The identification of some unique origins of environmental and health problems has been considerably enlightened through recently developed capabilities to analyse the definitive chemical species involved. This chapter involves speciation information and its use in the categorisation of chemicals in terms of their persistence, bioaccumulation, and toxicity. Such speciation data can be used to optimise the industrial processes thus reducing environmental impacts to within legislated environmental criteria (*1*).

Notable successes have already occurred in linking our environment with human health. From a new awareness which commenced with the publication of Carson's 'Silent Spring' in the 1960s, there has been increasing pressure to legislate and to enforce readily-biodegradability criteria to curtail the use of organic chemicals to those that disappear readily after use (*9,10*). Nowadays an enthusiastic citizens' lobby demands cessation of use of all chemicals unless certified as low hazard and more openness and transparency in "permission to use" decision making based upon risk assessments.

Multinational agencies such as the European Union (EU) and the Organisation for Economic Cooperation and Development (OECD) are now focussing upon new means of chemical analysis (not disturbing labile equilibrium) which identify and categorize hazards to the health of their citizens and thus lead to risk reduction management. Chemical speciation studies are one such promising approach (*11*).

Defining terms used.- The International Union of Pure and Applied Chemistry (IUPAC) defines chemical speciation as "the specific form of an element defined as to isotopic composition, electronic or oxidation state, and/or complex or molecular structure" (*12*). 'Speciation analysis' – the act of identifying and quantifying the species present – and 'the speciation of an element' which refers to the distribution of the element between different chemical species – follow on from the IUPAC definition.

Analyses of every individual species such that they summate to the total concentration of a chosen element are often impossible. However, the distribution of these species between different fractions is an achievable target; such fractionations are based upon different physical chemistry parameters such as molecular mass sizes, solubilities, partition into other solvents (often chosen to reflect the lipophilicity of cell membranes), charge densities, and readily ability to exchange ligands (its 'lability') or, alternatively, not to do so (its 'inertness').

Unfortunately, such fractionation researches often disturb delicate labile equilibria and so, in principle, computer simulation/modelling based upon the same labile equilibria can better reflect the species present at equilibrium with much greater accuracy. Such an approach is critically dependent upon the accuracy and completeness of the input data and of the model representing the prevailing equilibrium/steady state of the topic being researched. Thus, much time is spent verifying input data and validating the models (*13,14*).

Acceptable Biodegradability Criteria for Sustainable Development and Causal Environment – Health Relationship.

Real, or just perceived, chemical threats produce alarmist headlines, possibly because the lay-person and press regard these agents as invisible, unpronouceable, and man-made. Although society is safer now than at any time in its history and expected lifespans at birth are double those in the 1850s when drugs, the sanitary water and air era, and modern healthcare commenced, globally there are still 2 million children who die each year from contaminated water and one fifth of diseases in the developing world originate from environmental causes (*15*). The mismanaged media response to questions about safety such as arose recently for Measles, Mumps and Rubella vaccination safety has led governing organisations to demand cessation of use of persistent (P) organic chemicals in the environment which bioaccumulate (B) or are toxic (T) (*16,17*).

Thus, the serious challenges of organic agents entering the environment are based upon such PBT criteria. The UK acceptance concepts are illustrated in Table 1, there being many qualifying sub-criteria (*16*). Considering readily-biodegradability aspects (the converse of persistence), this refers to aqueous reactions in the environment into which the chemical is disposed. Typically these discharges enter rivers and drains. Thus, the OECD-definitions of this term rely upon naturally occurring bacteria to degrade the organic agent within a specified time (*11, 18, 19*). The criteria are not related to the intentional introduction of microbiologically selected bacteria having a specific affinity for the organic substrate; rather they assume only those naturally occurring bacteria which are ubiquitous. Further, it must be realised that even naturally occurring ground water bacteria require the prevailing conditions, such as temperature, ionic composition, lumination, *etc*, to support their biodegrading activities. Such natural readily-biodegradable criteria embody terms such as the percentage of agent biodegraded within a certain time (typically 10 days) after the initiation period has ended. In general, definitions are based upon CO_2 released, dissolved organic carbon, oxygen uptake, respirometry, *etc*. Typical criteria are exemplified in Table 1.

This realisation that environmental biodegradation is dependent upon commonly occurring bacteria has led to naturally-occurring entities such as amino-acids and sugars being included in the formulae of industrial ligands and also the careful selection of agents such that their degradation products are of the naturally occurring chiral isomer; these are more easily accepted into the host environment.

Table 1. ***PBT "first tier" criteria used by the UK expert committee on hazardous substances for substances of highest concern.***

- **P**ersistence, P, $t_{1/2}$ marine > 60 days, fresh water > 40 days or $t_{1/2}$ marine sediment > 180 days, freshwater sediment > 120 days.
- **B**ioaccumulation, B, log K_{ow} > 4.5 unless Bioconcentration factor (BCF) > 2000 where data are available. If experimental BCF < 2000, log K_{ow} does not apply.
- **T**oxicity, T, Acute Lethal (Effect) Concentration $L(E)C_{50}$ < 0.1 mg/litre or long term No Observable Effect Concentration, NOEC < 0.01 mg/litre OR category 1 or 2 carcinogen, mutagen or reprotoxin, and category 3 mutagens and reprotoxins.

Safety net = expert judgement of Advisory Committee on a case-by-case basis.

Key $t_{1/2}$ = half-life; logK_{ow} ≡ log of partition coefficient between oil and water.

Similarly, the increasingly lower levels of analytical equipment detection leads to media alarmism – "It has an analytical presence therefore it caused it!" rather than a more scientifically meaningful approach to proving the existence of a statistically valid causal link between chemical species and health effects.

A more cautious approach to assessment in terms of P, B, and T is needed for inorganic agents since the elements will never biodegrade *per se* but do change their bonding, and sometimes their oxidation states in order to give another species. The fact that such bonding exchanges are frequently labile in the environment, in humans, and in healthcare, has not been widely understood by non-chemistry scientists even though Sillén *et al* revealed its geological importance in the 1960s (*6*).

This chapter describes progress as seen from the viewpoint of permitted registered chemical usage and environmental health risks in the UK; this scenario closely reflects the EU sustainable environment position, and shows how chemical speciation answers many current problems (*11, 18,19*).

Categorising chelates in terms of their P, B, and T properties requires speciation information to optimise desirable effects of chemicals in numerous applications such as the manufacture of pulp for paper and in the foliar nutrition of crops. Simultaneously, the geosphere migration of metals through the environment is discussed in terms of speciation. Here, the vast amount of research which has already been achieved in the field of radioactive waste disposal species migration to biospheres is taken as a springboard for principles of risk management involving non-radioactive metals. Finally, the challenges of communicating speciation-based recommendations to decision makers are discussed (*20, 21*).

Establishing Chemical Speciation in the Presence of Chelating Agents.

Reliable risk assessment when metal ions and chelates are involved demands detailed chemical speciation knowledge. This is not without its challenges!

Commonly encountered problems include.-

- Concentrations of species are sometimes at parts per billion, and even lower, levels. Ebdon et al quote solid geological analytes as typically containing As, Cu, Ni, Cr, Hg and Sn at parts per hundred thousand to ppbillion levels, and biotissues containing Hg, As, and Se over the range ppm to ppb and the Hg and S contents in water samples being ppb to ppt! (*22*).

- There are literally thousands of individual species in interdependent competition through a vast equilibrium involving protons, metal ions and ligands.
- Perhaps more difficult to understand by non-inorganic solution chemists is the fact that most of these species are in labile equilibrium such that adding one probably disturbs most of the other species concentrations and compositions. Although there are a minority of inert metal-ligand reactions which are non-labile possibly restricted through the directional nature of the metal's bonds, it is the labile complexes which have consequences for the chosen analytical technique used since it must not disturb this delicate equilibrium balance (2, 3, 7).

Books and conferences have described analytical methods for chemical speciation analyses in situ (*20–23*). This paper uses chemical occurring in industry, in biochemistry, and in the environment using the scheme shown in Figure 2 (*13, 14, 24, 25*).

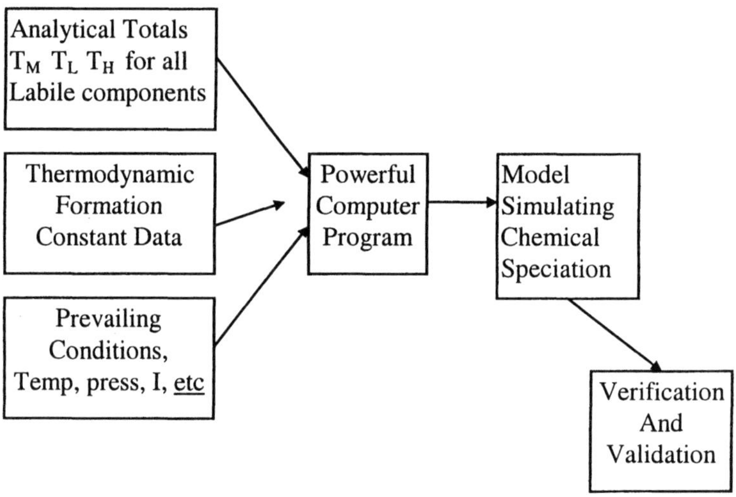

Figure 2. Fundamentals of computer speciation simulation. Analyses, as accurate as possible, of all the individual labile metals and each possible ligand are combined with solution data and formation constants for each possible reaction (in solution, involving precipitation/dissolution, and partitioning into other phases). The model is validated against carefully selected laboratory experiments (26).

There are many internationally verified computer programs and thermodynamic formation constant databases accepted by multinational scientific experts. This paper uses the JESS program and database (*13,14,26-33*). By means of illustration, the simple ligand iminodisuccinate^{4-} (IDS), if present in aqueous solution with six transition metal ions comes to equilibrium wherein the ligand is distributed as in Figure 3.

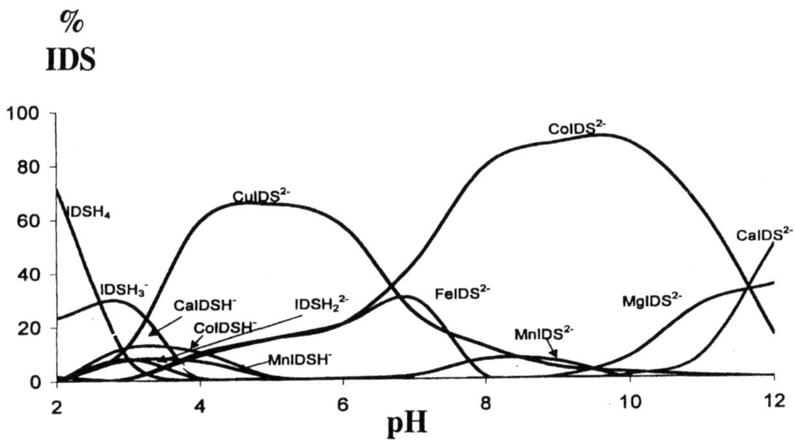

Figure 3. Chemical speciation plot for iminodisuccinate^{4-} equilibrating with Ca, Cu, Mn, Fe, Mn and Fe(II) ions at equimolar concentrations expressed as a percentage of the total chelate present.

Clearly, whichever species predominates is pH-, and concentration-, dependent and, by varying these parameters, the presence of the desired species can be optimised whilst simultaneously minimising interference from other species. This specific chelate, $(CH_2COO)_2.NH.(CH_2COO)_2^{4-}$, has four carboxylate donors and a secondary amine which cooperatively give the possibility of either five membered (N: and O^-) or eight membered (O^- and O^-) chelate rings; this combination of stereochemistries and donor groups specifically matches transition metal ions.

Biogeochemistry reactions are directly driven by a specific species rather than indirectly through the total amount of components present. From speciation data it is possible to optimise many features. When extended to the real life situation, involving dozens of ligands and metal ions, in biology and in the environment, it is relatively easy to model. Graphical speciation plots can reflect all species but would be too cluttered for clarity and so visual representation involves just the important active species which, hopefully, correlate with a causal relationship with the desired activity such as bio-response or environmental result.

Recent published examples include minimising the environmental impact statements (EIS) for EDTA, an agent which has made an outstanding contribution to chemistry over the last 60 years; approximately 20000 tonnes per annum are used. However, it is now falling victim to readily-biodegradability legislation *(7)*. Optimising the use of EDTA-replacement agents such as phosphates *etc*, the design of EDTA-based ligands which do readily biodegrade, the use of two or more ligands blended to optimise chemistries, minimising purchase prices and environmental costs in the pulp industry, the foliar feeding of plants using metal-ligand mixtures, and the risk assessment for radio-nuclides in the presence of ligands following the co-disposal of inorganic and organic wastes – all are viable ways to establish a sustainable future for our environment *(35-37)*.

Chelating Agents Challenging the Environment and their Chemical Speciation.

The biogeochemistry of chelating agents is mediated through their metal complexing. Pathways from chemical manufacturing plants to the environment and into humans are complex-, and protonation-, dependent and there are manifold routes – through effluents and emissions, through food and drink, through contamination, and as healthcare agents (most pharmaceuticals are ligands whose *modus operandi* involve metal seeking).

Although the majority of pollutant/contaminant chemical species differ at their point of introduction to naturally occurring species containing the same elements, because of lability, the geochemical/biochemical effects are usually through overloading substantially buffered natural processes frequently by accelerating the mobility of metal ions therein. In brief, the chelates move metal ions more rapidly to established sites where they overload the regular chemistry by exceeding the "healthy concentration window", to use Schwarz's term of 1970s (*38, 39*).

Provided that these influences are labile they are relatively easy to simulate. This permits possible changes in conditions many thousands of years hence to be estimated and risks to be managed. Separation of particulate material from soluble species can disturb the equilibrium data and give false conclusions. The scheme in Figure 4 shows how heavy metals in waters can be fractionated; each subfraction will have an individual hazard potential.

Size in nm	Species Type and Fraction Separated	Thermodynamic Behaviour	Ability to Exchange Metal Ions
	First	***Fraction***	
1-10	Free metal ions	Soluble	Labile equilibria
1-10	Low mol mass organic chelates	Soluble	Labile equilibria
1-10	High mol mass organic chelates	Soluble	Labile equilibria
	Second	***Fraction***	
10-100	Metal - humates	Colloidal	Labile equilibria
10-100	Metal - colloids	Colloidal	Labile equilibria
	Third	***Fraction***	
1000	Particulates	Solids	Mainly inert to exchange

Figure 4. Different fractions separated from a sample of metal ions equilibrating with chelates (natural and manufactured) and with soil. Note that the fractionation process may disturb the equilibria governing the first two fractions and could give a misleading result.

Whereas 25 years ago "officialdom" was accused of obdurate complacency about health effects from environmental sources, nowadays such officials are enthusiastic about openness and transparency in restraining erring industries which contaminate (*40*). Naturally, there are cries of 'job-losses' and 'feeding opposition from imported overseas chemicals' when faced with such official demands for toxicology, PBT, and emissions data. There have been noteworthy exceptions where genuine progress has been made in justifying and almost eliminating metal – ligand contaminants in the environment. For example, organic tetraethyl lead, a vehicle fuel additive that prevents engine knocking, has been researched, found to be considerably more of a health risk than inorganic lead because of lipophilicity and ability to compete with essential zinc biochemicals therein, and has almost gone out of production over the last two decades – from 99.9% of fuel using this additive down to 0.01%. (see Table 2). Substitute chemicals which biodegrade readily have been launched and, simultaneously, engines have been re-designed to run on lead-free fuel. The human symptoms of such overload are typical of zinc deficiency – nervousness, tired, run-down feeling, lack of ambition, frequent colds and other infections, *etc*. Most benefit from zinc supplementation (*41*).

Table 2. Decline in use of lead in road-transport fuels once public opinion and government will had been established. Data are for UK usage per annum (*41*).

Year	*Total leaded fuel production in million tonnes*	*% of fuel that was sold leaded*	*Lead emissions in tonnes*
1987	22.2	99.9	4283
1990	16.0	66.0	3102
1994	9.68	42.4	1871
1998	4.68	21.4	906
2000	0.005	0.03	1.04
2002	0.003	0.01	0.56

The changes in chemical and biodegradability activity occurring when metal ions meet ligands can be easily illustrated using EDTA as an example ligand (*42*). Ordinarily, EDTA does not biodegrade at all readily in nature and so environmental concentrations are rising.

A Gram-negative proteobacteria (BNC1. DSM 6780) has been isolated and successfully uses EDTA as its nitrogen and carbon sources and hence it biodegrades (*43-45*). The first stage in this process in the transport of the ligand, EDTA, into the supporting bacteria usually as a lower charge density species of metal ions complexed to EDTA anion; calcium and magnesium complexes are the usual means of achieving this penetrating species. Given the incorrect metal ion presence or imperfect free metal ion gradient (possibly because of pH effects or other competing metal ion equilibria – see IDS in Figure 3), the ligand species cannot enter and so even this specially derived EDTA biodegrading bacteria (BNC1) is inactive – such blocking occurs with metal ions iron(III), cobalt(II), cadmium, lead, nickel and copper(II) (*46*).

In theory, should one sequester the transition metal ion away from the EDTA by using a more specific and powerful ligand, then the EDTA contaminated sample could be cleaned up by biological degradation. This is only possible when the original Fe(III) EDTA chelate in not inert and thus most reluctant to release the iron to the other ligand.

Such readily biodegradation of EDTA in the more widespread global environment is not a practical solution since the bacteria therein do not have the unique properties of BNC1. Thus EDTA, in use since the 1930s, has progressively accumulated in certain environments and so its environmental impact was surveyed by the EU in the 1980s (*47*). Ubiquitous presence in aqueous environments was found and some solubilization of heavy toxic metal was suspected (*48-50*). The survey quantities and associated risk assessment data, so far, have not found that the agent has breached acceptable limits in respect of mutagenicity and carcinogenicity to public or workers, of environment damage, or of biomagnification.

However, higher emission regions such as beverage and dairy sites, pulp and paper making, circuit printing and photographic wastes need to be controlled and substitute agents which readily biodegrade must be found as soon as possible.

Management Decisions and Public Perception.

Multinational agencies such as the EU and the OECD are now focussing upon new means of chemical analysis and legislation which identifies and reduces hazards to the health of their citizens and enables and manages risk reduction.

The reported links between organic DDT in the environment and avian health commenced with the publication of 'Silent Spring' in the 1960s, even though it took more than 30 years to prove a causal link (*9*). Meanwhile, pressure to legislate using readily bio-degradability criteria upon organic chemicals has increased and produced new legislation.

Originating with decades of geochemistry research into radioactive waste metal migration from disposal vaults moving through to biospheres - a process which can be accelerated a million-fold by the presence of chelates - there is now a growing number of substitute replacement agents which do biodegrade, provided that conditions such as the presence of water, pH, ambient temperature, and other conditions are satisfied; these conditions for bacterial activity are usually avoided in disposal site selection and so the two aspects have to be resolved *(33).* The solution is the principle of no co-disposal of organics (chelates) and inorganics.

One such substitute agent is *S,S'* EDDS which degrades to form *l*-aspartic acid. Not surprisingly *S,S'* EDDS satisfies the OECD 301E criteria and is being successfully introduced in many areas formally employing EDTA (*7,34,51,*).

Although our ability to model this new science has improved substantially over the last two decades, our influence upon management and licensing authorities has been weak. There has been reluctant acceptance that biological and industrial activities are chemical speciation dependent rather than total amount dependent. Rather, many material purchasing decisions are related to "the more ligand used to eliminate a metal based problem, the better!" rather than to environmental cost!

Similarly, legislators and licensing inspectors place more trust in total amounts rather than individual species. Sales teams habitually use conditional constants to compare chemical efficiencies (they are only valid for the simple system at a fixed pH) rather than the more accurate and modern Speciation Efficiency Indices (SEI) (*52*).

Further, the ability not to become an environmental burden is best expressed as Readily Bio-degradability Indices (RBI). Sometimes cheaper treatment prices and lower environmental costs can best be achieved using blends of two or more (*8*). All of these concepts are easily simulated using chemical speciation analysis and then validated by laboratory experiments and pilot scale trials.

The public need more persuasion that not all chemicals, including actual chelating agents, do environmental damage. On the contrary, many chemicals have virtually doubled human life expectancy and quality from birth. More information for lay-persons, often using the type of management tools listed above such as SEI and RBI, will help to allay these fears and to ensure that the chemical industry continues to add value and longevity to the quality of life and the sustainable biogeosphere.

The industry has great influence, EU being the largest chemical producing region in the World having a trade surplus of 41 billion Euros in 1998 (*11*).

On the one hand, all of our nutrition and support comes from the environment but, on the other hand, the environment and all industrial chemicals therein are held responsible by many for causal links with diseases and other undesirable events. Speciation chemists can help to unscramble these phenomena and to identify real tort.

Some of the public insisting on absolutely safe drugs for therapy, are also de-crying the processes which ensure their safety. Most do not understand the meaning of the word "safe". The UK Chief Medical Officer's annual report for 1995 defined "safe" as a negligible risk being quantified as less than one in one million chance of an event happening *per annum*, there being no such thing as zero risk. Just as it is incorrect to blame all undesirable events upon chemicals, so too, it would be equally misleading to claim that there are no dangerous chemicals that have been, and that are still being, released upon an unsuspecting public.

For many years, safety cases and licences in the nuclear industry have had to be justified on the grounds of "worst possible case" scenarios. Risk of migration of a hazard (in this case the radioactive element) under the speciation influences of the geochemistry of the site and of any chemicals co-disposed with the radioactive waste material, has been a critical factor in reaching planning and development decisions.

The OECD, the EU, and the UK as members of both, have set up scientific and legislative devices to protect the *status quo* and to ensure that no further damage to the environment and to health arises from chemicals. The global production of chemicals is 400 million tonnes *per annum*. There are more than 100,000 chemicals registered in the EU market and which contribute to society's lifestyle as taken for granted today; we encounter about 1000 chemicals daily in everyday activities. It is a gigantic task to assess current, and retrospectively to back assess existing, hazardous substances.

However, this paper is not about this perception as the reader is referred elsewhere for improving communications (*20*). Rather, this chapter demonstrates how speciation is pivotal to decisions on continued safe use, to improved risk management, or, in the extreme, to banning an agent. The EU, the OECD and other multinational agencies have progressed along broadly similar lines and are currently harmonising their inter-state limits and criteria of concern.

Globally, a dozen chemical agents (predominantly pesticides) have been targeted for immediate substitution (Aldrin, Clordane, Dieldrin, Endrin, Heptachlor, Hexachlorobenzene, Mirex, Toxaphene/Camphechlor, Polychlorinated biphenyls, DDT, Dioxins/Furans.). Unfortunately of the 60 substitutes researched most are said to have hazardous properties (see www.pan-germany.org) and a third of them are potential chelates of relevance to this chapter.

In October 2003 the European Commission published its REACH proposals which are now before the European Parliament. REACH stands for Registration, Evaluation, and Authorisation of Chemicals and is aimed at bringing chemicals in use before 1981 - the "existing chemicals" - in line with "new chemicals" produced and registered through a variety of schemes since then (*54*). Some pivotal concepts and consequential facts about the proposals are listed in Table 3. Again, about one third of those organic chemicals coming under REACH are potentially chelates; so far less thought seems to have been given to inorganics and to the question of co-disposal of chelates and inorganics

Known bio-active agents specifically controlled by existing legislation are not included in the considerations above. since strict controls and licencing systems already exist for these materials. To quote an example of a country's specific legislative provision, the UK has enacted stringent acts for.-

Pesticides – the Plant Protection Products Regulations 1995, Control of Pesticides Regulations 1986, and Part III of the Food and Environmental Protection Act 1985.
Biocides – Biocidal Products Directive 2001
Food Contact – Plastics Directive 1998
Pharmaceuticals and Medicines – Medicines Act 1968
Veterinary Medicines – Medicines Act 1968

Table 4 lists web sites for the various bodies involved in setting criteria for existing and new chemicals in the UK. Essentially, a committee of experts, guided by criteria such as those given in Tables 1 and 3, gives

Table3. Facts used in the EU REACH Proposals Document

Covers the *use* or *manufacture* of chemicals more than limit per year.
Existing chemicals come under REACH limit at 1 tonne per year of manufacture or import. New chemical registration has been set at 10kg per year for some time.
100,106 existing chemicals come into this category; 1 in 3 are chelators.
Both the maintenance of human health and of the environment are objectives.
Industry is given greater responsibility to manage risks from chemicals and to provide safety information.
The streamlined and cost-effective system has used 6000 Internet consultees.
Over 40 existing regulations are being replaced.
Dossiers are to be deposited with the European Chemicals Agency.
After registration approximately 80% of agents will require no further data provision.
Animal tests in producer's dossier will be available to other organisations thus reducing the needs for more in-house testing.
CMR, PBT, and vPvB criteria for assessing agents will be used*.
If risks from PBT *etc* categories of hazards can be controlled, continued should be authorised.
Expected costs to the producers and to users are put at 2.3 and 2.8 billion Euros over 11 years.
Health benefits of the order of 50 billion Euros over 30 years are anticipated.
Some 12% of existing chemicals are anticipated to be withdrawn.

*<u>Key</u>. ***Carcinogenic, Mutagenic, Bio-accumulative, Persistent, Bio-accumulative, Toxic.** v=very persistent or bio-accumulative.*

scientific advice to the UK Chemical Stakeholder Forum. This forum takes advice from bodies having an interest in a chemical *eg* the environmentalists, the manufacturing associations, the professional societies, the employee unions, the consumer councils, *etc* and then gives considered advice to Government concerning bans, chemicals likely to cause serious or irreversible damage to the environment, reassurances, import/export controls, the need for new risk management measures, the necessity to generate new data concerning persistence, bio-accumulation and toxicity, the need for research into substitute agents, and means of disposing of wastes and un-used chemicals, *etc*.

Table 4. Web sites describing chemical hazard legislation in Europe

http://www.defra.gov.uk/environment/chemicals/ukpolicy.htm
http://www.defra.gov.uk/environment/chemicals/csf/criteria/index.htm
http://www.iccahpv.com/hpvchallenge/about.cfm
http://www.europa.eu.int/comm/environment/chemicals/index.htm
http://www.defra.gov.uk/environment/chemicals/glossary.htm
http://www.environment-agency.gov.uk
http://www.defra.gov.uk/environment/chemicals/csf/concern/index.htm
http://ecb.jrc.it/existing-chemicals/
http://www.defra.gov.uk/environment/chemicals/strategy/03.htm#6
http://www.defra.gov.uk/environment/chemicals/achs/index.htm
http://www.food.gov.uk/science/ouradvisors/toxicity/
http://ecb.eu.jrc.it
http://ecb.jrc.it/existing-chemicals
http://www.europa.eu.int/comm./environment/docum/01262_en.htm
http://www.iccahpv.com/
http://www.oecd.org/pdf/M00017000/M00017224.pdf
http://www.ospar.org/eng/html/welcome.html
http://www.irptc.unep.ch/pops/default.html
http:forum.europa.eu.int/Public/irc/env/wfd/library
http://www.defra.gov.uk/environment/chemicals/eufuture.htm
http://www.cia.org.uk/industry/confidence.htm
http:www.iccahpv.com/hpvchallenge/about.cfm
http://www.epa.gov/enviro/html/emci/chemref/index.html
http://www.coshh-essentials.org.uk

In the UK this two year old UK campaign has already resulted in manufacturers accelerating acquisition of safety data, their having such data independently refereed and reported, industry depositing such data with open access bodies charged with responsibility of sharing screening and toxicity data (as part of a larger international effort to reduce animal experiments), and manufacturers initiating research projects to find substitute agents and to optimise (which usually means to reduce) the exposure to the PBT chemicals concerned.

All of these new regulations which differ marginally between countries and legislatures (a) are being normalised and harmonised into the detail of the EU REACH systems, (b) are over and above existing strict regulations for control of known hazards such as radionuclides, pharmaceuticals, *etc* (see Table 3), (c) are reducing the need for some animal experiments, and (d) bring an openness and trust to the perception that chemicals cause all the undesirable aspects of life whereas so-called "natural" or "health" cures currently are perceived as having the moral high ground. Yet to be achieved is the suppression of cheaply-produced chemical imports from countries not having such stringent environment and health protection regimes in force; the environment is potentially damaged just as much whether the agent is made in a non-EU/OECD country or within those countries having modern environmental sustainability standards. This suggests that the import of cheap non-environmental-friendly manufactured chemicals will, eventually, become history.

Biodegradability of Chelating Agents.

Organic agents, of which chelating agents are a fraction, generally biodegrade. However, the drive for extended shelf lives, and water being the preferred industrial solvent for atmospheric and precautionary first-aid provision reasons, has polarised the choice of industrial chemicals to those that may not readily biodegrade.

Further, by definition, chelating agents complex metal ions and so their biodegradation and biogeochemistry behaviours are vastly modified compared with uncomplexed agent (*55*). A wide range of factors are being researched in this newly emerging field of the molecular design of biodegradation. Agents for the future will be permitted only if they biodegrade readily before they chelate metal ions or if their metal ion complexes can biodegrade (*55*).

An additional consideration is that no chemical is 100% pure and there are many costly examples of the impurity which contaminates a material causing tangible chronic damage. Thus, REACH registration limits at 10kg and 1 tonne per year may be an oversimplification.

Exploring, Explaining, Exploiting, and Optimising Processes using Speciation Knowledge.

The greater the amount of speciation knowledge about an industrial process, the greater the confidence level, the more opportunities there are for optimising (*ie* exploiting) the amounts of chemicals used, and the more skilled are our abilities to reduce environmental impacts to near zero.

Harrison has pointed out that lead emissions some 25 years ago were in a "form" (now termed "speciation") vastly different to natural species and so our knowledge of environmental fates and pathways learned from natural deposit migration over many centuries was largely inapplicable *(38)*. Since then methods have been researched and published for determining speciation *in situ* and by modelling the system.

In general, the metal species are spread across a wide range of states and species and as circumstances change so too does the speciation (see Figure 4. This labile interchange of species can be used to advantage in understanding and in manipulating chemical processes; a few, of many, examples are given below.-

Hydroponic plant nutrition.

Copper, zinc and lead uptake from aqueous hydroponic solutions into sunflowers has been reported by Tandy *et al* and the new chelating agent *S,S'* EDDS discourages normal uptake of Cu and Zn but encourages and enhances Pb and other toxic metal intakes. Through detailed separation into each metal species its influence upon the others has been recognised and so it is particularly important that any ligands present rapidly biodegrade (as does *S,S'* EDDS) before permanent damage to the plants occurs (*55*). Interestingly, the physical symptoms of the lead toxicity are reduced even though intake of the metal increases.

Pulp and paper.

Traditionally, the largest World usage of EDTA (totalling circa 20 000 tonnes per annum) has been to sequester adventitious metal ions in the pulp and paper processes and to produce a clear white product having no spots. Details of the process are given in reference *53*. Additional to the clear white paper quality, smoothness of product requires the magnesium to remain in the system in spite of a complexing ligand being introduced to complex iron, managanese, and copper ions – the worst offenders with respect to flaw forming initiation. Although paper is produced world-wide most of the fundamental chemical research occurs in Scandinavia which means that the amount of ligand(s) introduced has not been pulp factory site specific and so ligand overdosing frequently occurs at other sites (*53*).

Realising that doses of EDTA traditionally used had not been optimised, speciation simulation using thousand species models focussed on the salient questions and produced site optimised formulations which (a) fine-tuned the amount of replacement ligand substituted for EDTA (in practise *S,S'* EDDS), and (b) sequestered the maximum amount of problematic transition metal ions whilst leaving the magnesium ions uncomplexed. (c) Previously, EDTA, being such an avid complexer of Ca^{2+} was mainly lost to that metal ion in the feed water (the term used is Calcium Distraction) and only a few percent were remaining for spot prevention. The expense of this distraction added to the price of the paper as well as $Ca.EDTA^{2-}$ posing a serious environmental cost.

By introducing *S,S'* EDDS as a substitute for EDTA there was almost negligible calcium distraction and considerably less agent to go into the environment whence it was readily-biodegradable. This effective 'tuning up' of ligand doses is referred to as "more for less". Further, for some stages of the pulp making process, optimisation employed blends of EDTA and *S,S'* EDDS mixed together in order to achieve some calcium, and total transition metal ion, complexing at a reasonable price and within the degradability limits permitted by OECD regulations. Marketers sometimes advertise their products as "inherently biodegradable" but these do not satisfy the international standard for sustainability of our planet legislated for under their definition of "readily-biodegradable" (*18,19*)

Reference *53* reports these achievements in some detail and shows Speciation Efficiency Index and Readily-Biodegradability Index plots understandable by plant managers without their need to master the intricacies of labile equilibria and of complex chemistry.

It is axiomatic that none of these conclusions ought to be implemented without using pilot scale laboratory experiments to check that all the data have been verified and the model has been validated for each of the several stages of the pulp making process which involves complex chemistry. The price economies of modelling are obvious when one considers the price and environmental cost of doing trial and error dose gradations at pilot plants sometimes half a world distance from the design and research laboratories.

Foliar feeding of agricultural crops and advantages of ligand blends

A similar approach has been taken to the control of nutrient metal complexes used in foliar feeding of plants. In agriculture, trace element availability is crucial to achieving higher yields and better crops (*8*). Two pivotal ingredients are ferric, Fe(III), and phosphate ions, PO_4^{3-}, which, in addition to manganese, copper, molybdenum and zinc deficiencies, cause serious growth problems (*8*). Recently, foliar application of these agents by spraying onto the leaves of the plants has been shown to be far more efficient than fertilising through soil and roots. Chelating agents such as EDTA are presently added to foliar sprays (i) to prevent metal phosphate/oxide/hydroxide precipitation which blocks the fine spray jets and also (ii) to prevent leaf scorching by free ions.

With the phasing out of EDTA more environment-friendly alternatives such as *S,S'* EDDS are being introduced but Fe(III) is bound more strongly to EDTA than to EDDS; also the EDDS price is more expensive. Fortunately, it has been possible to use computer simulation modelling of a range of blends of mixtures of *S,S'* EDDS and EDTA such that precipitation is prevented, purchase cheapness optimised, and environmental readily-biodegradability legality upheld (*ie* 60% of chelating agents biodegrade within 10 days *etc*). It was established that The blend range 60:40 through to 70:30 EDDS:EDTA is as effective as the EDTA alone. (*8*).

It is noteworthy that there is no straight-forward slot-in replacement of EDDS for EDTA (which would have produced precipitates anyway!). Rather, carefully selected two ligand blends held the solution. The ligand supply and agricultural spray formulation industry has confirmed the absence of solids and field trials by the industry are in hand.

The possibility now arises for optimising the foliar supplies of the other micronutrients mentioned above and also of tailoring spray formulations to the particular challenges of a specific plant and soil.

Legislative Limitations and Compliance; Safer Use of Chelates.

In the 1980s the impacts of EDTA upon the environment were reviewed since it had a widespread presence in our aqueous environment and, ostensibly, solubilised heavy metals (*49*). EDTA and Na_4EDTA were termed priority substances under European Union regulation number 793/93 considering risk assessment from the German authorities. In general, exposure levels were below the acceptable limits but rising; two decades ago such levels were thought to have no known mutagenic or carcinogenic potentials.

At that time, "worse possible case" scenarios, predicted no effect concentrations exceeded predicted environmental concentrations (*ie* PNEC of 2.2 mg/litre > PEC values). Rendering the environment slightly alkaline was thought to aid EDTA biodegradation. However, EDTA effluent sites of manufacture and use have environmental water concentrations which exceed 2.2mg/litre. Risk reduction measures at industrial cleaning (dairy/beverage) sites, pulp and paper manufacturing, printed circuit board production and photographic processing locations are now under strict environmental controls and restrictions and for the Northern Atlantic the outfalls have to comply with OSPAR zero-emission agreements.

Similar studies are directed at surveying platinum, palladium and rhodium from motor car exhaust catalysts. Sampling and surveying commenced in 1997 (22). The influence of ligands upon the bioavailability of soil metals is not, however, being surveyed (47).

Life Without Chelates and Future Needs.

Subtle differences between apparently similar situations can determine the success or failure of a chelate solution to a chemical problem involving metal ions. Willett and Rittman identified slow kinetics of complexation between Fe(III) – ferric and EDTA as the step rendering EDTA as non-readily biodegradable (*42*). Fe(III) is not taken into bacterial cells where it ought to undergo destruction whereas $CaEDTA^{2-}$ and $MgEDTA^{2-}$ are well biodegraded after cellular intake. Thus, some metal chelates are readily biodegraded whereas others are not. Similarly, although a ligand/chelate may be readily soluble and thus vulnerable to being a substrate, if it is in insoluble metal complexed form or is insoluble because of crystal lattice effects, *etc*, it is effectively removed from the thermodynamic equation and will remain nonbiodegraded. This difference between solubility and dissolving power has challenged the tablet making pharmaceutical industry for many decades!

Water solubility is usually the opposite of lipophilicity but Connors *et al* have combined very water soluble $EDTA^{4-}$ with a lipophilic side chain (*57*) in which one of the four acetate groups has been replaced by an acyl group bearing a C_{12} hydrocarbon tail. This permits Zn^{2+} species to be complexed in a lipophilic form and has possibilities in complexing zinc and taking it through cell membranes to where it acts as an antimicrobial agent.

The two topics of 'molecular design of bio-degradability' and 'stimulating biodegradation' are attracting much research attention. The former refers to the design of the chemical without a knowledge of the specific bacteria assumed to be assigned with the task of biodegradation. Key features, selected from many considered, are listed in Table 5.

The topic of encouraging biodegradation is a hot issue.- Starting with the days when dwellings and factories had soak-away drains and septic tanks, the subject of environmental/sanitary engineering in Europe has matured and blossomed. Sometimes rain water run-off (gray water) is separated from sewage treatment; the commendable campaign for recycling of waste and of biodegradable plastic bags is hindered by the sometimes overzealous use of bactericides to clean-wipe infant skin, bathroom/washroom touch surfaces, kitchens, *etc*.

The biochemistry of sewage disposal and of chemical factory effluent handling is now well understood. Whether the processes are best optimised and well policed is often a matter of finance! However, the modernisation of sanitary lifestyles alluded to in the previous paragraph coupled with the use of supplemental fertilisers in agriculture/home gardens *etc* often means that

Table 5. Some important considerations embodied into the strategy for designing biodegradable organic agents.

Use the stereo isomer, or biodegrade into the isomer, in the orientation best suited to the innoculum used.
Products of biodegradation to be naturally occurring species.
Thermal stability of agent under conditions of use and storage.
Agent to be subject to biodegradation by as broad a range of bacteria as possible.
Use a fundamental knowledge of how bespoke bacteria for agent's biodegradation differ to those targeted bacteria encountered in drains and water courses.
To have similar physical and chemical properties to the non-biodegradable agent being substituted.
Optimise the use of two ligand blending to maximise chemical efficiency and to minimise environmental cost and purchase price.
Biodegraded products also to be further biodegradable.
Influence of trace metal ions upon chelates.
Co-disposal of wastes challenges.
World trade and import legislation.

the biodegradability activities in ponds and smaller rivers, where so much of the undegraded material and end-product metabolites accumulate, is far from optimum.

In the late previous century, the severe chemical pollution of some of Europe's major rivers or ports led to aquatic life becoming virtually extinct; there has been a miraculous recovery after the polluting ceased and processes such as aeration commenced in addition to hot spots being treated with appropriate "bespoke" bacterial bio-technology.

Monitoring water sources to identify un-biodegraded material and the subsequent spiking with selected bacteria such as those specifically designed to degrade ligands, is a mine field of ethics, protesters against mass medication/chemicals, *etc*, and is best avoided by substituting all known PBT products with replacement materials.

Yet further challenges originate from those who believe in hormesis effects – the concept that a small amount of potentially hazardous material such as the sun's rays – can be beneficial in building up resistance against more formidable amounts. Such concepts oppose the totally sterile, absence of all biodegrading bacteria, views of the ultraclean germicidal society.

Environmental biomonitoring.- Although political acceptance of biomonitoring has made spectacular progress over the last 40 years, to use tax revenue to fund and to police the monitoring of man's actions which irrevocably damage the environment is still a sensitive issue (*58*). Figure 5 shows that the influence of metals in the environment is greatly magnified through chelates to give individual species which specifically affect the biogeosphere. However, the 'nanny state' is reluctant to fund the data acquisition which permits geosphere and biosphere modellers to research such phenomena.

Monitoring is, however, the only alternative to re-inventing all science and technology using biodegradable, totally environmental sustainable, chemical products.

Metals→Chelates→Environmental Species→Magnified Pattern of Effects

Effects = **_circa_ x1000** **_circa_ x1000 000**

Figure 5. Scheme illustrating that a chelate may increase metal solubility a thousandfold and then these species migrate through the environment a thousand or a million times more rapidly than the metal.

Viewed from the environmental metal aspect, the three key terms are 'species concentrations', 'pathways', and 'political will'. Lead in the environment is the best overall example *(38,41)*. The naturally occurring species in geochemistry and the main threat from leaded gasoline are vastly dissimilar and the commercial hazard has no similarity to those from natural deposits. Whereas, lead production emits $PbSO_4$ and PbO. and twenty years ago, leaded gasoline emitted lead tetraethyl through evaporation and incomplete vehicle engine combustion (cold starts and slow running in car parks accounted for 1-6 % total emissions into the atmosphere), most vehicle exhaust emissions are PbBrCl and its salts combined with NH_4Cl because of halide scavengers in petrol. Each of these lead species hazards needs to be risk-managed differently. Health effects of halides are profoundly more serious than natural sulfate/oxide. All these species eventually reach the water table and are found distributed in the geosphere/biosphere and reactions are modified by the chelates (natural and released) therein.

Laboratory analyses involving separation methods such as ultrafiltration are prone to disturb such labile equilibria and to produce results unrepresentative of the prevailing geochemistry. Computer simulation of these systems, provided that it is well validated and verified with the real scenario, is speedy, easy to use, and can simulate changes in conditions once the parent model has been established (*38*).

Public pressure (which leads to political will) to reduce these emissions from gasoline additives (i) has been based upon total amounts emitted (Table 2), (ii) was based on an understanding that all lead species had brain, renal and skeletal effects, and (iii) once government will had been won over, resulted in a rapid reduction quoted in Table 2. Paradoxically, particles taken into the lungs through a burning cigarette are aerosol species, their absorption is most efficient, and thus maximises damage (so too for other heavy metal aerosols such as environmental cadmium, arsenic *etc*). However, many smokers are insufficiently convinced to cease usage.

Much further back in the food chain, chelates in the soil environment double or even treble tobacco shoot uptake and thence onto the weed being smoked or aerosoled (59). Wenger has given detailed speciation explanations of how the aqueous chemistry of heavy metals is influenced doubly by chelates (i) by desorbing such metals from solids in soil, and (ii) by facilitating plant uptake and translocation.

Localised effluent sites of heavy metal complexes can be specifically targetted with specific bacteria, *eg* LPM-410 (a pseudomonas sp.) and LPM-4 (a microbacterium sp.), which attack metal-EDTA species and works effectively in the laboratory provided that the EDTA bonds have low formation constants (*60*). The pH, climate, and environmental survival of such bacteria has yet to be tested in the field (*61*).

Conclusions

There is a finite threat that future generations may have to face the challenge of living without manufactured chelates as part of their everyday lifestyles! This would be a trajedy brought on, in part, by the misbelief that all human-made compounds are toxic and also the fact that relatively few are purchased direct across a consumer counter; rather, they are used by the manufacturing trade as intermediates and so rarely familiar from their names by the lay-public.

Further, the fact that they are in the environment does not necessarily imply a hazard. However, without a rudimentary knowledge of the power and influence of speciation, lay-persons tend to over-simplify and call for bans. Furthermore, the presence of a chelating agent and its metal complexing dictates that metal-chelate species may have a completely different character of toxicology compared with the uncomplexed ligand.

The present UK government was elected on a manifesto containing more green items than any other previous British parliament – so too for many EC country and US governments! Global environmental management and explanations of its complexity must be an integrated and iterative approach. Scientists are playing with powerful gods should they opt not to participate and must increase their drive to produce more public awareness of speciation and so to set the science into perspective!

References

1. Williams, D. R. *J. Bioinorganic Chemistry and Applications*, **2004** at press.
2. Duffield, J. R.; Williams, D. R. *Chemistry in Britain*, **1989,** 25, 375.
3. *Analysis Using Glass Electrodes*. Linder, P. W.; Torrington, R. G.; Williams, D. R. Open University Press: Milton Keynes, UK, 1984.
4. *The Biological Chemistry of the Elements*, Frausto da Silva, J. J. R.; Williams, R. J. P. Oxford University Press, Oxford, UK. Edn 2. 2001.
5. *Selective Toxicity*. Albert A.A. (1973) Chapman and Hall, London. Edn 5. 2001
6. Sillén, L. G. *Chemistry in Britain*, **1967**, 291-297.
7. Williams, D. R. *Chemistry in Britain*, **1998**, 48-50.

8. Davidge J.; Williams D.R. *Inorganica Chimica Acta,* **2003,** *356*, 109-113
9. Carson, R. *Silent Spring*, Penguin and Houghton Mifflin Publ, New York. 1962.
10. Selborne, J. (Chair), *Annual Report of the UK Chemical Stakeholders Forum*; Department for Environment, Food, and Rural Affairs, Her Majesty's Stationary Office: London 2003. http://www.defra.gov.uk/environment/chemicals/csf/annrpt0102/index.htm
11. *Strategy for a Future Chemicals Policy.* White Paper, Commission of the European Communities, COM 88, 27 Feb 2001. Brussels, 2001. http://europa.eu.int/comm/environment/chemicals/0188-en.pdf
12. Templeton D. M.; Ariese F.; Cornelis R,. Danielsson L-G.; Muntau H, van Leeuwen H.P.; Lobinski, R. *IUPAC Recommendations*; *Pure and Applied Chemistry*, **2002**; *72*, 143.
13. May P.M; Murray K; *Talanta*, **1991** 38, 1409-1418, 1419-1428; **1993**, 40, 819-825.
14. Fillela M; Bugarin M. G.; May P. M. Analyst, **2001**, 126, 2093-2000.
15. Beckett, M. *Speech to Environmental Centre*, South Africa, 2002, 13 March. http://www.defra.gov.uk/corporate/minutes/speeches/mb1403.
16. Weber, C. *The Stockholm Convention (POPS Convention) for the Global Elimination of Extremely Dangerous Pollutants*, PAN, Hamburg, Germany, 2001.
17. *Plant J. (Chair); Williams D. R. et al* Department for the Environment, Food and Rural Affairs, *Annual Report of Advisory Committee on Hazardous Substances*, Her Majesty's Stationary Office, London, UK. http://wwwdefra.gov.uk. 2003.
18. OECD Report. *Testing of Chemicals. Section 3. Degradation and Accumulation.* OECD, Paris, France, 1996
19. *OECD Guidelines for the Testing of Chemicals, Proposal for a New guideline*, OECD, Paris, France 311, October, 2001. http://www.oecd.org/dataoecd/44/40/2741534.pdf
20. Williams, D. R, *What is Safe? The Risks of Living in a Nuclear Age.* Royal Society of Chemistry, Cambridge, 1998.
21. Williams, D. R. *J. Inorg. Biochemistry.* **2000,** *79,* 275-283.
22. Ebdon, L; Pitts, L; Cornelis, R; Crews, H.; Donard, O. F. X.; Quevauviller, Ph.; *Trace Element Speciation for Environment, Food, and Health,* Royal Society of Chemistry, Cambridge, UK. 2001.
23. *Elemental Speciation – New Approaches for Trace Element Analysis;* Caruso, J. A., Ed.; Elsevier, Amsterdam, Holland 1999.
24. Quevaullier, Ph. *Method Performance Studies for Speciation Analysis.* Royal Society of Chemistry, Cambridge, UK. 1998.

25. *Technical Guidance Document in Support of Commission Directive 93/67/EC on Risk Assessment for New Substances* and *Commission Regulation (EC) No 1488/94 on Risk Assessment for Existing Substances,* 4 parts. Office for Official Publications of the European Communities, Luxenbourg, 1996.
26. May P. M.; Murray K. *Journal of Chemical Engineering Data.* **2001**, 46, 1035-1040.
27. Beck, M. T.; Nagypal, I. (Williams, D. R.; Sci. Ed.) *Chemistry of Complex Equilibria.* 1990, Horwood Publ, Chichester. UK.
28. Popov, K. I. and Wanner, H. *Biogeochemistry of Chelating Agents*, Eds VanBriesen, J. M. and Nowack, B., American Chemical Society, 2004. This volume.
29. Motekaitis, R. J.; Martell, A. E. *Determination and Use of Stability Constants.* VCH Publishers, Weinheim, Germany.1988.
30. Powell, K. J. Ed. *IUPAC Stability Constants Database..* Academic Software, Ottley, UK. 2002.
31. Martell, A. E.; Smith, R. M.; Motekaitis. R. J.*; NIST Critically Selected Stability Constants of Metal Complexes Database.* Version 4.0. Texas A&M University, USA. 1997.
32. *HATCHES Thermodynamics Database*, Harwell.UK. see Tweed, C. J.; Cross. J. E.; Ewart. F. T. Radiochimica Acta.1991, 52/53, 421-422.
33. Duffield, J. R.; Williams, D. R.; *Chemical Society Reviews.* 1986, **15,** 291-307.
34. Neil, R.A.; Rose. N. J. *Inorg. Chem.* **1968**, *7(11),* 2405-2413. and US Patent 3158635 (1964).
35. Jones, P. W.; Williams, D.R.; *Inorganica Chimica Acta*, **2002**, *339*, 41-50.
36. *Inorganic and Organic Chemical Co-disposal Ban as of July 2004* http://defraweb/environment/waste/hazforum/index.htm
37. *Speciation of Fission and Activation Products in the Environment;* Editors, Bulman. J. A. and Cooper J. R.; Elsevier, London UK. 1985.
38. Harrison, R. M.; and Laxen, D. P. H.; *Chemistry in Britain*, **1980**, 16-320.
39. Schwarz, K. *Clinical Chemistry and Toxicology of Metals*; Elsevier, North Holland. 1977
40. Fell, G. S. *Chemistry in Britain*, **1980,** *15*, 323-326.
41. Snoddy, M. Environment Agency Data presented to Advisory Committee on Hazardous Substances, 28 April, 2003. mark.snoddy@environment-agency.gov.uk
42. Willet, A. I.; Rittman, B. E. *Proc. Amer Chem. Soc Conference*, New York, **2003**, 588-592.

43. Nőrtemann, B. *Applied and Environmental Microbiology*, **1992**, *58(2)*, 671-676.
44. Henneken, L.; Klűner, T.; Nőrtemann, B. Hempel, D. C. *Applied Microbiology and Biotechnology*; **1998**, *73*, 144-152.
45. Klűner, T.; Nőrtemann, B. *Applied Microbiol and Biotechnology*, **1998**, *49*, 194-201.
46. Hemeken, L.; Nortemann, B.; Hempel, D. C. *Applied Microbiology and Biotechnology;* **1995**, *44*, 190-197.
47. German Federal Institute for Occupational Safety and Health Notification Unit; *Risk Assessment of EDTA*, CAS 60-00-04. 2003;
48. German Federal Institute for Occupational Safety and Health Notification Unit) *Risk Assessment of Na_4EDTA*. CAS 64-02-8. 2003
49. van der Steen, A. T. M. *Proc. Amer Chem. Soc Conference*, New York, 2003, 595-597.
50. Grundler, O. van der Steen, A. T. M. Wilmot, J. *Biogeochemistry of Chelating Agents,* Eds VanBriesen, J. M. and Nowack, B., American Chemical Society, 2004; at press. This volume
51. Kezerian, C. Ramsey W.; US Patent 3158635; 1964.
52. On the State of the Public Health 1995. Chief Medical Officer, Department of Health, HMSO, London. 1996. ISBN 011-321989-X
53. Jones P. W.; and Williams D. R. *International. Journal of Environmental Analytical Chemistry;* **2001**; *81*, 73-88.
54. *EU Legislation for Registration, Evaluation, and Authorisation of Chemicals.* http://europa.eu.int/comm/enterprise/chemicals/chempol/whitepaper/reach.htm. Brussels, 2003
55. Tandy S.; Nowack B.; Wenger K.; Gupta S.; Schulin R. *Proc. Amer Chem. Soc Conference*, New York, **2003**; 612-615.
56. L, Cornelis, R. Crews, H, Donard, O. F. X., and Quevauviller, Ph. *Trace Element Speciation for Environment, Food, and Health*, Royal Society of Chemistry, Cambridge. UK. 2001; 176-187.
57. Connors T. F.; Berta J.; Nascimbeni B.; Case F.; D'Ambrogio B. *Proc. Amer Chem. Soc Conference*, New York, **2003**; 516-521.
58. Emons, H.; in Ebdon, L, Pitts, L, Cornelis, R. Crews, H, Donard, O. F. X., and Quevauviller, Ph.; *Trace Element Speciation for Environment, Food, and Health*, Royal Society of Chemistry, Cambridge. UK. 2001; 188 – 195.
59. Wenger, K.; Gupta, S. K.; Schulin, R. *Proc. Amer Chem. Soc Conference,* New York; **2003**; 621 – 625.
60. Chistyakova T. I.; Dedyukhina E. G.; Satroutdinov A. D., Belikova V. L.; Eroshin V. K. *Proc. Amer Chem. Soc Conference*, New York, **2003**; 666 – 669.
61. van Ginkel, C. G. *Proc. Amer. Chem. Soc.* Conference, New York. 2003, 655 – 658.

Chapter 3

Stability Constants Data Sources: Critical Evaluation and Application for Environmental Speciation

Konstantin I. Popov[1] and Hans Wanner[2]

[1]Physical and Colloid Chemistry Department, Moscow State University of Food Technologies, Volokolamskoye sh., 11, 125080 Moscow, Russia (ki-popov@mtu-net.ru)
[2]Swiss Federal Nuclear Safety Inspectorate, CH–5232 Villigen, Switzerland (hans.wanner@hsk.ch)

Numerous computerized chemical speciation models based on thermodynamic principles are nowadays available to estimate pollutant behavior in the environment. However, these models are sensitive to the reliability of their thermodynamic databases. The quality, accuracy, conflicts and diversity of published stability constants are discussed along with a comparative analysis of critically evaluated thermodynamic data sources. Guidelines that a user would employ in selecting data from the different compilations are proposed. Some applications of chemical speciations for prediction of chelating agent remediation activity in a contaminated soil washing test are considered.

Metal chelation is of great importance in soils because it increases the solubility of metal ions and affects many important chemical and biological processes. It has influence on the availability and mobility of numerous plant nutrients (*1-3*). Besides, the formation of chelated complexes drastically changes the toxicity and sorption ability of both metal ion and ligand as well as the biodegradation rate of complexant *(3-7)*. At the same time the chelating agents used in soil decontamination technologies could have a negative impact on land *(3, 8-20)*.

The toxicities and environmental behavior of pollutants strongly depend on the distribution of metals among the ligands present. The prediction of metal chelate equilibria in soils is based on the use of stability constants in chemical chemical speciation calculations *(1, 2, 5, 10, 12, 19)*. Recently, numerous computer chemical speciation programs, *e.g.*, JESS, MINEQL, WinSGW, MINTEQA2, SPECIES, PHREEQE, GEOCHEM, GEMS-PSI, EQ3/6 are widely available. Figure 1 demonstrates such a speciation for aluminium(III)-EDTA system in a ground water.

Such a speciation gives an opportunity to identify the dominating chemical forms of metal in the solution as well as the pH range of complex formation, and to estimate the relative maintenance of these forms. Moreover, the speciation programs give the log[M] values of free metal ion at any pH, and therefore are able to estimate the masking ability of a ligand.

Unfortunately, there are still abundant recent examples where changes in chelating agent response in the environment are interpreted without due consideration of the role of metal speciation *(5)*. The corresponding research groups still work on the empirical level *(3, 8, 9, 11-18, 20)*. Partly, this situation arises due to a high sensitivity of computer models to the quality of their thermodynamic databases. Another reason for this is associated with an inadequate description of chemical equilibria in a particular system.

It is worthwhile to mention that the chemical speciation diagrams of ligand (cation) forms versus pH in aquatic chemistry have a different meaning being compared with an environmental (geochemical) speciation. The latter indicates the percentage of metal M (ligand L) bound to different soil phases in order to differentiate soluble phase, the exchangeable cations, the carbonate bound *etc*. Thus it likely describes the ***physical*** speciation of a metal *(5)*. The former demonstrates the distribution of a cation (ligand) among the particular ***chemical*** species, both soluble and solid, *e.g.* ML, ML_2, M, M(OH)L, $M(OH)_n^{solid}$ *etc* (electric charges are omitted) .

In soil chemistry, equilibrium constant should correspond to an ionic strength 0.01 mol/L according to 0.003 mol/L $CaCl_2$ solution, which approximates that of many well-drained soils *(1)*. Numerous stability constants data compilations in book and software forms are recently available embracing more than 400 000 equilibrium constants; see for example *(21-32)*. Although most of these data correspond to ionic strengths 0.1-1.0 mol/L, the 0.01 mol/L values can be calculated from the former ones using SIT *(33)* or other approximations. The Davies equation may be used for corrections at ionic strength below 0.1 mol/L. For the prediction of relative leaching ability of pollutant by different completing agents in soils, the data obtained at I=0.1 mol/L are also suitable.

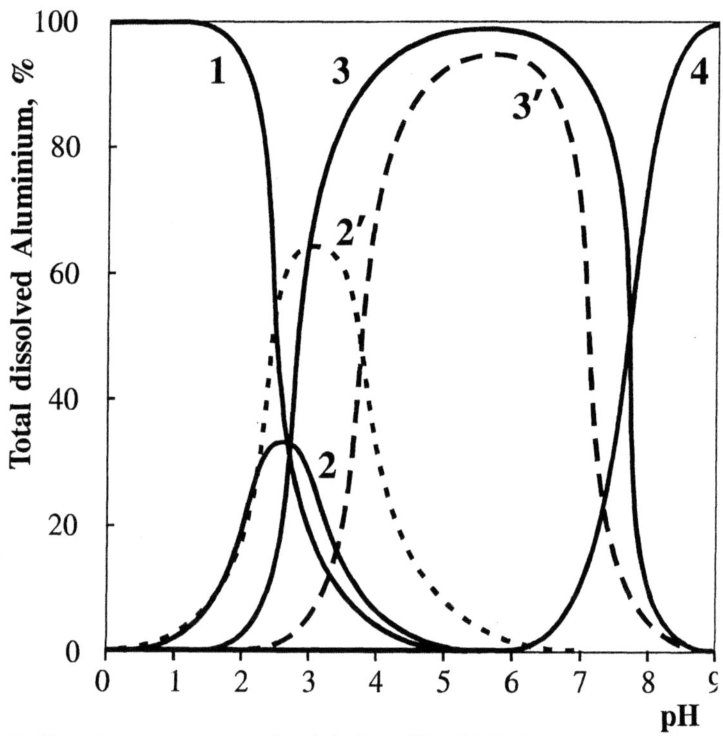

Figure 1. Aluminum speciation for 0.005 mol/L $Al(NO_3)_3$ aqueous solutions in presence of 0.005 mol/L EDTA at 25 °C

(Solid curves indicate the speciation done for $\log\beta_{Aledta}$ = 16.7. The dominating species are Al^{3+} (1), [AlHedta] (2), $[Aledta]^-$ (3) and $[Al(OH)_4]^-$ (4). The dashed lines demonstrate the speciation of the same system if $\log\beta_{Aledta}$ is reduced from 16.7 to 15.3, while all other constants are the same: AlHedta (2′), $Aledta^-$ (3′); the corresponding curves for Al^{3+} and $Al(OH)_4^-$ are not listed)

The purpose of this review is to characterize and to compare the different thermodynamic data sources and to provide some practical guidelines on how to avoid speciation-related errors and subsequent faulty interpretations.

Stability Constants Diversity

Although true equilibrium in soils is hardly possible due to the long equilibration time, the stability constants based approach appears to be quite adequate and fruitful if the competing equilibria, which include also numerous nonchelated species, are considered *(1)*. The main problem here is mostly associated with the choice of correct stability constants among those published. Indeed, Table I shows that conflicting data are quite abundant in the open literature, and discrepancies of as many as ten orders of magnitude may arise. This drastic data diversity arises mostly from the inadequacy of the experimental method applied and the lack of thermodynamic rigor in treating the equilibria under the study. At first glance the data contradiction could leave an impression that reliable measurement of stability constants is hardly possible. However, an experimental benchmarking undertaken within seven experienced research groups for the nickel - glycine system resulted in an excellent agreement of the measured values, cf. Table II *(34)*.

Table I. Comparison of EDTA and Sulfate stability constants, measured by independent research groups at 20-25 °C, in 0.1 M KNO_3

EDTA			*Sulfate*		
Cation	Research group	$\log\beta_{ML}$	Cation	Research Group	$\log\beta_{ML}$
	1	25.10		1	1.53
Fe(III)	2	24.95	Fe(III)	2	1.65
	3	25.10		3	2.36
	4	25.10		4	2.66
	$\Delta\log\beta_{ML}\max = 0.15$			$\Delta\log\beta_{ML}\max = 1.13$	
	1	23.40		1	1.78
Cr(III)	2	23.1	In(III)	2	1.79
	3	12.8		3	2.0
	$\Delta\log\beta_{ML}\max = 10.6$			$\Delta\log\beta_{ML}\max = 0.22$	
Th(IV)	1	21.17	Th(IV)	1	3.32
	2	25.3		2	3.28
	$\Delta\log\beta_{ML}\max = 4.1$			$\Delta\log\beta_{ML}\max = 0.04$	
	1	8.0			
Sr(II)	2	8.53			
	3	8.60		1	1.34
	4	8.63	Ca(II)	2	1.40
	5	8.70		3	1.49
	6	8.8		4	1.54
	$\Delta\log\beta_{ML}\max = 0.8$			$\Delta\log\beta_{ML}\max = 0.20$	

NOTE: $\Delta\log\beta_{ML}\max$ indicates the maximal diversity of logarithmic values among the research groups; SOURCE: Data from IUPAC Data base *(32)*.

It can be concluded that reliable experimental procedures and calculation methods exist for the characterization of all type of complexes *(37)*.

However, the selection of proper values requires profound knowledge of the chemistry of complex equilibria.

Table II. Nickel – Glycine Project; 25 °C, 1.0 mol/L NaCl

Research Group	*log* β_{HL}	*log* β_{H2L}	*log* β_{NiL}	*log* β_{NiL2}	*log* β_{NiL3}
1	9.629	12.036	5.80	10.588	14.308
2	9.67	12.14	5.53	10.26	13.59
3	9.654	12.067	5.625	10.357	13.75
4	9.656	12.076	5.625	10.398	13.911
5	9.652	12.109	5.60	10.325	13.65
6	9.659	12.071	5.625	10.381	13.805
7	-	-	5.66	10.43	14.08
Mean	9.659	12.083	5.638	10.391	13.922
S.d.	0.011	0.031	0.071	0.089	0.191
Δ max	0.041	0.104	0.27	0.328	0.558

SOURCE: Data from Reference 34, Copyright 1987 IUPAC

Critical and Noncritical Stability Constants Compilations

This situation stimulated the IUPAC Commission on Equilibrium Data to start publication of a continuous series "Critical Surveys of Stability Constants of Metal Complexes" in 1975 with the aim to evaluate the most reliable equilibrium constants from the available literature data *(35)*. Each survey was prepared by an expert, working actively in the field of thermodynamics of complexes, and commented on by the members of the Commission on Equilibria Data. However, it obviously follows from the nature of the data considered, that even the recommended values cannot be regarded as "final" ones and further research may change its rank, embracing the four categories as defined by IUPAC: recommended, tentative, doubtful and rejected. The authors' responsibility was to state clearly why certain data are regarded as reliable and why other data are rejected. At the same time, this best up-to-date IUPAC evaluation procedure still remains somewhat arbitrary because it relies largely on "expert opinions".

According to the first IUPAC regulations, the data could be recommended (***R***) if the results of at least two independent research groups are in good agreement; if the surveyor has no doubt as to the adequacy of the applied experimental and calculation procedure; if the consideration of the activity-concentration relation is correct, and the standard state is unambiguous. The given error of such a constant must be no less than ±0.05 logarithmic units. Data could be regarded as tentative (***T***) if all the conditions mentioned for R category are fulfilled, except the first one; or if the surveyor observes some deviation from the necessary rigorousness, but this probably caused no serious mistakes. The given error of such a constant must not exceed ±0.2 logarithmic units.

Data should be considered as doubtful (***D***) if the surveyor suspects some errors in the evaluation of the constants, which are nevertheless of semi-quantitative value. The probable error of such a constant should not exceed ±0.2 logarithmic units. Data determined by either an inadequate method, or under undefined conditions should be rejected (***Rj***). The same catergory is given if any serious objection is found in the evaluation.

The reliability of the data is normally revealed by the care with which the reaction conditions and measurements are controlled and described. Papers deficient in specifying the measurement conditions (*e.g.*, temperature, ionic strength, nature of supporting electrolyte), the purity of reagents, the details of the calibration procedure and the consideration of possible side reactions, are eliminated from the evaluation. In addition, the adequacy of the applied method is thoroughly analyzed. The agreement of data from different sources is a valuable verification tool, but is not considered of main priority. Cases are known where several published data showed excellent agreement but were nevertheless rejected *(55)*. Moreover, sometimes one and the same value given by one research group is recommended, and by another one is rejected due to an inadequate data presentation *(51)*.

As can be seen from Table II, even the highly experienced laboratories fail to reach the ***R*** level of agreement for Ni-complexes. Moreover, the standard deviation value for NiL_3 is beyond the error threshold for Tentative data. Indeed, the increasing complexity of particular species leads normally to higher errors. This could be clearly illustrated by a comparison of nickel NiL, NiL_2 and NiL_3 complexes. Generally, the protonation constants of a ligand H_iL ($\log\beta_{HiL}$) are more accurate than the formation constants of the corresponding complex ML_i because the calculation of the latter one includes inevitably the experimental values of the former species. Thus a report on ML complex stability constant of H_iL with an accuracy better than the standard error of $\log\beta_{HiL}$ indicates likely the reproducibility of experimental data, but not the real range of confidence. For example, for IDA and MIDA the precision of stability constants should be approximately as high as that found for Ni(II)-glycine *(34)*. For EDTA, DTPA, DOTA, TETA and TTHA it is significantly lower. For example an evaluation of expected uncertainty of the constants for $[(UO_2)_2dtpa]^-$, measured by direct potentiometric titration, gave a value of 0.36 logarithmic units *(36)*.

In general, the constants of low stability complexes reveal a better accuracy and agreement than those of high stability complexes. This can be explained by the fact that the stability of "weak" complexes can usually be determined directly in one or two steps, while for the "strong" complexes indirect methods are necessary requiring the use of competing ligand or cation

(37, 38). For example, easily hydrolyzed cations such as Al(III), In(III), Tl(III), Bi(III), Zr(IV), Th(IV), Pd(II) *etc.* are strongly complexed by complexones. K_{ML} values for many of these metals range between 10^{30} and 10^{40}. In these cases the direct potentiometric or spectrophotometric titrations cannot be readily used and measurements with competitive ligands, *e.g.* 2,2′,2″-triaminotriethylamine (tren), N,N′-di-(2-hydroxybenzyl)-diaminoethane-N,N′-diacetic acid (HBED), or cations (Hg^{2+}, Ga^{3+}, Cu^{2+}) are required. This introduces additional systematic errors, associated with protonation and complex formation of the competing ligands and cations. Similar problems can arise due to fairly slow kinetics of ligand-ligand displacement reactions, *e.g.* for Fe(III) complexonates equilibrium in most cases is established in a few days *(38)*.

In the late 1990-ies the IUPAC Commission adopted "milder" and more flexible requirements for evaluation. The four categories have been replaced by only two: recommended (***R***) and provisional (***P***), which approximately correspond to the ***R*** and ***D*** levels of the former ones: for ***R***-level s.d. ≤0.05 for H-complexes (*e.g.* H+L) and ≤0.1 for metal-complexes (M+L, H+ML or M+HL); for ***P***-level: 0.05<s.d.≤0.2 for H-complexes and 0.1< s.d. ≤0.2 for metal-complexes. Within the last 30 years eighteen IUPAC critical reviews *(38-58)* have been published, cf. Table III. Besides, an independent high quality critical evaluation of hydrolysis constants published up to 1974 was undertaken and is widely recognized *(27)*.

In 1986, the Nuclear Energy Agency (NEA) of the Organisation for Economic Co-operation and Development (OECD) undertook the development of a critically evaluated chemical thermodynamic data base for a number of elements of interest to nuclear technology, in particular to the performance assessment of geological repositories for spent nuclear fuel and radioactive waste (*59, 60*). This project applies rigorous evaluation techniques such as a uniform method to treat ionic strength dependencies and uncertainties. The recommended data refer to the standard state and to 25 °C, and the data analysis and review process is made entirely transparent in the review reports. Five such reviews have been published to date on U, Np, Pu, Am and Tc (*61-65*), and critical reviews on Ni, Zr, Se, Fe, Th, Sn, Mo, and selected organic ligands are being carried out at present.

Such evaluations are very valuable, insofar as they pass rather substantial multilevel expertise, give the full reference list, a substantial characterization of data quality and rather clear reasons for recommendation or rejection of a particular constant. Indeed, they can be treated as the most reliable thermodynamic data source and used as milestones for other data comparison. A clear advantage of the IUPAC and the NEA approaches is also associated with the evaluation transparency. Unfortunately, the preparation of these reviews appeared to be extremely time consuming. As can be seen from Table III, recently the rate of IUPAC data evaluation is lagging far behind to rate of original thermodynamic data publication. Thus the ligands considered in *(38-58, 61-65)* represent only a small fraction of those described in the literature and offer totally *ca.* 4000-5000 values. The IUPAC surveys also leave many decisions unresolved. In particular, the evaluation of environmentally important systems involving sulphate, phosphate and carbonate anions and environmentally common cations has been started only recently *(58)*.

A somewhat alternative approach to create a critical and unique compilation of metal complex equilibrium constants was developed at Texas A & M University *(26, 66)*. The authors have produced six volumes of critical stability constants, which cover the publications that have appeared in the literature through 1985 *(26)*. Later all the compiled critical data with some additional constants have been placed on computer disks and are now available as an upgraded extended version: NIST Data base *(30, 31)*. Due to elimination of the time consuming external expertise of the selected data and to the much more flexible and less rigid selection criteria than in IUPAC series, the NIST Data base recently covers the 47,950 stability constants of around 4500 ligands and provides the users with one value for each aqueous compile at standard conditions.

The basic criteria employed for selecting "Critical Stability Constants" are generally the same as in the IUPAC Series as far as an essential experiment control is concerned. But unlike IUPAC, the NIST reviewers selection may be guided by a comparison with values obtained from other metal ions with the same ligand and with values obtained from the same metal ion with other ligands. Besides, when there is poor agreement between published values, and comparison with other metal ions and ligands does not suggest the best value, the results of more experienced research groups who have supplied reliable values for other ligands are selected. Moreover, when such assurance is lacking, the preference could sometimes be given to values reported by an investigator who has published other demonstrably reliable values obtained by the same methods *(66)*.

Another important difference between IUPAC and NIST critical evaluations is associated with single investigators. The values reported by only one investigator are included in the NIST Data base unless there is some reason to doubt their validity *(66)*. In the IUPAC data series such data could be treated as provisional (***P***-level), but not as recommended. The NIST Data base normally does not indicate the data category, although the data listed there in brackets could be treated as doubtful. The references in *(26,30)* are not linked to any particular stability constants, so it is not possible to check on individual values; many do not give any stability constants.

The most comprehensive computerized stability constants database – IUPAC Data base has recently been developed by IUPAC and Academic Software *(32)* (cf. Table IV). It embraces most of the papers published in the

Table III. IUPAC Critical Evaluations of Stability Constants 1977-2004

Year	*Ligand (Metal)*	*Authors/Reference*
1977	EDTA	G. Anderegg *(39)*
1978	1,10-phenanthroline, 2,2′-Bipyridyl and related compounds	W.A.E. McBryde *(40)*
1979	8-hydroxyquinoline	J.Stary,Y.Zolotov, O.M. Petrukhin *(41)*
1980	Fluoride	A.M.Bond, G.T. Hefter *(42)*
1982	NTA	G.Anderegg *(43)*
1982	Acetylacetone	J.Stary, J.O.Lilenzin *(44)*
1983	Indium	D.G.Tuck *(45)*
1984	Histidine, Phenylalanine, Tyrosine, L-DOPA, Tryptophan	L.D.Pettit *(46)*
1984	Ethylenediamine	P.Paoletti *(47)*
1987	Cyano complexes	M.T.Beck *(48)*
1991	Nucleotide complexes	R.M.Smith, A.E.Martell, Y.Chen *(49)*
1991	Glycine	T.Kiss, I.Sovago, A.Gergely *(50)*
1993	Alanine, Valine, Leucine	I.Sovago, T.Kiss, A.Gergely *(51)*
1995	Cysteine, Cystine, Methionine, Serine, Threonine, Asparagine, Glutamine.	G.Berthon *(52)*
1996	Agrinine, Lysine, Ornithine	O. Yamauchi, A.Odani *(53)*
1997	Salicylic acid and other aromatic hydroxycarboxylic acids	L.H.J.Lajunen, R.Portanova, J.Piispanen, M.Tolazzi *(54)*
2001	Phosphonic acids	K.Popov, H.Rönkkömäki, L.H.J.Lajunen *(55)*
2003	Crown ethers	R.Delgado, F. Arnaud-Neu, S. Chaves *(56)*
2003	Aliphatic hydroxo-acids	L.H.J.Lajunen, R.Portanova *(57)*
In press	IDA, MIDA, DTPA, TTHA, DOTA, NOTA, TETA	G.Anderegg, R.Delgado, F.Arnaud -Neu, J.Felcman, K.Popov *(38)*
In prep.	Cu(II),Zn(II),Cd(II), Hg(II), Pb(II) with OH^-, Cl^-, CO_3^{2-}, PO_4^{3-}, SO_4^{2-}	P.L. Brown, R.H. Byrne, T. Gajda, G. Hefter, K.J.Powell, S. Sjöberg, H. Wanner *(58)*

NOTE: MIDA – (Methylimino)diacetic acid; TTHA – 3,6,9,12-tetrakis(carboxymethyl)-3,6,9,12-tetraazatetradecanedioic acid; NOTA-2,2',2"-(1,4,7-triazonane-1,4,7-triyl)triacetic acid; TETA – 2,2',2",2'"-(1,4,8,11- Tetraazacyclotetradecane-1,4,8,11-tetrayl)tetraacetic acid; DOTA - 2,2',2",2'"-(1,4,7,10-Tetraazacyclododecane-1,4,7,10-tetrayl)tetraacetic acid twentieth century in the field of metal complex equilibria (stability constants, enthalpies, entropies and free energies) for both aqueous and non-aqueous media. The recent upgraded versions include the data through 2003.

The data sources include also the Royal Society Compilations *(21-24)* and all the IUPAC critical data series *(39-57)* except those given in *(26,30)*. Along with non-critical data, the IUPAC critically evaluated ***R***-values are also indicated. The user can therefore make an independent comparison of data from original publications with those recommended by IUPAC. Unfortunately, the ***T*** and ***P*** levels of critically evaluated data are not indicated. In all over 400 000 stability constants for over 8570 ligands are now present in *(32)*. The IUPAC Data base has a built-in speciation software (SPECIES), a possibility to make temperature and ionic strength corrections and many other useful options. Among these, there is a very useful option to select the highest log*K*, lowest log*K* and log*K* range.

Table IV. Comparison of Databases: general overview

Characteristics	*NIST (31)*	*IUPAC Data base (32)*	*JESS (67)*
Number of Constants	47,951	over 400,000	188,352
Number of Ligands	4,500	8,570	over 3,000
Reference year range	Not indicated	1877-2003	Not indicated
Number of References	17,600	21,588	11,500 [a]
logβ values	one for each M-L combination	All published data	All published data + recommended one for each M-L combination
Data	Critical	Both critical and noncritical	Both critical and noncritical
Non aqueous solutions	No	Yes	Yes
Mixed solvents	No	Yes	Yes
IUPAC Critical Data	Not indicated	Indicated	Not indicated

[a] Recently an updated version has over 17,000 references

Another critical data compilation, Joint Expert Speciation System (JESS), represents some alternative to the NIST Data base *(31)*. JESS *(67)* offers the critically assessed constants, although the evaluation process is not entirely uniform. Mainly, the consistency between thermodynamic parameters for the same reaction and for other equivalent linear combinations of reactions is considered. Also the reputation of the source is taken into account when there is no other data for comparison. It is the job of the JESS code to take all this information and to produce from it a thermodynamically consistent calculation. Importantly, these decisions to eliminate data remain clearly marked in the database. An expert system makes assessment, embodied only in the weight given to each datum, shown as 'Wgt' with a scale ranging from 0 (wrong values) to 9 (most reliable constants). By using a process that is entirely automatic from the (inconsistent) database onward, JESS is able to identify the effects of the many assumptions that are involved in achieving thermodynamic consistency. Recently JESS is one of the largest data bases used directly in speciation calculations. Although the JESS system reveals really good results *(68)*, the validity of approach developed in JESS has to be proved in some circle tests and critical data comparison to be run in future.

Besides the comprehensive and international data bases mentioned above, there are some application specific data bases *(70-72)*. Obviously, these have considerably smaller data bases. Some concentrate on elements relevant to radioactive waste management such as Al, Am, Eu, Fe, Np, Ni, Pd, Pu, Se, Si, Tc, Th, Sn, U and Zr *(70)*, others have the specialty of considering natural organic chelators (fulvic and humic acids) *(69)*. In the field of implemented stability constants, MINTEQA2/PRODEFA2 *(69)* gives priority to NIST data *(30)* with very few values taken from IUPAC Database and original papers. In contrast, the Nagra/PSI data base *(70)* relies to a large extent on the values recommended so far by the NEA *(61-65)*, and in addition performs critical reviews of original experimental papers. The Nagra/PSI data base exists in two formats, one of them consisting of equilibrium constants which can be used by conventional speciation modelling codes, the other consisting of SUPCRT-based *(73)* Gibbs energies of formation of all contained chemical species. The latter has the advantage to be tailor-made for the use in Gibbs energy minimization (GEM) codes. GEM codes facilitate the calculation of multi-component systems and can thus be easily used to calculate chemical adsorption and to model solid solutions *(74)*. It should be mentioned that MINEQL+ *(72)* has some definition errors, which are analysed in *(5)*.

Critically selected Stability Constants Comparison

The large number of critical stability data sources justifies some comparison. A comprehensive comparison of data presented in *(30, 31, 67)* with those in the IUPAC Critical Data series is hardly possible in the framework of the current chapter, but some observations for selected chelating agents such as amino acids and phosphonates can be presented. The NIST data base became recently the most widely accepted one, and the corresponding data are frequently used blindly without any reasonable doubt. Therefore, a special emphasis on the data quality of particularly this data base is done in a present paper.

As can be seen from Table V, the numerical values of the stability constants selected by NIST and JESS agree in general with those proposed by IUPAC. The NIST, JESS and IUPAC reviewers similarly recognized the data on Bi(III)-TTHA and Fe(III)-glycine as doubtful, and reveal an excellent agreement for H, Zn(II)-glycine.

Table V. Comparison of Critical Data Quality for Amino Acids ($\log\beta_{ML}$; 20-25 °C, I= 0.1-0.2 mol/L) [a]

Ligand	*Cation*	*NIST (30)*	*IUPAC critical review* [b, c]	*JESS (67)* [c,d]
Glycine	H^+	9.58	9.60(***R***)	9.62 (9)
	Mg^{2+}	1.66	Doubtful	1.9 (4)
	Sr^{2+}	0.6	Doubtful	0.4 (2)
	Fe^{3+}	Doubtful	Doubtful	10.4(2)
	Zn^{2+}	4.96	5.03 (***T***)	5.1(8)
	Hg^{2+}	10.3	Doubtful	4.1(2)
	Cd^{2+}	4.25	4.28 (***T***)	4.2(6)
DOTA	K^+	1.6	1.6 (***R***)	1.4 (3)
	Be^{2+}	13.2	Doubtful	13.4(2)
	Cu^{2+}	22.7	22.3 (***R***)	22.2(3)
	Fe^{3+}	29.4	Doubtful	29.0(2)
TTHA	Na^+	Missing	1.0 (***P***)	Missing
	Fe^{3+}	26.8	Doubtful	27.3 (2)
	Ag^+	8.7	8.7(***P***)	8.7 (2)
	Tl^+	4.91	4.9 (***P***)	Missing
	Zn^{2+}	17.9	18.1(***P***)	18.1(2)
	Cd^{2+}	18.9	18.7(***P***)	18.58 (3)
	Hg^{2+}	26.9	28.1(***P***)	25.2 (2)
	Al^{3+}	21.1	Doubtful	19.76 (2)
	Bi^{3+}	Doubtful	Doubtful	17.3 (2)

[a] Some data refer to I = 1 mol/L and I = 0.5 mol/L; [b] Sources: glycine *(50)*, DOTA, TTHA *(38)*; the DOTA and TTHA have been taken for comparison for the reason that the corresponding critical evaluation is not yet published and therefore could not be used by *(30,67)*; [c] figures in brackets indicate critical evaluation rank; [d] critical evaluation rank ranging from 0 to 9

Meanwhile the Hg(II)-TTHA stability constants indicate a rather large disagreement within the three evaluations. The reasons for preference of the value of 28.1 over 26.9 and 25.2 by IUPAC are discussed in *(38)*. However, the IUPAC and JESS reviewers appear to be more critical and rigorous in general relative to NIST, as several constants recommended by *(30)* were considered as doubtful by IUPAC and JESS, *i.e.*, those for the glycine complexes of Mg^{2+}, Sr^{2+}, Hg^{2+}, the DOTA complexes of Be^{2+} and Fe^{3+}, and the TTHA complexes of Fe^{3+} and Al^{3+}. JESS reveals even more rigor (and/or caution) treating most of glycine and all TTHA, DOTA constants as doubtful ('Wgt' = 2-3).

At the same time, a comparison of Glycine complexes evaluation done in *(30)* and in *(67)* for Zn(II), Cd(II), Hg(II) group, indicates an evident error in NIST evaluation: recommendation of anomalously high value for Hg(II)-Glycine complex. Meanwhile both IUPAC and JESS reviewers estimate the corresponding measurements as doubtful, and JESS offers a more realistic value. On the whole the agreement between IUPAC and JESS evaluations is much better than with NIST.

For phosphonates, NIST has recommended some values, while IUPAC has rejected them, cf. Table VI. Indeed, the phosphonate anion of NTPH and

Table VI. Comparison of critical data quality for phosphonates, $\log\beta_{ML}$ (20-25 °C, I= 0.1 mol/L)

Ligand	*Cation*	*NIST (30)*	*IUPAC critical reviews (55)*
IDMP	H^+	$\log\beta_{HL}$ = 10.79 Recommended	$\log\beta_{HL}$ = 10.79 Rejected
NTMP	H^+	$\log\beta_{HL}$ = (12.4) Doubtful	$\log\beta_{HL}$ = (12.4) Rejected
EDTMP	H^+	$\log\beta_{HL}$ = (13.0) Doubtful	$\log\beta_{HL}$ = (13.0) Rejected
DTPMP	H^+	$\log\beta_{HL}$ = (12.0) Doubtful	$\log\beta_{HL}$ = (12.0) Rejected
	Mg^{2+}	$\log\beta_{ML}$ = 6.4	Rejected
	Ca^{2+}	$\log\beta_{ML}$ = 7.1	Rejected
	Co^{2+}	$\log\beta_{ML}$ = 11.1	Rejected
	Zn^{2+}	$\log\beta_{ML}$ = 15.7	Rejected
	Cu^{2+}	$\log\beta_{ML}$ = 19.5	Rejected
	Ni^{2+}	$\log\beta_{ML}$ = 13.4	Rejected

EDTPH is protonated at a very high pH (pK >> 12). Therefore the glass electrode technique used in publications recommended in *(30)* could not be adequate and was rejected by IUPAC reviewers *(55)*. Later re-estimation of these constants by NMR technique *(75,76)* has proved the validity of this decision. For DTPPH it is essential to verify the purity of the ligand. However, this had not been done in any of the publications. This was the reason why the IUPAC reviewers rejected all the data, while NIST recommends values that refer to the complexing ability of a mixture of DTPPH and its underphosphorilated derivatives, rather than to pure DTPPH.

Another example of an obvious reviewer's error` is given in Table VII. The formation constant of K^+ ionic pair with hydroxyl cannot be lower than that of Na^+. Therefore the data proposed by Baes and Mesmer *(27)* clearly give the correct trend ($\beta_{KOH} < \beta_{NaOH}$), unlike the NIST compilation, which suggests the reverse order: $\beta_{KOH} > \beta_{NaOH}$ *(30)*.

Table VII. Comparison of critical data quality for alkali metal hydroxocomplexes

Cation	logβ*(MOH)*	
	(30) ***I***= 0.1 mol/L, 25 °C	*(27)* ***I***=1.0 mol/L, 25 °C
Na^+	-0.1	-0.5
K^+	0.0	-0.8

The examples given above demonstrate that the values recommended in the NIST Data base *(30)* are not free of errors and have to be used with caution in spite of the fact that they are called "critical" by their authors. The structure of the NIST data base provides inability of the user to correlate the recommended constants with original references from which they have been taken. Actually errors are unavoidable part of data bases. The sources of errors are numerous: besides the issues of expert errors, the transcription of data from original papers and other sources to the data base may lead to misprints. In fact, any new version of data bases reports on substantial corrections *(30, 31, 32)*. Another source of corrections arises due to new experimental studies requiring a continuous updating of the older data. The NEA has recognized this need and decided to carry out periodical critical updates of its databases. The first update on uranium, neptunium, plutonium, americium and technetium thermodynamics *(65)* has been published recently and supersedes the earlier critical reviews on these elements *(61-64)*. The users of such data are thus strongly discouraged from using such compilations blindly and without a reasonable control. Generally, a potential drawback of any critical compilation is the occasional exclusion of important equilibrium constants and complexes due to low quality of the source data.

Some impression on the agreement degree of some specific data compilations is presented by the data given in Table VIII. Generally, these data reveal a reasonable agreement within the experimental errors.

Chelating Agents Speciations

The problems stated above should be accounted for in environmental modeling. It is reasonable to note that IUPAC has developed electronic tutorials *(77)* for environmental applications of stability constants as an independent supplement to the IUPAC Data base. This is a valuable introduction to the field for an inexperienced user.

Table VIII. Comparison of Critical Data for Hydroxocomplexes ($\log\beta_{MOH}$; 25 °C, I = 0 mol/L)

Cation	*NIST (30)*	*NEA (65)*	***Nagra/PSI** (70)*
Fe^{3+}	11.81	-	-
Al^{3+}	9.0	-	-
Am^{3+}	7.5	6.8 ± 0.5	6.7 ± 0.3
Th^{4+}	10.8	-	11.6 ± 0.5
Pu^{4+}	- [a]	14.6 ± 0.2	13.22 ± 0.60
U^{4+}	13.48 ± 0.1	13.46 ± 0.06	13.46 ± 0.06
UO_2^{2+}	8.18 ± 0.1	8.75 ± 0.24	8.8 ± 0.3

[a] NOTE: The data are missing only for this particular ionic strength (I = 0)

Although there are some compilations that present the selected stability constants values (*26, 27, 30, 31, 38-52, 61-64, 67, 68*), for a proper environmental speciation a careful comparison of data from different sources including original publications is still strongly recommended. In some cases the knowledge of a critical value is not needed, and even the scientists who are not sufficiently expert in the field of complex equilibrium could estimate the degree of confidence by trying speciation with the conflicting values.

The following case study illustrates how conflicting data can be qualified and judged: the speciation of 0.005 mM Al(III) in a ground water in the presence of an equimolar content of EDTA at pH 6 is to be modelled. For $[Aledta]^-$ (ML complex) the IUPAC full database *(32)* gives five different $\log\beta_{ML}$ values, measured at I= 0.1 - 0.2 M and 20-25 °C: 15.3, 16.01, 16.11, 16.13, 16.7 with a maximal difference of 1.4 logarithmic units (IUPAC critical evaluation *(39)* qualifies $\log\beta_{AlL}$ =16.7 as "tentative"). The pK values for AlHedta are in better agreement, ranging within 2.0; 2.63 and 2.77. Taking the IUPAC recommended pK values for EDTA *(39),* the Al(III) hydrolysis constants which include the species $AlOH^{2+}$, $Al(OH)_2^+$, $Al_3(OH)_6^{3+}$, $Al_{13}(OH)_{32}^{7+}$, $Al(OH)_3$, $Al(OH)_4^-$ and a solubility product of $Al(OH)_3$ from *(77),* as well as an intermediate [AlHedta] pK value (2.63), one can easily construct a speciation diagram with the extreme ML formation constants: 15.3 and 16.7. Figure 1 represents such a speciation (solid lines) for $\log\beta_{AlL}$ =16.7 done with a software

'Species' *(77, 32)* (Al(III) complexes, such as $[Al(OH)edta]^{2-}$ and $[Al(OH)_2edta]^{3-}$ are omitted here for simplicity).

As can be seen, almost 100% of aluminum at pH 6 is bound by EDTA in the soluble complex $[Aledta]^-$. Therefore EDTA dissolves solid (colloidal) $Al(OH)_3$, which dominates in aluminum solutions in fresh water at this pH almost completely. The 1.4 logarithmic units change of stability constant from 16.7 to 15.3 (dashed lines, the corresponding curves for Al^{3+} and $Al(OH)_4^-$ are not indicated) results in no drastic variation of particle distribution. At pH 6 almost 98% of aluminum is bound by EDTA as $[Aledta]^-$ species. Thus the conflict in the Al-EDTA stability constants, which appears substantial at first glance, is absolutely not important in this particular case, and the large data uncertainty does not affect the final conclusion on the role of EDTA in aluminum complexation.

A significant change takes place only within pH 7-8, but aluminum in any case remains soluble under these conditions, either as $[Aledta]^-$ or as $Al(OH)_4^-$ particles. The reduction of the stability of $[Aledta]^-$ after lowering its formation constant is compensated by the relative increase of the AlHedta content. Therefore, the sum of the $[Aledta]^-$ and AlHedta species covers almost the same pH range, despite the 1.4 log units difference in the constant for $[Aledta]^-$. In order for the integral stability field of the Al-EDTA complexes to decrease substantially, a significantly larger difference in the mentioned constant would be required.

Turning back to Table I, it is not too difficult to declare the $\log\beta_{CrEDTA}$ value of 12.8 as an outlier, in view of the good agreement of the other values and the comparison with Fe(III)-EDTA, which should be of the same order of magnitude as the Cr(III) complexes. Ideally, however, the reviewer should find out why the value is so different. Perhaps it is only an erroneous evaluation which can be corrected, so that the corrected value can be used in the analysis. In this respect, another reason for discrimination is the very slow chromium equilibration time that frequently leads to an underestimation of the stability constant. Thus the user is left with the two $\log\beta_{CrEDTA}$ values of 23.40 and 23.1, which are in rather good agreement.

The use of stability constants and chemical speciation is a valuable tool for the understanding of the role of chelating agents in soils, and for the prediction of their leaching ability. Table IX presents a comparison of complexone chelating and leaching ability. It combines the results of batch soil washing experiments with the metal speciation calculated by us with the program 'Species' *(77)*.

It is generally assumed that those complexants that have the highest affinity towards pollutants (highest $\log\beta_{ML}$) provide the best extraction. Indeed, for the Pb-NTA and Pb-EDTA systems this assumption seems to be valid, but for the EDTA/DTPA pair the ***reversed*** order is observed. The situation becomes clearer when the speciation is modelled and equilibrium concentrations of free metal ions are compared. In Table IX a qualitative correlation of -log[M] and pollutant removal is observed: the lower free metal ion equilibrium concentrations correspond to a better extraction. Indeed, such a correlation is observed for EDTA/DTPA comparative leaching ability towards Pb(II), and Zn(II), cf. Table IX. This proves the validity of simple speciation model for a prediction of relative chelating agent leaching ability. At the same time, for the extraction of Cd(II) the reversed order still remains for EDTA and DTPA chelators, although within the error of the stability constants measurement. Probably, more species and equilibrium constants should be accounted in order to obtain a better agreement.

Table IX. Pollutant Removal from Soils with Chelating Agents for pH 6 and Calculated Free Metal Ion Equilibrium Concentrations [a]

M	*L*	*M Removal, %*	*pH*	*Ref.*	$\log\beta_{ML}$ [b]	*-log[M]* [a, b] pH 6
Pb	NTA	30	6.0	*(17)*	11.4	5.35
	EDTA	80			18.3	8.32
Pb	NTA	5	6.0	*(3)*	11.4	5.35
	EDTA	65			18.3	8.32
Pb	EDTA	42	6.2	*(16)*	18.3	8.32
	DTPA	9			18.9	7.30
Cd	EDTA	76		*(16)*	16.62	7.5
	DTPA	40			19.3	7.6
Pb	EDTA	99	4-8[c]	*(15)*	18.3	8.32
	DTPA	96			18.9	7.30
Zn	EDTA	90	4-8[c]	*(15)*	16.68	7.57
	DTPA	75			18.6	7.23

NOTE: [a] the total [M] and [L] in speciation experiments are assumed to constitute 0.001 mol/l (-log[M]=3.0); [b] calculations are done with stability constants data taken from IUPAC critical evaluations: NTA *(44)*; EDTA *(39);* DTPA *(38)*; [c] pollutant extraction was pH independent *(15)*

The differences in leaching tests depend on absolute metal content, chelating agent concentration, type of soil, contaminant location, contaminant interaction with soil ingredients, contact time and many other parameters.

Therefore, the results of independent research groups can be significantly different (see NTA/EDTA comparative tests in Table IX presented in *(3, 17)*). It should be noted that the variations of leaching conditions might strongly influence the free metal ion concentration and therefore the remediation results. For example, if the ligand-to-metal ratio is changed from 1:1 to 2:1, then NTA reduces free Pb(II) concentration to $\log[Pb^{2+}] = -7.69$, while EDTA reduces it to $\log[Pb^{2+}] = -13.6$. Thus the difference in relative remediation ability becomes more pronounced.

Besides the issues of proper selection of reliable stability constants among the published values and an adequate approach to the prediction of batch leaching experiment results, the problem of missing data is also important. It is most probable that for a large number of complexes that may be formed in the environment, experimental data are still missing. In particular, this concerns ternary species such as mixed hydroxy-carbonate complexes. Moreover, for some of them, direct measurement of their stability is impossible and estimated values must be employed *(78)*. At the same time, in more complicated cases the competing interaction of complexants with main soil background cations should be taken into account.

The theoretical and empirical methods of stability constants estimation, including the use of linear free energy relationships, are now proposed for the natural amino acids and their synthetic analogues *(78, 79)*, hydroxo complexes *(28)* crown ethers *(80)*, natural organic matter *(69)* and some other ligands *(67)*. A viable procedure consisting of extrapolations and interpolations based on the most reliable data shows that stability constants for some species may be estimated at a precision of about 0.1 of a log unit *(78)*.

Conclusions

Stability constants based speciations represent a powerful and effective tool of metal and chelating agent behaviour and toxicity prediction for soils, sediments and groundwater. Even a rather simplified model can give an excellent description of a ligand's relative affinity towards a target pollutant, and thus of the decontamination processes.

Databases which have been recently reviewed, and which can be regarded as comprehensive, high-quality, and state-of-the-art data sets for specific applications *(e.g., 58, 65, 67, 70)*, can be used readily for applied speciation modelling. Unfortunately, the number of such data sets is still limited, and in most cases a user who wishes to make quantitative use of stability constants is left with the task of analysing the system equilibria and composing the data set relevant to the particular problem to be solved. For this task, the IUPAC Data base *(32)* should always be used as a start because it contains the largest collection of references with equilibrium constants and allows the user to assess the data diversity for a particular system. The next step includes the examination of original papers in order to exclude the possible transcription errors from the papers to databases. Then the best constants for the conditions used have to be selected. This is not a straightforward task as it requires specialized knowledge of the particular system. Existing critical evaluations such as the IUPAC "Critical Surveys of Stability Constants" *(38-52)*, the NEA critical compilations *(61-65)*, the monograph on metal ion hydrolysis by Baes and Mesmer *(27)*, the JESS expert system *(67)*, or the compilations of selected stability constants in NIST Data base *(26, 30, 31)* may be useful in solving the problem of conflicting data. Since there is no guarantee that these sources are free of misprints and errors, it is recommended to compare the values from different critical compilations. In addition, new experimental evidence may outdate older compilations, and therefore a check on new literature on the system of interest is essential. Although large amounts of effort have been invested in the review of thermodynamic data in the past, errors cannot *a priori* be excluded, and even the 'critical' data have to be treated with care.

Equilibrium constants can be adjusted for temperature changes if ΔH values are known (many are reported in *(32, 31, 67)*) using the van't Hoff equation, available as a part of *(32)*. Changes in stability constants resulting from ionic strength variations can be calculated by using the Davies equation (included in *(32)*) at low ionic strengths. The SIT equation described in the Appendices of *(61-65)* and contained also in *(32, 33)* allows corrections beyond $I = 0.1$ M if the ion interaction parameters are known. (Alternative approaches are discussed in *(81)*). For sea water some specific corrections for high pressure are necessary *(82)*.

Acknowledgements

We are grateful to Robert Byrne, Peter Campbell, Peter May and Leslie Pettit for valuable suggestions and help.

References

1. Lindsay, W. L. *Chemical Equilibria in Soils;* Wiley: New York, Chichester, Brisbane, Toronto. 1979.
2. Bohn, H. L.; McNeal, B. L.; O'Connor G. A. *Soil Chemistry.* Wiley: New York, Chichester, Brisbane, Toronto. 1979.
3. Campbell, P. In *Metal speciation and Bioavailability in Aquatic Systems*; Tessier, A., Turner, D. R. Eds.; Wiley: New York, 1995; *Metal Speciation and Contamination of Soil*; Allen, H. E.; Huang, C. P.; Bailey, G. W.; Bowers A. R. Eds.; Lewis Publishers: Boca Raton *e.a.*, 1995.
4. Francis, A. J.; Dodge, C. J.; Gillow, J. B. *Letters to Nature.* **1992**, *356*, 140.

5. Twiss, M. R.; Errecalde, O.; Fortin, C.; Campbell, P. G. C.; Jumarie, C.; Denizeau, F.; Berkelaar, E.; van Rees, K. *Chem. Spec. Bioavail.* **2001**, *13*, 9.
6. Cacheris, W. P.; Quay, S. C.; Rocklage, S. M. *Magn. Res. Imaging*. **1990**, *8*, 467; Means, J. L.; Crerar, D. A.; Doguid, J. O. *Science*. **1978**, *200*, 1477.
7. Romney, E. M.; Wallace, A.; Mueller, R. T.; Cha J. W.; Wood, R. A. *Soil Sci.* **1981,** *132*, 104; Wallace, A.; Mueller, R. T.; Romney, E. M. *Soil Sci.* **1981,** *132*,120.; Wallace, A.; Mueller, R. T.; Romney, E. M.; Soufi, S. M. *Soil Sci.* **1981,** *132*, 114.
8. Heil, D. M.; Samani, Z.; Hanson, A. T.; Rudd, B. *Water, Air Soil Pollut.* **1999,** *113*, 77.
9. Quevauviller, Ph.; Lachica, M.; Barahona, E.; Gomez, A.; Rauret, G.; Ure, A.; Muntau, H. *Fresenius J. Anal. Chem.* **1998,** *360*, 505.
10. Harnlem, B. J.; Vane, L. M.; Sayles, G. D. *Wat. Res.* **1999**, *13*, 951.
11. Garcia-Delgado, R. A.; Rodriguez-Maroto, J. M.; Gomez-Lahoz, C.; Vereda-Alonso, C.; Garcia-Herruzo, F. *Sep. Sci. Technol.* **1998**, *33*,867.
12. Byegard, J.; Skarnemark, G.; Skalberg, M. *J.Radioanal. Nucl. Chem.* **1999**, *241*, 281.
13. Kedziorek, M. A. M.; Dupuy, A.; Bourg, A. C. M.; Compere, F. *Environ. Sci. Technol.* **1998**, *32*, 1609.
14. Lu, N.; Kung, K. S.; Mason, C. F. V.; Triay, I. R.; Cotter, C. R.; Pappas A. J.; Pappas, M. E. G. *Environ. Sci. Technol.* **1998**, *32*, 370.
15. Wasay, S. A.; Barrington, S. F.; Tokunaga, S. *Environ. Technol.* **1998**, *19*, 369.
16. Hornburg, V.; Bruemmer, G. *Z. Pflanzenernaehr. Bodenk.* **1993**, *156*, 467.
17. Elliott, H.; Brown, G. *Water, Air Soil Pollut.* **1989**, *45*, 361.
18. Kayser, A.; Wenger, K.; Keller, A.; Attinger, W.; Felix, H. R.; Gupta, S. K.; Schulin, R. *Environ. Sci. Technol.* **2000**, *34*, 1778.
19. Wu, J.; Hsu F. C.; Cunningham. S. D. *Environ. Sci. Technol.* **1999** *33*, 1898.
20. Yeung, A. T.; Hsu, C.; Menon R. M. *J. Geotechn. Engineering*. **1996**, *122*, 666.
21. *Stability Constants, Ligands*; Schwarzenbach, G.; Sillen, L. G., Eds., The Chemical Society: London, 1957.
22. *Stability Constants. Part. II. Organic Ligands;* Sillen, L. G.; Martell, A. E., Eds., Special Publication No. 17, The Chemical Society: London, 1964.
23. *Stability Constants*; Sillen, L. G.; Hogfeldt, A. E.; Martell, A. E.; Smith, R. M., Eds., Supplement 1, Special Publication No. 25. The Chemical Society: London, 1971.
24. *Stability Constants of Metal-ion Complexes. Part B: Organic Ligands*; Perrin, D. D. Ed., Pergamon: Oxford, 1979.
25. *Stability Constants of Metal-ion Complexes. Part A: Inorganic Ligands.* Hogfeldt, E. Ed., Pergamon: Oxford, 1982.

26. Martell, A. E.; Smith, R. M. *Critical Stability Constants*; Plenum Press: New York, Vols.1-6, 1974, 1975, 1977, 1976, 1982, 1989.
27. Baes, C. F.; Mesmer, R. E. *The Hydrolysis of Cations*; Wiley: New York, 1976.
28. Barnum, D. W. *Inorg. Chem.* **1983**, *22*, 2297.
29. Kiss, T. In *Handbook of Metal-Ligand Interactions in Biological Fluids*. G. Berthon, G. Ed.; Marcel Dekker: New York, **1995,** 717-724; Rizkalla, E. N. *Coord. Chem. Rev.* **1983**, *5*, 223.
30. Martell, A. E.; Smith, R. M.; Motekaitis, R. J. NIST *Critically Selected Stability Constants of Metal Complexes Database*. Version 4.0. 1997. Texas A&M University, US.
31. NIST Standard Reference Database 46. *Critically Selected Stability Constants of Metal Complexes Database*. Compiled by Smith, R. M.; Martell, A. E.; Motekaitis, R. J. Version 7.0 for Windows. 2003. US National Institute of Standards and Technology Standard Reference Data Program; Gaithersburg, MD 20899.
32. *Stability Constants Database and Mini-SCDatabase*. IUPAC and Academic Software. Version 5.3. 2003. Sourby Old Farm, Timble, Otley, Yorks. UK; scdbase@acadsoft.co.uk.
33. *Ionic Strength Corrections for Stability Constants* using *Specific Interaction Theory (SIT)* for Windows 9x, NT, 2000, and xp, Pettit, L. D. Academic Software, IUPAC, 2003. (supplied as attachment to *(32)* and available on the IUPAC site www.iupac.org and www.acadsoft.co.uk.); Grenthe, I.; Plyasunov, A. V.; Spahiu, K. In *Modelling in Aqueous Chemistry*; Grenthe, I.; Puigdomenesh, I. Eds.; Organisation for Economic Co-operation and Development: Paris, 1997.
34. Bottari, A.; Ostacoli, G.; Paoletti, P.; Pettit, L. D.; Sammartano, S. *Pure Appl. Chem.* **1987**, *59*, 1721.
35. Critical Surveys of Stability Constants of Metal Complexes – Guidelines for prospective Authors. *Coordination Chemistry Rev.* **1975**, *17*, 358.
36. Overvoll, P. A.; Lund, W. *Anal. Chim. Acta* **1982**, *143*, 153.
37. Beck, M.T.; Nagypal, I. *Chemistry of Complex Equilibria*. John Wiley: New York, 1990; Motekaitis, R. J.; Martell, A. E. *The Determination and Use of Stability Constants*, VCH: New York, 1988; Martell, A. E.; Motekaitis, R. J. *Coord. Chem. Rev.* **1990**, *100*, 323.
38. Anderegg, G.; Arnaud-Neu, F.; Delgado, R.; Felcman, J.; Popov, K. *Pure Appl. Chem.* **2004**, accepted for publication.
39. Anderegg, G. *Critical survey of stability constants of EDTA complexes*. IUPAC chemical data series. No.14. Pergamon Press: Oxford *e.a.*, UK, 1977.
40. McBryde, W. A. E. *A critical Review of equilibrium data for proton and metal complexes of 1,10-phenanthroline, 2,2'-bipyridyl, and related compounds*. Pergamon Press: Oxford, England, 1977.

41. Stary, J.; Zolotov, Yu. A.; Petrukhin, O. M. *Critical evaluation of equilibrium Constants involving 8-hydroxyquinoline and its metal chelates.* Pergamon Press: Oxford, England, 1979.
42. Bond, A. M.; Hefter, G. T. *Critical survey of stability constants and related thermodynamic data of fluoride complexes in aqueous solutions.* Pergamon Press: Oxford, England, 1980.
43. Anderegg, G. *Pure Appl. Chem.* **1982**, *54*, 2693.
44. Stary, J.; Liljenzin, J. O. *Pure Appl. Chem.* **1982**, *54*, 2557.
45. Tuck, D. G. *Pure Appl. Chem.* **1983**, *55*, 1477.
46. Pettit, L. D. *Pure Appl. Chem.* **1984**, *56*, 247.
47. Paoletti, P. *Pure Appl. Chem.* **1984**, *56*, 491.
48. Beck, M. T. *Pure Appl. Chem.* **1987**, *59*, 1703.
49. Smith, R. M.; Martell, A. E.; Chen, Y. *Pure Appl. Chem.* **1991**, *63*, 1015.
50. Kiss, T.; Sovago, I. ; Gergely, A. *Pure Appl. Chem.* **1991**, *63*, 597.
51. Sovago, I.; Kiss, T.; Gergely, A. *Pure Appl. Chem.* **1993**, *65*, 1029.
52. Berthon, G. *Pure Appl. Chem.* **1995**, *67*, 1117.
53. Yamauchi, O.; Odani, A. *Pure Appl. Chem.* **1996**, *68*, 469.
54. Lajunen, L. H. J. ; Portanova, R.; Piispanen J. ; Tolazzi, M. *Pure Appl. Chem.* **1997**, *69*, 329.
55. Popov, K.; Rönkkömäki, H.; Lajunen, L. H. J. *Pure Appl. Chem.* **2001**, *73*,1641.
56. Arnaud-Neu, F.; Delgado, R.; Chaves, S. *Pure Appl. Chem.* **2003**, *75*, 71.
57. Portanova, R.; Lajunen, L. H. J.; Tolazzi, M.; Piispanen, J. *Pure Appl. Chem.* **2003**, *75*, 495.
58. Brown, P. L.; Byrne, R. H.; Gajda, T.; Hefter, G.; Powell, K. J.; Sjöberg, S.; Wanner, H. *Pure Appl. Chem.*, in press.
59. Wanner, H. *Radiochim. Acta*, **1988**, *44/45*, 325.
60. Mompean, F. J.; Wanner, H. *Radiochim. Acta*, **2003**, *91*, 617.
61. Grenthe, I.; Fuger, J.; Konings, R. J. M.; Lemire, R. J.; Muller, A. B.; Nguyen-Trung, C.; Wanner, H. *Chemical Thermodynamics of Uranium*, Amsterdam: North-Holland, 1992.
62. Silva, R. J.; Bidoglio, G.; Rand, M. H.; Robouch, P. B.; Wanner, H.; Puigdomenech, I. *Chemical Thermodynamics of Americium*, Amsterdam: Elsevier, 1995.
63. Rard, J. A.; Rand, M. H.; Anderegg, G.; Wanner, H.; Puigdomenech, I. *Chemical Thermodynamics of Americium,* Amsterdam: Elsevier, 1999.
64. Lemire, R. J.; Fuger, J.; Nitsche, H.; Potter, P.; Rand, M. H.; Rydberg, J.; Spahiu, K.; Sullivan, J. C.; Ullman, W. J.; Vitorge, P.; Wanner, H. *Chemical Thermodynamics of Neptunium and Plutonium*, Amsterdam: North-Holland, 2001.
65. Guillaumont, R.; Fanghänel, T.; Fuger, J.; Grenthe, I.; Neck, V.; Palmer, D. A.; Rand, M. H. *Update on the Chemical Thermodynamics of Uranium,*

Neptunium, Plutonium, Americium and Technetium, Amsterdam: Elsevier, 2003.

66. Smith, R. M.; Martell, A. E. The Selection of Critical Stability Constants. In *Chemical Equilibrium and Reaction Models*. Soil Sci. Soc. of America Special Publication 42, **1995,** p. 7.
67. May, P. M.; Murray, K. *Talanta*, **1991**, *38*, 1419; May, P. M; Murray, K. *J. Chem. Eng. Data*, **2001**, *46*, 1035; see also website 'jess.murdoch.edu.au' .
68. Filella, M; May, P. M. *Geochim. Cosmochim. Acta*, **2003**, *67*, 4013.
69. MINTEQA2/PRODEFA2, *A Geochemical Assessment Model for Environmental Systems: User Manual Supplement for Version 4.0*; Prepared by HydroGeoLogic, Inc. Herdon, Virginia and Allison Geoscience Consultants, Inc. Flowery Branch, Georgia; 1999.
70. Hummel, W.; Berner, U.; Curti, E.; Pearson, F. J.; Thoenen, T. Nagra/PSI Chemical Thermodynamic Data Base 01/01, Universal Publishers, USA, 2002.
71. Sehmel, G.A. Cyanid and Antimony Thermodynamic Database for the Aqueous Species and Solids for EPA-MINTEQ Geochemical Code, Report prepared by Battelle Pacific Northwest Laboratory for U.S. Environmental Protection Agency, 1989, Athens, Georgia. USA; Nordstrom, D. K.; Plummer, L. N .; Wigley, T. M. L.; Wolery, T. J. ; Ball, J. W. ; Jenne, E. A.; Bassett, R. L. E. A. Comparison of computerized chemical models for equilibrium calculations in aqueous systems. In *Chemical Modeling in Aqueous Systems - Speciation, sorption, solubility & kinetics*. ACS Symposium Series No. 93. Jenne, E. A. Ed. 1979. American Chemical Society: Washington, DC. 857-892; Florence, T. M.; Batley, G. E. *CRC Crit. Rev. Anal. Chem.* 1980, 219.
72. Schecher, W. D. *Thermodynamic data used in MINEQL+ version 4.5 Environmental Research Software*, **2001**, Hallowell, ME, USA.
73. Shock, E. L., Sassani, D. C., Willis, M., Sverjensky, D. A. *Geochim. Cosmochim. Acta* **1997**, *61*, 907 (also http://geopig.asu.edu/supcrt_data.html).
74. Thoenen, T., Kulik, D. PSI Techmical Report TM-44-03-04, Paul Scherrer Institute, Villigen, Switzerland, 2003 (also http: //les.web.psi.ch/ Software /GEMS-PSI/index.html).
75. Popov, A.; Rönkkömäki, H.; Lajunen, L. H. J.; Vendilo, A.; Popov, K. *Inorg. Chim. Acta* **2003**, *353*, 1.
76. Popov, K.; Popov, A.; Rönkkömäki, H.; Vendilo, A.; Lajunen, L. H. J. *J. Solut. Chem.* **2002,** *31*, 511.
77. *Solution Equilibria: Principles and Applications* [Windows 95, 98] Academic Software and K. J. Powell. Release 1.04. **2000**. (SolEq). Interactive, problem-oriented softbook. *Contributing Authors*: Byrne, R.; Kiss, T.; Lövgren, L.; May, P. M.; Orindo, C. O.; Pettit, L. D.; Popov, K. I.; Powell, K. J.; Ramette, R. W.; Sjöberg, S.; Town, R. M.; Ohman, L. O.
78. Smith, R. M.; Martell, A. E.; Motekaitis, R. J. *Inorg. Chim. Acta.* **1985**, *99*, 207.

79. Smith, R. M.; Motekaitis, R. J.; Martell, A. E. *Inorg. Chim. Acta.* **1985**, ***103***, 73.
80. Varnek, A. A.; Wipff, G.; Solov'ev, V. P.; Solotnov, A. F. *J. Chem. Inf. Comput. Sci.* **2002**, *42*, 812.
81. May, P. M. *J. Chem. Soc. Chem. Commun.* **2000**, 1265.
82. Byrne, R. H. *Pure Appl. Chem.* **1996,** *68*, 1639; Byrne, R. H.; Laurie, S. H. *Pure Appl. Chem.* **1999,** *71*, 871.

Analytical Methods

Chapter 4

Analysis of Aminopolycarboxylates and Organophosphonates

Carsten K. Schmidt and Heinz-Jürgen Brauch

Chemical Analysis Department, DVGW-Water Technology Center (TZW), Karlsruher Strasse 84, 76139 Karlsruhe, Germany

A number of analytical assays exist for the analytical determination of aminopolycarboxylates and organophosphonates. Relevant assays include gas and liquid chromatography, capillary electrophoresis, electrochemical methods, spectrophotometry, atomic absorption spectrometry, and titrimetric assays. The method to be selected depends on the individual problem. The following discussion is intended as an overview of the available methods and as a short guide.

Introduction

Aminopolycarboxylates such as ethylenediaminetetraacetic acid (EDTA), nitrilotriacetic acid (NTA), and diethylenetriaminepentaacetic acid (DTPA) and organophosphonates such as 1-hydroxy-ethanediphosphonic acid (HEDP), nitrilotris(methylenephosphonic acid) (NTMP), ethylenediaminetetra(methylenephosphonic acid) (EDTMP), and diethylenetriaminepenta(methylenephosphonic acid) (DTPMP) find wide applicability as chelating agents in many industrial processes and products worldwide (1-5). Recently, some alternative aminopolycarboxylates such as 1,3-propylenediaminetetraacetic acid (1,3-PDTA), ß-alaninediacetic acid (ß-ADA), and methylglycinediacetic acid (MGDA) were established (1,6-8). The chemical structures of these compounds are depicted in Figure 1. Aminopolycarboxylates and organophosphonates form stable and water-soluble complexes with multivalent metal ions and restrict them from playing their normal chemical role, an aspect that makes them attractive for a variety of applications (e.g. in industrial cleaners, photochemicals, textile and paper industries, cosmetics and pharmaceuticals). As a consequence, these compounds are used in huge quantities worldwide. The global consumption of aminopolycarboxylates and organophosphonates amounts to roughly 200,000 and 60,000 tons per year, respectively (5,6).

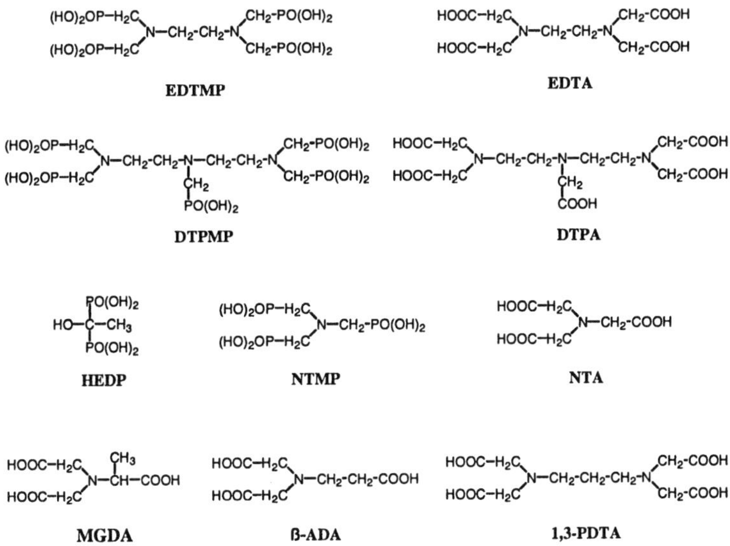

Figure 1. Structural formulas of important organophosphonates and aminopolycarboxylates.

Due to their high polarity and predominantly low degradability these compounds enter our creeks, rivers, and lakes, mainly via industrial and domestic wastewaters. Typical aminopolycarboxylate concentrations in European rivers are in the range of 0-60 μg/L (9). Aminopolycarboxylate concentrations that can be typically expected in runoffs of industrial sewage treatment plants are in the low to middle mg/L-range (9). Pollution levels of municipal wastewater effluents are usually around two orders of magnitude lower (in the middle μg/L-range) (9). Knowledge on concentrations of organophosphonates in the aquatic environment is scarce, mainly due to a lack of adequate sensitive analytical methods. However, determination of organophosphonates in influents and effluents of some Swiss wastewater treatment plants revealed concentrations in the range of

0-70 μg/L and 20-970 μg/L, respectively (10). Environmental organophosphonate levels in surface waters are estimated to be in the middle ng/L- (4) or very low μg/L-range (3).

Aminopolycarboxylates and organophosphonates occur in natural waters predominantly in the form of metal complexes (5,11,12). Several studies revealed that environmental fate, microbial degradability, toxicology and ecotoxicology of complexing agents depend significantly on their actual speciation (5,9,11,12). Thus, adequate analytical assays have to be chosen according to the water matrix to be analyzed and the respective research problem. Even though they are often used for the same purpose, aminopolycarboxylates and organophosphonates partially differ in their physico-chemical properties and require different analytical approaches. Therefore, analytical assays for their determination are described separately.

Aminopolycarboxylates

Gas Chromatography

Gas chromatography is available since many decades and has become a fully developed separation technique. However, it requires that the analytes are sufficiently volatile or vaporizable, while aminopolycarboxylic acids lack this property. Nevertheless, also aminopolycarboxylic acids are applicable to gas chromatographic analysis, when they are converted into another chemical form (derivatization). For this reason, in particular to its high sensitivity, gas chromatography has become a valuable and common tool in the analysis of aminopolycarboxylic acids (13).

The sample preparation of aminopolycarboxylic acids involves their conversion to readily volatile alkyl ester derivatives, like methyl (14-22), ethyl (23-26), n-propyl (27-34), isopropyl (35-38) or n-butyl esters (27,39-47). Detailed studies have demonstrated that often the derivatization to propyl or butyl esters yields better results than derivatization to methyl esters (27,29,33,35). Other authors have discussed the use of silylating agents (48,49). The derivatization step makes sample preparation tedious and time-consuming. When aminopolycarboxylic acids exist as metal complexes, their derivatization is more difficult; therefore, metal complexes must be decomposed prior to derivatization by decreasing the pH value causing a conversion of the aminopolycarboxylates into their free acid forms. Therefore, gas chromatographic methods are unfortunately not suitable for the determination of individual aminopolycarboxylate-metal species and determine only the integral content of all existing species of a specific aminopolycarboxylic acid.

In most cases, derivatization is preceded by a concentration step to increase the sensitivity of the overall process. In principle, two approaches exist: (i) concentration of the aminopolycarboxylic acids as anions on an anion exchanger (e.g. SAX) at pH 2-3, elution with formic acid and evaporation of the eluate at 100-110 °C to dryness (16,31,32,36-38,41,43,46,50), (ii) simple evaporation of the acidified aqueous sample to dryness at 100-115 °C (23,24,27,29,30,33,35-38,42,47). More detailed studies have demonstrated that simple evaporation yields better results than the first approach as long as suitable internal standards are included from the beginning (34,36-38). Both approaches imperatively require the complete removal of all water traces for the following esterification. After derivatization followed by water addition, the resulting aminopolycarboxylic acid alkyl esters can be extracted from the aqueous phase and purified by liquid-liquid extraction with an organic solvent like n-hexane (30,35,41,42), toluene (16,23,24,33), MTBE (36-38) or methylene chloride (27). Usually, the final injection into a gas chromatograph is preceded by drying the extract over Na_2SO_4 and reduction of the sample volume by partial evaporation.

While detection in earlier studies often involved the use of flame ionization detectors (FID) (18,20-22,25,26,32,43-45), they have been almost completely replaced in current studies by mass-selective (MSD) (17,27,28,31,34-40,42,47) or nitrogen-sensitive (NPD) detectors (16,23,24,29-31,33,35,41,42,46,51-53) that have become state-of-the-art due to their increased sensitivity. In case of GC-MSD or GC-NPD detection, the detection limits are usually in the low μg/L-range. Most published applications involve the analysis of all types of aqueous solutions (drinking water, surface water, groundwater, wastewater) (15-17,21,24-26,28,39-41,45), but also of foods (22), sediments and fish (17), and the determination of the contamination of DTPA with EDTA (20).

An international standard procedure has been published on the analysis of aqueous samples; it allows the simultaneous determination of EDTA, DTPA, NTA, ß-ADA, 1,3-PDTA and MGDA (54,55). Initially, the water sample containing formaldehyde as a preservative is evaporated to dryness, first without additive, then after addition of hydrochloric or formic acid. A subsequent esterification step converts the complexing agents into their n-propyl, isopropyl or butyl esters. Finally, water is added to the reaction mixture, and the formed esters are extracted from the mixture with n-hexane. Finally, they are separated by gas chromatography and quantified with a nitrogen-sensitive or preferably mass-spectrometric detector. In the case of nitrogen-sensitive detection, heptanoic and/or octadecanoic acid nitrile is used as control standard for gas chromatography; a suitable control standard for mass-spectrometric detection is

1-chlorotetradecane. 1,2-diaminopropane-N,N,N',N'-tetraacetic acid (DPTA) is used as an internal standard for the overall procedure. As an option, also ^{13}C-labeled standards are suitable for gas chromatography with mass-spectrometric detection (36,37). The method can be applied to the analysis of drinking water, surface water, groundwater and wastewater in a concentration range of 0.5 to 200 µg/L, utilizing sample volumes from 50 to 100 mL. When wastewater is analyzed, the sample volume should be reduced to e.g. 5 mL or 10 mL to minimize matrix effects. In particular elevated salt concentrations (> 2 g/L NaCl) interfere with the method due to their negative impact on the complete drying of the evaporated sample. The presence of calcium ions results in falsely low EDTA levels at concentrations of 200 mg/L Ca^{2+} and higher (54,55). In contrast to manganese and copper, also increased iron (> 10 mg/L) and bismut (> 100 µg/L) contents may cause falsely lower concentrations to a certain extent (21,24,54,56).

Liquid Chromatography

Liquid chromatography (LC) represents another option for the determination of aminopolycarboxylates. Compared to gas chromatography, it offers the major benefits that no extraction of the aqueous samples is required, and that no derivatization of the analytes for increased volatility is necessary. Basically, the samples can be directly applied to the separating column.

Liquid chromatography has been applied to the determination of aminopolycarboxylates in many different sample matrices (13), including drinking water (57,58), surface water (57-70), wastewater (35,59,71-76), foods (14,77,78), sediments (62), medical products (79-83), blood (84), fertilizers and micronutrients (85-88), and chemical purification and reaction solutions (89,90).

The LC separation of aminopolycarboxylic acids is performed on reversed-phase columns (57-59,62-65,67,69,70,72-76,78-83,85-97) or by ion chromatography (IC) on an anion exchange column (35,60,61,66,68,71,84,97-100). The reversed-phase technique is usually based on ion-pair chromatography (IPC) (13). Both separation techniques are related to many problems and interferences, in particular when high inorganic cation or anion levels are present (56,58,59,63,71,74,93,101). However, methods have been developed to overcome such problems, and a number of authors have reported on successful and robust liquid chromatographic determinations in various matrices.

Detection is normally performed with a UV detector. For this purpose, by addition of an excess of metal ions prior to (56-58,71,74,78,84,85,90,95), during (92,93,96) or after (61,72,94,102) the chromatographic run, the aminopolycarboxylic acids to be analyzed are converted into a highly stable defined metal complex with favorable UV absorption characteristics. Most frequently, Fe(III) ions (35,58,60,62-64,71-74,76,79,80,82,85,87,88,90,92,96) or Cu(II) ions (61,78,81,83,84,86,91,94) are used. In addition to UV detection, some studies were performed using electrochemical (59,61,68,98), fluorescence (69,70,103,104), and mass-spectrometric detection (in part involving a suppressor module) (84,97,99,105) and also ICP-MS coupling (106-108).

A DIN (German Institute for Standardization, Berlin) standard is available for the detection of NTA, EDTA and DTPA in water and wastewater (101). It requires no concentration step. First, the aminopolycarboxylic acids are converted into their Fe(III) complexes and then are separated by ion-pair chromatography on a reversed-phase column. The mobile phase is a tetrabutylammonium hydrogen sulfate or tetrabutylammonium hydroxide solution acidified with nitric acid. Detection is performed by direct UV/Vis measurement at 260 nm. The working range is 0.1 to 20 mg/L (101). This example clearly demonstrates the major problem related to the liquid-chromatographic determination of aminopolycarboxylic acids, i.e. poor sensitivity. Typical detection limits are in the low mg/L-range (13). Standard approaches to increase sensitivity include sample concentration, large injection volumes and the use of more sensitive detectors. Some methods taking these aspects into account are rewarded by lower detection limits down to 1 μg/L (56,57,62,96,99) and therefore represent alternatives to gas chromatography, making them suitable for the determination of aminopolycarboxylic acids in surface water. Sample enrichment options include sample evaporation or concentration in a rotary evaporator (57), anion exchange chromatography (96) and ion-pair extraction (56), i.e. methods that have already been partially discussed in the gas chromatography section.

In contrast to gas chromatography, liquid chromatography in principle also allows the determination of the aminopolycarboxylate speciation, i.e. the identity of the bound metal ion (61,68,69,72,97,107-109). To some extent, also gel permeation chromatography allows the fractionation of aminopolycarboxylate species (110,111). However, these approaches suffer from insufficient sensitivity, because each aminopolycarboxylate signal is split to many different species and consequently into several peaks. Furthermore, sample concentration would influence the complex equilibrium and change speciation. Therefore, when the determination of the speciation is intended, any recomplexation with Fe(III) or Cu(II), in particular needed for UV detection, can only be performed after the chromatographic run. Published methods have demonstrated that even more sensitive detectors like mass-spectrometry and fluorescence detectors will not improve the detection limit below 0.1-10 μM (69,72,97). Consequently, these methods are suitable for wastewater analysis, but they are not sensitive enough when it comes to the determination of aminopolycarboxylate species in surface waters.

Currently, a promising method for the determination of aminopolycarboxylate speciation at trace levels has been published. It is based on the coupling of ion chromatography with ICP/MS involving on-line sample enrichment via column switching. This method allows the determination of nM levels of aminopolycarboxylate metal complexes (NTA, EDTA and DTPA) and thus the determination of the aminopolycarboxylate speciation in surface waters (107,108).

However, also relatively simple means are available for the determination of Ni(II)-EDTA- and Fe(III)-EDTA species in water samples. The Fe(III) portion of the total EDTA speciation can be determined by illumination of the water sample, resulting in the complete photochemical decomposition of Fe(III)-EDTA (112,113). Even though also other metal-EDTA complexes (in particular Mn(II)-EDTA) turned out to be slightly photodegradable, experience has demonstrated that their impact on the determination of Fe(III)-EDTA via its photodegradation is rather low for natural waters. Corresponding reaction rates of interfering metal-EDTA species are either too slow or they account only for a negligible fraction of the total EDTA speciation. The determination of Ni(II)-EDTA is possible with a comparative method based on the conversion of the species into the Fe(III)-EDTA complex. The recomplexation of Ni(II)-EDTA to Fe(III)-EDTA is relatively slow and necessitates heating of the sample to 90 °C for approximately 3 hours. Omitting the heating step prevents the conversion of Ni(II)-EDTA to Fe(III)-EDTA (12,62,95,114).

Finally, it should be also mentioned that EDTA and DTPA are applied for the separation of metal ions in metal chelate analytics. They are either added to the samples during sample preparation or to the eluent as additives (13,109,115).

Capillary Electrophoresis (CE)

In recent years, capillary electrophoresis (CE) has evolved to an alternative separation technique besides HPLC allowing the determination of the speciation of complexing agents. The principle of capillary electrophoresis is based on the different migration speeds of electrically charged particles in an electric field; its particular benefit is its high separation efficiency at short analysis times. Complexing agents are traditionally used in particular as an electrolyte additive for the modification of the ion mobility of metal cations (116-129). The first direct assay has been published by O`Keeffe et al. in 1995 (130). So far, the analysis of free aminopolycarboxylic acids and their metal complexes has been predominantly performed by capillary zone electrophoresis (CZE) techniques

(130-150), but also the application of micellar electrokinetic capillary chromatography (MEKC) has been described (134).

Electrophoretic assays for aminopolycarboxylates include all possible variations like the determination of free acids (130,132,133,138,144,145), the determination following conversion of all existing complex species into a single defined metal complex (130,134,137,140,141,143,146-150) and also the differentiated determination of individual metal complex species (131,133-136,138,139,142,145). Most studies are limited to the investigation of the aminopolycarboxylic acids EDTA (130,133-143,145,147-150), DTPA (130-134,146-149), and NTA (130,133,134,138,144,147,148). Published application ranges include biological fluids (140,141), fixing baths (139), detergent formulations (147), wastewater (131,143), aqueous solutions in a desulfurization process (144), surface water (132,143), radioactive waste solutions (145,148), mayonnaise (137), and culture media (138).

The detection predominantly involves UV detectors, occasionally, laser-induced fluorescence (LIF) detectors (149), electrochemical detectors (132) and mass spectrometers (140,141,143) are used. Depending on the type of complex, the UV detection of the metal complexes is mostly performed in the wavelength range 200-290 nm (130,131,134-137,139,145-148,150). Methods allowing both the simultaneous determination of free acids and metal complexes species comprise the (less selective) direct UV detection at 185 nm (133,138) or the indirect UV detection (142,144). An elegant option is the in-line coupling of UV detection with potentiometric detection based on a metal copper electrode (132); the UV trace identifies the metal complexes, while the free acids are determined by potentiometry. In potentiometric detection, the electrode potential depends on the concentration of free copper ions at the electrode surface; it decreases, when the complexing agents pass the electrode. Buchberger and Mülleder demonstrated the possibility to determine the free DTPA acid portion after selective complexation with Fe(III) utilizing the pH dependence of the complex stabilities (131). The detection limits for these techniques are usually in the low mg/L-range. Sensitivity can be considerably increased by mass-spectrometric detection. Sheppard and Henion reported on a detection limit of 150 ng EDTA/L when using CE-ion spray-MS/MS coupling (143). LIF detection after conversion of EDTA or DTPA into ternary complexes with 8-hydroxyquinoline-5-sulfonic acid (HQS) and lutetium or thorium allowed to achieve detection limits in the low μg/L-range (149). Improved sensitivities in the low μg/L-range were also reported by Zhu et al. using a large-volume sample stacking technique in connection with UV detection (150,151).

The addition of cationic polymers (135,136) or micelle-forming agents (134) to the electrolyte superimposes the electrophoretic separation with a corresponding ion-exchange or a chromatographic reveresed-phase mechanism, offering improved separation options for individual metal complexes.

Sample preparation procedures described for individual applications often involve only a dilution step or filtration of the sample solution. Depending on the procedure, this step is preceded by a derivatization step for the conversion of individual complex species into a single well-defined metal complex. Only Sheppard and Henion reported on a more complex sample preparation technique utilizing an anion exchanger based solid-phase extraction of the Ni(II)-EDTA complex, previously prepared by derivatization, followed by an evaporation step (140,141,143). An interesting sample preparation variant is the method for the determination of EDTA in mayonnaise by Kvasnicka and Mikova that allows to remove a part of the matrix by a preceding on-line-coupled capillary isotachophoresis (137).

Capillary electrophoresis is a comparably new technique (equipment is commercially available only since around 1988); and its distribution is still only limited. However, to the opinion of the authors, its potential in environmental applications appears as being capable of further development, in particular due to the increasing availability of mass spectrometry and electrochemical detectors. The approaches discussed in the present review appear promising, and it can be expected that CE might develop as a true alternative to ion chromatography, in particular regarding the determination of individual metal complex species. With respect to the analysis of complexing agents in water samples, the discussed CE techniques based on UV detection demonstrate sufficient sensitivity for the analysis of wastewater samples. It even appears that there is still some upside potential for increased detection sensitivity by means of an improved sample preparation. The coupling of CE with mass spectrometry appears to be an attractive option for the determination of metal complex species in surface waters.

Electrochemical Methods

A number of electrochemical methods for the determination of aminopolycarboxylic acids have been developed in the past. In addition to potentiometry, in particular voltammetric and polarographic methods have been used (13,152,153), the latter in the following techniques: direct current polarography, differential pulse polarography, square-wave polarography and inverse polarography (stripping techniques). Sillanpää and Sihvonen (13) and also

Kaiser (152) have published excellent review articles on this matter, in particular with respect to earlier methods. The predominant number of methods refers to the determination of EDTA and NTA, only a small portion takes DTPA into account.

One major disadvantage of the discussed methods is their poor selectivity and robustness. All compounds and parameters that might influence the electrochemical process (e.g. humic substances, detergents, other complexing agents, increased salt levels, etc.) should be taken into account for calibration. Therefore, unknown environmental samples are particularly critical and require thorough validation. Consequently, environmental samples often require a more complex samplc preparation aiming at the reduction of interfering components (154). Furthermore, due to the poor selectivity, the simultaneous determination of different complexing agents in a single sample is complicated.

A well validated DIN (German Institute for Standardization, Berlin) standard is available for the simultaneous detection of EDTA and NTA in polluted waters and wastewaters by differential pulse polarography (DPP) (153-155). The determination is performed after conversion of all species into the corresponding Bi(III) complex. The limit of quantitation of this method is 0.1 mg/L. A recently published differential pulse polarography method, however, based on a Eu(III) complex, demonstrated a detection limit of 0.1 mg/L EDTA (156). Potentiometric methods usually demonstrate a somewhat lower detection sensitivity in the low mg/L-range (13,157). Although the direct determination of the bismuth complex by differential pulse polarography according to the DIN standard is feasible, it lacks the sensitivity required for the levels found in surface waters. As expected, the determination of the bismuth complex by differential pulse polarography utilizing the anodic-stripping technique (DPASV) results in significantly lower detection limits of about 0.1 μg/L (30,158-160).

Due to the poor selectivity, more recent papers propose to use electrochemical methods as a summation parameter in screening tests (30,159,161,162). An "index of bismuth complexation" established by DPASV can be used for the determination of complexing agents in drinking water and groundwater (30,159), while a "copper complexation index" determined by DPP might be applied to the analysis of highly polluted waters (161,162).

Spectrophotometry

Also a number of spectrophotometric methods have been published on the determination of aminopolycarboxylates (13,152,163-165). In principle, they can be classified into four groups: (i) measurement of the absorption of a metal-aminopolycarboxylate complex, (ii) measurement of the absorption of a non-complexed metal remaining after addition of a known amount of metal to an aminopolycarboxylate sample solution, (iii) measurement of the absorption of a colored or fluorescent metal complex in the presence of aminopolycarboxylates or (iv) measurement of the absorption of a colored or fluorescent ternary complex based on a metal-aminopolycarboxylate with another organic ligand. The first method involved complexes with Cr(III) (166,167), Co(III) (168,169), Fe(III) (170,171), Ni(II) (172), Cu(II) (173,174) and phosphomolybdic acid (175,176), the second metals like Cu(II) (177), Ni(II) (178) and Fe(III) (163). The third method was performed with the Bi(III)-pyrocatechol violet complex (179,180), Fe(II)-bis(2,4,6-tripyridyl-s-triazine) (163,181), Ca(II)-fluo-3 (181), Fe(III)-thiocyanate (182), Fe(II)-1,10-phenathroline complex (34), Cu(II)-chromo-azurol-S (183), Zn(II)-zincon (184), zirkonium-xylenol orange (102), neocuproine (185), bromopyrogallol red (186), Ti(IV)-p-carboxyphenyl fluorone (165) and 4-aminoantipyrene (187). The following ternary EDTA complexes have been used in papers on the fourth method (188): 4,7-diphenyl-1,10-phenanthroline-Fe(III)-EDTA (189), alizarin red-S-Zr(IV)-EDTA (190), salicylic acid-Tb(III)-EDTA (191) and 8-hydroxyquinoline-5-sulfonate-Lu(III)-EDTA (103).

Usually, spectrophotometric methods for the determination of aminopolycarboxylic acids only demonstrate little sensitivity in the low mg/L-range, and they are prone to many interferences. Due to their comparably poor selectivity, these methods are mostly only suitable for the determination of the general complexing capacity of a sample. Only methods based on ternary EDTA complexes are characterized by some selectivity combined with high sensitivity. For example, the detection limit of the alizarin red-S-Zr(IV)-EDTA method is 3.4 μg/L EDTA.

Atomic Absorption Spectrometry

In particular some earlier studies on the analysis of aminopolycarboxylates involve an indirect detection of the complexing agent by atomic absorption spectrometry (AAS) of the complexed metal ion (102,184,192-195). For example, in EDTA analysis, EDTA is first converted to its Cu(II)-EDTA complex; then the complexed copper is determined by AAS. The EDTA concentration in the sample is calculated assuming a 1:1 binding relationship between EDTA and copper. Major problems related to this approach are in particular the separation of Cu(II)-EDTA from non-complexed copper ions in the solution and interferences by other metal ions having an impact on the generation of the Cu(II)-EDTA complex (184,192,193). Two approaches were published on the isolation of the Cu(II)-EDTA complex. The first is based on the precipitation of excess copper ions as $Cu(OH)_2$ at pH 10. After careful removal of the precipitate by membrane filtration, the still dissolved copper complexed by EDTA can be determined by AAS (184). Alternatively, the conversion of EDTA into its copper complex can be performed with a cation exchanger in the Cu^{2+} form. The water sample is flushed through the cation exchanger, and EDTA contained in the sample abstracts an equivalent Cu^{2+} amount from the ion exchanger, and the resulting solution containing Cu(II)-EDTA complexes is then analyzed via AAS (192,193).

Depending on the method, these approaches offer detection limits down to the low mg/L-range. However, although such techniques are comparably fast and simple, they suffer from poor selectivity when the sample also contains other complexing agents. Therefore, the usefulness of AAS methods is limited when it comes to wastewaters and natural water samples; they only allow the determination of a general complexing capacity.

Titration

Also a number of titrimetric methods have been published on the determination of aminopolycarboxylates (13,196-204). However, they all are characterized by the same drawbacks as the spectrophotometric methods. Therefore, considerable concessions are to be made with respect to sensitivity (usually in the low mg/L-range) and selectivity. In addition, matrix-containing samples often cause problems in end-point recognition. EDTA can be determined by potentiometric titration with a $FeCl_3$ solution prepared by titration of an aqueous EDTA sample buffered to pH 2.5 with a solution of known $FeCl_3$ content. Measurement is performed with a platinum-calomel electrode (102). Also the titration with $ZnSO_4$ solution and eriochrome black T as the indicator has been published. The titration is stopped when transition in the indicator range occurs. For easier detection of the color change, methyl red can be added (205).

Summary

Aminopolycarboxylic acids can be determined in different sample types by gas chromatography, liquid chromatography, capillary electrophoresis, electrochemical techniques, spectrophotometry, atom absorption spectrometry and titration. Gas chromatography (GC-MS and GC-NPD) allows the sensitive and reliable identification of the complexing agents. However, it involves cumbersome sample preparation and does not allow the determination of the aminopolycarboxylate speciation in the sample. GC can be applied to drinking waters, surface waters and wastewaters. Alternatively, the determination in these matrices can be performed by liquid chromatography (IC, IPC). Liquid chromatography techniques also allow the separation and quantification of individual aminopolycarboxylate-metal species. However, the detection limit that is required for surface water analysis (i.e. low µg/L-range) can be only achieved via coupling of ion chromatography and ICP/MS. A number of aminopolycarboxylate-metal species can also be differentiated by capillary electrophoresis (CE); the sensitivity required for surface waters is achieved by ion spray-MS/MS coupling. Polarography, spectrophotometry, atom absorption spectrometry and electrochemical methods are prone to interferences, in particular in case of complex matrices. As these techniques demonstrate relatively poor selectivity, they are usually only suitable for the determination of the general complexing capacity of samples.

Organophosphonates

Gas Chromatography

Unlike aminpolycarboxylates, organophosphonates can be converted into their alkyl ester derivatives with great difficulty only (206). In order to get the totally methylated organophosphonates, Klinger et al. tried to derivatize them with different methylation reagents such as methyl alcohol/sulfuric acid, trimethylanilinium hydroxide, and trimethyl orthoformate. Although reaction conditions like temperature, time and added quantity of methylation reagent were varied, all attempts were in vain (207). It was found that diazomethane is the only reagent that yields the totally methylated organophosphonic acids (206). However, trying to measure the derivatives by GC/MS revealed that only methylated HEDP was volatile enough for successful analysis (206). HEDP was determined by GC/MS and GC/FID also after conversion to the trimethylsilyl derivative (208).

Liquid Chromatography

Anion Exchange Chromatography

A number of analytical assays for the determination of organophosphonates have been developed based on anion-exchange chromatography. With regard to the compound spectrum under consideration, published methods include the analysis of single organophosphonates (107,209-214) and also the differentiated determination of organophosphonate mixtures (66,215-226). Published applications include water samples (environmental and spiked drinking water) (107,217,225), purity studies of technical products (212,216,220) and detergent powders (222,223), biological samples (211,213), and pharmaceutical dosage formulations (210).

Separations were carried out on anion exchange columns from various suppliers including Dionex AS7 (66,210,212,213,215-219,223,226), Dionex AS11 (107,222), Hamilton PRP-X 100 (209,211,214), and Waters IC-Pak HR phases (210,221). Strongly acidic organophosphonates show a strong affinity to stationary phases of anion exchangers, since they provide in their totally dissociated state up to ten negative charges. To assure a reasonable run time, the effective charge of the analytes can be decreased by sufficient protonation at lower pH values making the pH of the mobile phase, as well as the concentration of the displacing anion (e.g. nitrate), an important parameter in controlling the retention of organophosphonates during anion-exchange chromatography (218,221). It was repeatedly reported that organophosphonates with more than two phosphonate groups could not be eluted from anion exchangers by common slightly acidic carbonate/bicarbonate gradients (212,225,227). As a consequence, most often strongly acidic mobile phases based on HNO_3 (66,213,214,215,218,221,223,226) or mixtures of HNO_3 and KNO_3 (210) were used. However, other researchers also reported on adequate separations based on slightly acidic [trimesic acid (209) or mixtures of EDTA and KCl (216,217,224)] or even basic eluents [NaOH (222), mixtures of NaOH and $NaNO_3$ (211), or basic NH_4NO_3 solutions (107)]. Excellent separations of HEDP, NTMP, EDTMP and DTPMP were described by Weiß and Hägele (66), Fitchett and Woodruff (215), Wong et al. (221), and Vaeth et al. (216).

Detection of the analytes is often performed with a UV detector. However, the lack of characteristic strong UV signals of organophosphonates necessitates major efforts when using this detection technique. UV-based detection approaches for organophosphonates can be classified into three groups: (i) post-column complexation of the phosphonate with Fe(III), (ii) post-column oxidation of the phosphonate to phosphate or (iii) indirect photometric detection.

In methods of the first kind, ferric nitrate in perchloric acid is used to form a complex directly with organophosphonates eluting from the column. The formed Fe(III) complexes provide favorable UV absorption characteristics and can be easily detected at 300-330 nm (66,212,215,219,223,226).

The second method is based on oxidative decomposition of the separated organophosphonates with ammonium persulphate into orthophosphate at elevated temperatures followed by addition of molybdenum reagent to generate either phosphomolybdate (detected at 820 nm) (213) or molybdenum blue (detected at 660 nm) (224) or molybdovanadophosphoric acid (detected at 410 nm) (216,217,219,226).

The third method requires a mobile phase species which absorbs highly at a wavelength that the organophosphonate does not. Analyte detection is then performed by measuring the decrease in absorption of the mobile phase due to the replacement of the UV-absorbing species by organophosphonates. As adequate displacing ions for organophosphonate determinations, nitric acid (210) and 1,3,5-benzenetricarboxylic acid (trimesate ion) (209) have been reported. A few reports also describe an indirect fluorescence detection method (211,214,226). In these methods the organophosphonate is combined post-column with a highly fluorescent Al(III)-morin complex solution and fluorescence decreases because of the formation of the non-fluorescent Al(III)-phosphonate complex. This decrease is correlated to the amount of organophosphonate present in the sample.

Furthermore, some studies were performed using conductivity (225), refractive index (219,221), pulsed amperometric (222) and element specific detection of phosphorus with an inductively coupled plasma (ICP) detector (228). Recently, the latter approach was also used in combination with mass spectrometry (ICP-MS) (107).

Typical detection limits of these methods are in the low to middle mg/L-range. Lower detection limits in the low to middle μg/L-range can be achieved by employing some kind of sample enrichment in combination with the most sensitive detection approaches including ICP-MS (107), UV-based detection of phosphate following post-column oxidation of the organophosphonate (213,217), indirect fluorescence detection with Al(III)-morin (211) and UV detection following conversion of the organophosphonate into the Fe(III) complex (215,223). These detection limits are sufficient for wastewater samples, however, the sensitivity that is required for organophosphonate surface water analysis (i.e. sub-μg/L-range) is predominantly not met by these methods.

Separations based on anion-exchange chromatography are prone of many interferences. Many metal cations cause, by complex formation with the organophosphonates, a decrease in sensitivity of the analytical assay, leading in the worst case to the total suppression of the signal, to shifted retention times or to a loss of separation quality due to tailing peaks. Strong interferences are particularly caused by Al(III), Fe(III), Mn(II), Ni(II) and In(III), whereas Bi(III), Ca(II), Cu(II), Mg(II) and Zn(II) interfere only slightly (217,223). However, these interferences can be avoided by use of strongly acidic mobile phases or by addition of adequate sequestrants such as EDTA or triethanolamine (217,224). Felber et al. reported on an oxine-based liquid-liquid extraction of Al(III) from aqueous samples by utilizing the good solubility of the Al(III)-oxine complex in chloroform (223). Furthermore, critical metal cations can be eliminated by cation exchange cartridges (217,224). Depending on the method, also inorganic anions such as chloride, phosphate and sulfate might interfere with the separation or obstruct relevant analyte signals (66,210,226). However, since organophosphonates are strongly retained on anion-exchange cartridges at high pH, they can be readily separated from weakly retained inorganic and organic anions by sequential elution (210,211,226).

Ion-pair Chromatography

A few ion-pair high-performance liquid chromatography assays have also appeared in the literature. Organophosphonate separations were achieved on reversed-phase columns after addition of different ion-pairing reagents such as tetraethylammonium acetate (TEA-Ac) (227), tetrabutylammonium bromide (TBA-Br) (229-233) or tetrahexylammonium bromide (THxA-Br) (226) to mobile phases based on water/methanol (226,229-231) or water/acetonitrile mixtures (227,232,233) at either basic (232,233) or acidic (226,229-231) conditions. Detection is normally performed with a UV detector, mostly based on conversion of the organophosphonates into adequate UV-active metal complexes prior to the chromatographic run (230-233). In analogy to methods developed for aminopolycarboxylates, Reichert tried, however, only with limited success, to form Fe(III) and Cu(II) organophosphonate complexes during the run by addition of an excess of Fe(III) or Cu(II) to the mobile phase (226). Other methods involve the application of mass spectrometry detection (in combination with a suppressor module) utilizing electrospray ionization in the negative mode (227) or flame photometric phosphorus-selective detection (229,234). The latter approach was introduced by Julin et al. (235) and is based on the flame photometric measurement of phosphorus from the molecular light emission of HPO.

Nowack described an excellent ion-pair HPLC method for the determination of the parent compounds HEDP, NTMP, EDTMP and DTPMP in natural waters following precolumn formation of the Fe(III) complexes (232), and Chester demonstrated the successful separation of DTPMP and EDTMP (229).

Furthermore, approaches based on ion-pair chromatography are principally applicable for organophosphonate speciation determination. Li et al. presented the separation of Cu(II)-EDTMP, Fe(III)-EDTMP, Pb(II)-EDTMP and EDTMP acid (230,231). However, ion-pair chromatography of organophosphonates is very susceptible to interferences by metal ions and it is crucial to assure stable and defined organophosphonate species during analysis. Especially, pH dependencies of complex stabilities and possible metal exchange reactions have to be taken into account. Approaches that disregarded these aspects were consequently less successful (226) or even failed completely when analyzing natural water samples (227).

Nowack avoided such drawbacks by conversion of all organophosphonates in the corresponding Fe(III) complexes prior to the chromatographic separation (232,233). In a first step, water samples are passed through cation-exchange columns in the H^+-form (Dionex OnGuard H) to remove calcium and magnesium, which interfere with the complexation of Fe(III) by the phosphonates. Then, the pH is brought to about 3 and Fe(III)-organophosphonate complexes are formed by addition of an excess of ferric nitrate. Since the subsequent chromatographic analysis in this assay is performed at basic pH conditions, excess Fe(III) has to be masked by adding an adequate amount of NTA to the sample. Otherwise, uncomplexed Fe(III) would precipitate and organophosphonates would adsorb onto the oxides formed (232,233). This analytical assay turned out to be relatively robust, however, Cr(III) forms very strong and kinetically inert complexes with organophosphonates and suppresses the complexation reaction of NTMP, EDTMP and DTPMP by Fe(III). Furthermore, the conversion of the Cu(II)-HEDP into the Fe(III)-HEDP complex is slightly hindered (232). Depending on the compound, the application of this method allows organophosphonate detections in the range of 15-100 μg/L without any enrichment step. However, NTMP, EDTMP and DTPMP can be preconcentrated by adsorption onto calcium carbonate, allowing detection limits in the low μg/L-range. Similar detection limits can be obtained by the flame photometric detection approach (229,235). However, these detectors were self-constructed and are not commercially available.

Surprisingly, mass spectrometric detection has received only little attention in organophosphonate analysis, most probably due to general problems encountered by interfacing ion-pair or ion chromatography and mass spectrometry. Organophosphonates are typically separated using mobile phases with high ionic strength, that cause adduct formation, suppression of analyte ionization, elevated backgrounds in the MS chromatogram and strong incrustations in the vacuum interface. In such cases, the use of ion-exchange devices (suppressors), that are positioned between the chromatographic column and the MS interface, is often of significant benefit. Suppressors can be used to remove involatile eluent constituents such as tetraalkylammonium or sodium cations by cation-exchange against H^+.

Knepper et al. reported on an ion-pair chromatography coupled to electrospray mass spectrometry in the negative mode using a suppressor module to remove non-volatile mobile phase constituents (227). It was demonstrated by analyzing a spiked sample of deionized water, that organophosphonates survived the applied electrospray process and that singly and doubly negatively charged molecular ions represented the most intense signals in the mass spectrum. Furthermore, no salt adducts of the molecular ion were observed. However, in the chromatogram of spiked surface water, no phosphonate signals could be detected. This was partially attributed to apolar matrix constituents (e.g. humic substances) suppressing the signal, since a clean-up of the spiked surface water sample via RP-C18-based solid phase extraction improved the signal considerably (227). In the applied procedure, however, matrix components most probably also changed organophosphonate speciation and molecular complex masses, thereby causing a general restriction of the overall assay performance. Even though no detection limits were reported for this procedure, presented experiments with the direct injection into the mass spectrometer via integrated syringe pump lead one to assume that detection limits in the low μg/L-range might be achievable under optimized conditions (7,236).

Klinger tried to analyze organophosphonates by particle-beam mass spectrometry based on electron impact (EI) ionization, but the expected EI spectra could not be observed, most probably due to the low vapor pressure of the target compounds (207) and their highly anionic character making the molecular ion radical ionization process more difficult. Therefore, the organophosphonates were transformed into their methyl ester derivatives using diazomethane. This approach, however, requires the complete removal of all water traces prior to the derivatization step, for which reason the water sample was evaporated to dryness first. Separation of the totally methylated organophosphonates was then performed on a diol column with a gradient mobile phase containing isopropyl alcohol and n-hexane. Starting with a sample volume of 500 mL water, limits of determination were about 10 μg/L. Similar sensitivities were obtained following replacement of the particle-beam interface by a thermospray interface (207). The method could, however, in either case be satisfactorily applied to demineralized water only, since the derivatization step turned out to be negatively influenced by the major cations and anions of the water matrix when analyzing drinking water samples (207).

Options for Sample Preconcentration

Since detection limits necessary for organophosphonate determination in surface waters are difficult to achieve by direct injection onto the liquid chromatograph, sample enrichment is an important option for sensitivity improvement of the overall assay. Methods for the preconcentration of organophosphonates from aqueous samples were intensively investigated (207,217,225). The major concepts are simple sample evaporation (206,207,225,227), enrichment by adsorption on collector precipitates (metal hydroxides, carbonates and phosphates) (217,232) and enrichment utilizing adsorbent materials (adsorption resins and exchange celluloses) (217). However, all sample preparation procedures suffer from some drawbacks and most of them are complex, trouble-prone and laborious. Furthermore, up to now, no convincing concept is available for the simultaneous enrichment of all relevant organophosphonates in major use (HEDP, NTMP, EDTMP, DTPMP) from environmental aqueous samples.

Klinger demonstrated that organophosphonates are stable in aqueous solutions at 80 °C for at least 3 hours (207). Consequently, the easiest way to concentrate a sample would be partial or complete water evaporation. Indeed, several authors reported on quite successful enrichments of spiked demineralized water samples by gentle evaporation of the water at 60 °C using for instance a rotary evaporator. So far, however, transfer of this approach to drinking or even surface water was always less successful and somehow connected with difficulties. The main problem is caused by calcareous and saline residues to which organophosphonates naturally show relatively high adsorption affinities (206,207,225,227). Such residues can be avoided by passing the sample prior to the evaporation step over a cation exchanger in the H^+ form (e.g. Bio-Rad AG MP-50) (207). Anions and organophosphonates pass the column and are enriched in the following evaporation step yielding a highly acidic and anion-rich sample concentrate. Despite intensive circumvention attempts (207,225), such sample properties often turned out to be completely incompatible with the subsequently applied analytical method (207,225,227). Nonetheless, to the opinion of the author, this preconcentration approach might be well compatible to other assays reported in the literature (e.g. conversion of organophosphonates into their Fe(III) complexes as described by Nowack (232)).

Organophosphonates can also be preconcentrated by adsorption onto a number of insoluble metal carbonates [Ca(II)] (217,232), hydroxides [Ca(II), Fe(III), Al(III), Bi(III)] (217,237) or phosphates [Ca(II)] (213,238-240). Assays of this kind can be carried out either as suspensions with the preformed salt or by coprecipitation. After centrifugation or filtration the insoluble salt or precipitate can be redissolved in acids (213,217,232,237-239,241). Frigge and Jackwerth demonstrated the successful enrichment of HEDP, NTMP and EDTMP by coprecipitation with bismuth hydroxide (217). If Bi(III) cations, which are present in relatively high concentrations after dissolution of the collector precipitate with sulfuric acid, interfere with the subsequent analytical assay, they can be separated by ion exchange using a strongly acidic cation-exchange resin (217). With this approach, recovery rates between 76 and > 95 % were obtained. However, significantly lower recoveries were found in the presence of 100 mg/L chloride, which was attributed to the formation of bismuth oxide chloride instead of the „hydroxide“ and its different surface and adsorption properties (217). Nowack reported on successful preconcentration of NTMP, EDTMP and DTPMP from natural waters or wastewaters using freshly precipitated $CaCO_3$ providing recoveries of 95-102 % at the 1 μM level (232).

Another option is the use of anion-exchange resins [Bio-Rad AG1-X2 (207), Bio-Rad BioRex5 (207), Riedel-de Haën VI9 (217), Waters IC Pak A (221), Dionex AG11 (107), Dionex AG4A (225), Varian Bond Elut SAX (213)] or celluloses [Whatman DE32 (217)]. Organophosphonates are typically well adsorbed on these materials so that elution has to performed with strong eluents like mineral acids that cause occasionally some difficulties in the subsequent analytical steps. Elution with organic acids like formic or acetic acid turned out to be unsuccessful (207,227). Some authors reported on interferences with cations (217) and anions (213,225) present in natural waters. However, according to the work of Frigge and Jackwerth (217), cationic interferences can be partially overcome by adding a competitive complexing agent like EDTA to the sample. An elegant option is sample enrichment based on anion-exchangers coupled on-line to adequate chromatographic systems (107,221,225).

Titration and Spectrophotometry

Analytical methods that were initially used for the determination of organophosphonates include a number of titrimetric and spectrophotometric assays. In early titrimetric methods organophosphonates were titrated as acids (211). Later, similar to aminopolycarboxylates, a series of spectrotitration assays was established based on the measurement of absorption changes of colored or luminescent metal complexes in the presence of organophosphonates. Published methods utilize the Fe(II)-1,10-phenanthroline complex (242), Fe(III)-azide (242), Fe(III)-thiocyanate (242), Fe(III)-sulfosaclicylate (204), Eu(III)-beta-diketones (204), and Th(IV)-diaminocyclohexanetetraacetate (237,240). For organophosphonate determination in boiler and cooler water, Sloat and Buck converted organophosphonates at acidic pH into their corresponding Fe(III) complexes by addition of a known amount of Fe(III) in excess (243). Samples were then passed through a cation exchange column to remove uncomplexed Fe(III). Fe(III) chelated by organophosphonates passes through the column and can be determined by addition of a strong reducing agent (bisulfite and hydrosulfite) and 1,10-phenanthrolin as indicator. The amount of iron present is stoichiometrically related to the concentration of organophosphonate present. However, the methods mentioned above provide detection limits in the low mg/L-range and are thus not sensitive enough for the determination of organophosphonates at trace levels.

An interesting method to determine organophosphonates in natural waters is based on the spectrophotometric determination of phosphate as phosphomolybdate after oxidation of the organophosphonate (204,207,219,241,244-246). However, applied without further sample preparation, this assay does not allow the distinction between phosphonates, inorganic phosphate and other organic phosphorus sources. Inorganic phosphate can be eliminated prior to the oxidation step as the insoluble triethylamine-phosphomolybdate complex (241). The sum of phosphonates can also be determined more selectively by fractionation of the phosphorus-containing substances in the sample (247). This method takes advantage of the resistance of phosphonates to phosphatase action and acid hydrolysis. A comparative analysis of the sample following complete oxidation and following partial conversion of the phosphorus-containing sample constituents by phosphatase and acid hydrolysis gives a measure of the phosphonates present in the sample (247). Typical detection limits of these assays are in the low to middle μg/L-range (207,241,247).

Furthermore, several kinetic methods have been developed for the sensitive determination of organophosphonates. Such assays are typically based on either inhibiting or activating effects of organophosphonates on catalyst ions in distinct redox reactions (204). Any change in catalyst ion concentration in the system causes a change in the reaction rate that can be recorded spectrophotometrically. The following reaction systems were described: (i) inhibition of the copper(II)ethylenediamine catalysis on the oxidation of hydroquinone by hydrogen peroxide at pH 8.7 (204); (ii) activation of manganese catalysis on the oxidation of o-phenylenediamine by hydrogen peroxide at pH 5.5 (204); (iii) inhibition of copper catalysis on the oxidation of hydroquinone by hydrogen peroxide with ammonium fluoride as activator at pH 7.3 (248); (iv) inhibition of chromium catalysis on the oxidation of o-dianisidine by hydrogen peroxide at pH 3.0 (248); (v) inhibition of manganese catalysis on the oxidation of o-dianisidine by periodate (204). Kinetic methods are characterized by high sensitivity with detection limits in the low µg/L-range. However, they are not very selective, since the presence of other complexones in the sample will also exhibit an inhibiting or activating effect on the activity of catalyst ions (204).

Other Methods

Taulli reported on a separation of HEDP and NTMP from aqueous media based on thin-layer chromatography (249). For detection, organophosphonates were converted to phosphate by a sodium hypochlorite solution followed by subsequent molybdate and sulfite solutions and observation of the blue coloration (249).

Structure and identity of the organophosphonates under consideration can be unequivocally determined by ^{31}P nuclear magnetic resonance spectroscopy (212,220,250-252). At high enough concentrations this technique can also be used for organophosphonate quantification. However, due to the low sensitivity of the ^{31}P nucleus, achievable detection limits are not practically relevant for environmental analytics (66).

Some electroanalytical methods have also been described. Perosa et al. reported on polarography-based titrations of HEDP and EDTMP with Pb(II) cations as titrant (253). Similar to aminopolycarboxylates, NTMP, DTPMP, EDTMP and HEDP can be determined by differential pulse polarography (DPP) (161,162). However, these methods are characterized by limited selectivity and sensitivity (in the low mg/L-range).

Furthermore, analysis of organophosphonates has been performed by some electrophoretic approaches (254-257). Shamsi and Danielson reported on a separation of HEDP, NTMP, EDTMP and DTPMP based on capillary zone electrophoresis with negative polarity, where ribonucleotides and adenosine monophosphonate are used as electrolytes and detection is performed by indirect photometric detection (254). Koudelková and Jedináková-Křížová studied the rhenium complexation by HEDP utilizing capillary zone electrophoresis with negative polarity and direct UV detection (257). HEDP, NTMP, EDTMP and DTPMP were also separated via isotachophoresis with conductivity detection (256,258). Reported detection limits of these assays were in the middle mg/L-range.

Summary

Analysis of organophosphonates is challenging due to their highly ionic character, low reactivity and lack of strong UV chromophores. Furthermore, these compounds easily form chelates with metal ions and show a high adsorption affinity to calcareous and saline residues, making analytical procedures complicated and laborious. The most useful and sensitive analytical assays include methods based on liquid chromatography, methods based on mineralization and subsequent determination of the formed phosphate, and kinetic methods based on spectrophotometry. However, most of these approaches have detection limits above the expected natural concentrations or suffer from interferences. Optimized methods based on either anion exchange or ion-pair chromatography allow, at best, for the determination of organophosphonates in wastewaters or highly polluted surface waters. Thus, one of the major limitations in the occurrence and fate assessment of organophosphonates in the aquatic environment is still the lack of an adequate method for reliable determination of these compounds at concentrations in the very low μg/L-range. However, optimization of sample preconcentration procedures and the use of negative ion electrospray (tandem) mass spectrometry and ICP-MS detectors in combination with liquid chromatography or capillary electrophoresis are still promising options for further improvement of organophosphonate analytics.

References

1. Bucheli-Witschel, M.; Egli, T. *FEMS Microbiol. Rev.* **2001,** *25,* 69-106.
2. Wolf, K.; Gilbert, P. A. In *The Handbook of Environmental Chemistry,* Hutzinger, O., Ed.; Springer: Berlin, 1992; Vol. 3/F, pp 243-259.
3. Gledhill, W. E.; Feijtel, T. C. J. In *The Handbook of Environmental Chemistry,* Hutzinger, O., Ed.; Springer: Berlin, 1992; Vol. 3/F, pp 261-285.
4. Jaworska, J.; Van Genderen-Takken, H.; Hanstveit, A.; Van de Plassche, E.; Feijtel, T. *Chemosphere* **2002,** *47,* 655-665.
5. Nowack, B. *Water Res.* **2003,** *37*(11), 2533-2546.
6. Jäger, H.-U.; Schul, W. *Muench. Beitr. Abwasser-, Fisch, Flussbiol.* **2001,** *54,* 207-226.
7. Knepper, T. P. *TRAC* **2003,** *22*(10), 708-724.
8. Potthoff-Karl, B. *SÖFW J.* **1994,** *120*(2-3), 104-109.
9. Schmidt, C. K.; Fleig, M.; Sacher, F.; Brauch, H.-J. *Environ. Pollut.* **2004,** *131,* 107-124.
10. Nowack, B. *Water Res.* **1998,** *32*(4), 1271-1279.

11. Sillanpää, M.; Orama, M.; Rämö, J.; Oikari, A. *Sci. Total Environ.* **2001,** *267,* 23-31.
12. Nowack, B. *Environ. Sci.Technol.* **2002,** *36,* 4009-4016.
13. Sillanpää, M.; Sihvonen, M.-L. *Talanta* **1997,** *44,* 1487-1497.
14. Retho, C.; Diep, L. *Zeitsch. Lebensm.Unters. Forsch.* **1989,** *188*(3), 223-226.
15. Pietsch, J.; Schmidt, W.; Sacher, F.; Fichtner, S.; Brauch, H.-J. *Fresenius J. Anal. Chem.* **1995,** *353*(1), 75-82.
16. Otteneder, H.; Schleser, B. *Lebensmittelchemie* **1992,** *46* 87-90.
17. Nishikawa, Y.; Okumura, T. *J. Chromatogr. A* **1995,** *690*(1), 109-118.
18. Rudling, L. *Water Res.* **1972,** *6*(7), 871-876.
19. Murray, D.; Povoledo, D. *J. Fish. Res. Bd. Can.* **1971,** *28*(7), 1043-1047.
20. Blank, M. L.; Snyder, F. *J. Chromatogr.* **1979,** *170*(2), 379-383.
21. Cassidy, R. M.; Harpur, R.; Elchuk, S. *J. Chromatogr.* **1980,** *190*(1), 188-192.
22. Williams, D. T. *J. AOAC* **1974,** *57*(6), 1383-1385.
23. Sillanpää, M.; Sorvari, J.; Sihvonen, M.-L. *Chromatographia* **1996,** *42*(9/10), 578-582.
24. Tuulos-Tikka, S.; Sillanpää, M.; Rämö, J. *Intern. J. Environ. Anal. Chem.* **2000,** *77*(3), 221-232.
25. Gardiner, J. *Analyst* **1977,** *102*(1211), 120-123.
26. Ribick, M. A.; Jemal, M.; Cohen, A. I. *J. Pharm. Biomed. Anal.* **1987,** *5*(7), 687-694.
27. Raksit, A. *J. AOAC Int.* **2002,** *85*(1), 50-55.
28. Wanke, T.; Eberle, S. H. *Acta hydrochim. hydrobiol.* **1992,** *20*(4), 192-196.
29. Reichert, J. K.; Linckens, A. H. M. *Environ. Tech. Lett.* **1980,** *1,* 42-49.
30. Bauer, K.-H.; Bonin, H.; Fink, A.; Aimene, A.; Schön, P.; Weber, L. *Vom Wasser* **1997,** *88,* 351-364.
31. Lee, H.-B.; Peart, T. E.; Kaiser, K. L. E. *J. Chromatogr. A* **1996,** *738,* 91-99.
32. Chau, Y. K.; Fox, M. E. *J. Chromatogr. Sci.* **1971,** *9*(5), 271-275.
33. Sillanpää, M.; Vickackaite, V.; Rämö, J.; Niinistö, L. *Analyst* **1998,** *123,* 2161-2165.
34. Meißner, G.; Ried, A.; Stork, G. *gwf Wasser Abwasser* **1997,** *138*(11), 564-569.
35. Randt, C.; Wittlinger, R.; Merz, W. *Fresenius J. Anal. Chem.* **1993,** *346*(6-9), 728-731.
36. Soßdorf, D.; Brenner-Weiß, G.; Kreckel, P.; Ternes, T.; Wilken, R.-D. *Vom Wasser* **2000,** *94,* 121-134.
37. Soßdorf, D.; Dürner, B.; Henatsch, J.; Wilken, R.-D. *Vom Wasser* **2001,** *96,* 211-222.

38. Ternes, T. A.; Stumpf, M.; Steinbrecher, T.; Brenner-Weiß, G.; Haberer, K. *Vom Wasser* **1996,** *87,* 275-290.
39. Nguyen, D. L.; Bruchet, A.; Aprino, P. *J. High Res. Chromatogr.* **1994,** *17*(3), 153-159.
40. Dietz, F. *gwf Wasser Abwasser* **1987,** *128*(5), 286-288.
41. Brauch, H.-J.; Schullerer, S. *Vom Wasser* **1987,** *69,* 155-164.
42. Lindner, K.; Knepper, T. P.; Karrenbrock, F.; Rörden, O.; Brauch, H.-J.; Lange, F. T.; Sacher, F. *IAWR-Rheinthemen, Band 1. Erfassung und Identifizierung von trinkwassergängigen Einzelsubstanzen in Abwässern und im Rhein;* IAWR, 1996.
43. Aue, W. A.; Hastings, C. R.; Gerhardt, K.; Pierce, J.; Hill, H. H.; Moseman, R. F. *J. Chromatogr.* **1972,** *72,* 259-267.
44. Warren, C. B.; Malec, E. J. *J. Chromatogr.* **1972,** *64*(2), 219-237.
45. Sniegoski, P. J.; Venezky, D. L. *J. Chromatogr. Sci.* **1974,** *12*(6), 359-361.
46. Games, L. M.; Staubach, J. A.; Kappeler, T. U. *Tenside Det. Surf.* **1981,** *18*(5), 262-265.
47. Sacher, F.; Lochow, E.; Brauch, H.-J. *Vom Wasser* **1998,** *90,* 31-41.
48. Stolzberg, R. J.; Hume, D. N. *Anal. Chem.* **1977,** *49*(3), 374-378.
49. Taylor, J.; Zielinski, W.; Marienthal, E.; Durst, R.; Burke, R. *Interferences from TMS-esters of NTA with TMS-esters of lauric and myristic acid*; EPA-report R2-72-057; US-Environmental Protection Agency: 1972.
50. Rudling, L. *Water Res.* **1971,** *5*(10), 831-838.
51. Williams, D. T.; Benoit, F.; Muzika, K.; Ogrady, R. *J. Chromatogr.* **1977,** *136*(3), 423-427.
52. Schaffner, C.; Giger, W. *J. Chromatogr.* **1984,** *312,* 413-421.
53. Schürch, S.; Dübendörfer, G. *Mitt. Gebiete Lebensm. Hyg.* **1989,** *80,* 324-334.
54. DIN 38413-10. 2000.
55. EN ISO 16588. 2002.
56. Neitzel, P.; Nestler, W.; Dehnert, J.; Haupt, A. *Vom Wasser* **1997,** *88,* 33-48.
57. Bergers, P. J. M.; de Groot, A. C. *Water Res.* **1994,** *28*(3), 639-642.
58. Loyaux-Lawniczak, S.; Douch, J.; Behra, P. *Fresenius J. Anal. Chem.* **1999,** *364,* 727-731.
59. Dai, J.; Helz, R. G. *Anal. Chem.* **1988,** *60,* 301-305.
60. Harmsen, J.; van den Toorn, A. *J. Chromatogr.* **1982,** *249,* 379-384.
61. Buchberger, W.; Haddad, P. R.; Alexander, P. W. *J. Chromatogr.* **1991,** *558*(1), 181-186.
62. Nowack, B.; Kari, F. G.; Hilger, S. U.; Sigg, L. *Anal. Chem.* **1996,** *68,* 561-566.
63. Yamaguchi, A.; Rajput, A. R.; Ohzeki, K.; Kambara, T. *Bull. Chem. Soc. Japan* **1983,** *56*(9), 2621-2623.

64. Yamaguchi, A.; Toda, A.; Ohzeki, K.; Kambara, T. *Bull. Chem. Soc. Japan* **1983,** *56*(10), 2949-2951.
65. Vora, M. M.; Wukovnig, S.; Finn, R. D.; Emran, A. M.; Boothe, T. E.; Kothari, P. J. *J. Chromatogr.* **1986,** *369*(1), 187-192.
66. Weiss, J.; Hägele, G. *Fresenius Z. Anal. Chem.* **1987,** *328*(1-2), 46-50.
67. Knox, J. H.; Shibukawa, M. *J. Chromatogr.* **1991,** *545*(1), 123-134.
68. Taylor, D. L.; Jardine, P. M. *J. Environ. Qual.* **1995,** *24*(4), 789-792.
69. Ye, L.; Lucy, C. A. *Anal. Chem.* **1995,** *67,* 2534-2538.
70. Ye, L.; Lucy, C. A. *J. Chromatogr. A* **1996,** *739,* 307-315.
71. Randt, C.; Klein, J.; Merz, W. *Vom Wasser* **1995,** *84,* 61-67.
72. Bedsworth, W. W.; Sedlak, D. L. *J. Chromatogr. A* **2001,** *905,* 157-162.
73. Richardson, D. E.; Ash, G. H.; Harden, P. E. *J. Chromatogr. A* **1994,** *688,* 47-53.
74. Venezky, D. L.; Rudzinski, W. E. *Anal. Chem.* **1984,** *56,* 315-317.
75. Jen, J. F.; Chen, C. S. *Anal. Chim. Acta* **1992,** *270*(1), 55-61.
76. Sillanpää, M.; Raimo, K. B.; Silhoven, M. L. *Anal. Chim. Acta* **1995,** *303*(2-3), 187-192.
77. Perfetti, G. A.; Warner, C. R. *J. AOAC* **1979,** *62*(5), 1092-1095.
78. de Jong, J.; van Polanen, A.; Driessen, J. J. M. *J. Chromatogr.* **1991,** *553*(1-2), 243-248.
79. Inman, E. L.; Clemens, R. L.; Olson, B. A. *J. Pharm. Biomed. Anal.* **1990,** *8*(6), 513-520.
80. Stalberg, O.; Arvidsson, T. *J. Chromatogr. A* **1994,** *684*(2), 213-219.
81. Kord, A. S.; Tumanova, I.; Matier, W. L. *J. Pharm. Biomed. Anal.* **1995,** *13*(4-5), 575-580.
82. Tran, G.; Chen, C.; Miller, R. B. *J. Liquid Chromatogr. Rel. Tech.* **1996,** *19*(9), 1499-1508.
83. Pistos, C.; Parissi-Poulou, M. *J. Pharm. Biomed. Anal.* **2002,** *28,* 1073-1079.
84. Miller, M. L.; McCord, B. R.; Martz R.; Budowle B. *J. Anal. Toxicol.* **1997,** *21,* 521-528.
85. Van de Gucht, I. *J. Chromatogr. A* **1994,** *671*(1-2), 359-365.
86. Göttlicher, U.; Siegfried, R.; Birke, H. *Fresenius J. Anal. Chem.* **1995,** *352*(3-4), 398-400.
87. Lucena, J. J.; Barak, P.; Hernández-Apaolaza, L. *J. Chromatogr. A* **1996,** *727*(2), 253-264.
88. Hernández-Apaolaza, L.; Barak, P.; Lucena, J. J. *J. Chromatogr. A* **1997,** *789,* 453-460.
89. Cassidy, R. M.; Elchuk, S. *Anal. Chem.* **1985,** *57*(3), 615-620.
90. Virtaphoja, J. *Pulp Paper Can.* **1998,** *99*(10), 33-35.
91. Unger, M.; Mainka, E.; König, W. *Fresenius Z. Anal. Chem.* **1987,** *329*(1), 50-54.

92. Huber, W. *Acta hydrochim. hydrobiol.* **1992,** *20*(1), 6-8.
93. Chinnik, C. C. T. *Analyst* **1981,** *106* 1203-1207.
94. Deacon, M.; Smyth, M. R.; Tuinstra, L. G. M. T. *J. Chromatogr. A* **1994,** *659*(2), 349-357.
95. Nirel, P. M.; Pardo, P.-E.; Landry, J.-C.; Revaclier, R. *Water Res.* **1998,** *32*(12), 3615-3620.
96. Geschke, R.; Zehringer, M. *Fresenius J. Anal. Chem.* **1997,** *357,* 773-776.
97. Collins, R. N.; Onisko, B.; McLaughlin, J.; Merrington, G. *Environ. Sci. Technol.* **2001,** *35,* 2589-2593.
98. Buchberger, W.; Haddad, P. R.; Alexander, P. W. *J. Chromatogr.* **1991,** *546*(1-2), 311-315.
99. Bauer, K.-H.; Knepper, T. P.; Maes, A.; Schatz, V.; Voihsel, M. *J. Chromatogr. A* **1999,** *837,* 117-128.
100. Metrohm *IC Application Note* **2002** No. S-125.
101. DIN 38413-8. 2000.
102. Erlmann, W.; Foery, M.; Jantschke, H. *Galvanotechnik* **1990,** *81*(4), 1249-1258.
103. Lucy, C. A.; Ye, L. *Anal. Chem.* **1995,** *67,* 79-82.
104. Kumar, K.; Sukumaran, K. V.; Tweedle, M. F. *Anal. Chem.* **1994,** *66*(2), 295-299.
105. Baron, D.; Hering, J. G. *J. Environ. Qual.* **1998,** *27,* 844-850.
106. Schoppenthau, J.; Dunemann, L. *Fresenius J. Anal. Chem.* **1994,** *349*(12), 794-799.
107. Ammann, A. A. *J. Chromatogr. A* **2002,** *947,* 205-216.
108. Ammann, A. A. *Anal. Bioanal. Chem.* **2002,** *372,* 448-452.
109. Hajós, P.; Révész, G.; Horváth, O.; Peear, J.; Sarzanini, C. *J. Chromatogr. Sci.* **1996,** *34,* 291-299.
110. Deacon, M.; Smyth, M. R.; Tuinstra, L. G. M. T. *J. Chromatogr. A* **1993,** *657,* 69-76.
111. Deguchi, T. *J. Chromatogr.* **1976,** *120*(1), 159-170.
112. Kari, F. G.; Giger, W. *Environ. Sci. Technol.* **1995,** *29,* 2814-2827.
113. Kari, F. G.; Giger, W. *Water Res.* **1996,** *30*(1), 122-134.
114. Bedsworth, W. W.; Sedlak, D. L. *Environ. Sci. Technol.* **1999,** *33,* 926-931.
115. Sarzanini, C.; Mentasti, E. *J. Chromatogr. A* **1997,** *789,* 301-321.
116. Baraj, B.; Martinez, M.; Sastre, A.; Aguilar, M. *J. Chromatogr. A* **1995,** *695*(1), 103-112.
117. Blatny, P.; Kvasnicka, F.; Kenndler, E. *J. Chromatogr. A* **1997,** *757*(1-2), 297-302.
118. Conradi, S.; Vogt, C.; Wittrisch, H.; Knobloch, G.; Werner, G. *J. Chromatogr. A* **1996,** *745*(1-2), 103-110.

119. Fukushi, K.; Takeda, S.; Wakida, S.; Higashi, K.; Hiiro, K. *J. Chromatogr. A* **1997,** *759*(1-2), 211-216.
120. Haumann, I.; Bächmann, K. *J. Chromatogr. A* **1995,** *717*(1-2), 385-391.
121. Jung, G. Y.; Kim, Y. S.; Lim, H. B. *Anal. Sci.* **1997,** *13*(3), 463-467.
122. Liu, W.; Lee, H. K. *J. Chromatogr. A* **1998,** *796*(2), 385-395.
123. Motomizu, S.; Oshima, M.; Matsuda, S.; Obata, Y.; Tanaka, H. *Anal. Sci.* **1992,** *8,* 619-625.
124. Semenova, O. P.; Timerbaev, A. R.; Gagstädter, R.; Bonn, G. K. *J. High Res. Chromatogr.* **1997,** *19*(3), 177-179.
125. Timerbaev, A. R.; Andrei, R.; Semenova, O. P.; Olga, P.; Bonn, G. K. *Analyst - Letchworth* **1994,** *119*(12), 2795-2800.
126. Wang, T.; Li, S. F. Y. *J. Chromatogr. A* **1995,** *707*(2), 343-354.
127. Weston, A.; Brown, P. R.; Jandik, P.; Jones, W. R.; Heckenberg, A. L. *J. Chromatogr.* **1992,** *593,* 289-295.
128. Liu, W.; Lee H. K. *J. Chromatogr. A* **1999,** *834,* 45-63.
129. Timerbaev, A. R.; Semenova, O. P.; Petrukhin O. M. *J. Chromatogr. A* **2002,** *943,* 263-274.
130. O'Keeffe, M.; Dunemann, L.; Theobald, A.; Svehla, G. *Anal. Chim. Acta* **1995,** *306,* 91-97.
131. Buchberger, W.; Mülleder, S. *Mikrochim. Acta* **1995,** *119,* 103-111.
132. Buchberger, W.; Aichhorn, D.; Niessner, G.; Haddad, P. R.; Bogan, D. *Monatshefte Chem.* **1998,** *129* 811-816.
133. Bürgisser, C. S.; Stone, A. T. *Environ. Sci. Technol.* **1997,** *31,* 2656-2664.
134. Harvey, S. D. *J. Chromatogr.* **1996,** *736,* 333-340.
135. Krokhin, O. V.; Xu, W.; Hoshino, H.; Shpigun, O. A.; Yotsuyanagi, T. *Chem. Lett.* **1996,** *12,* 1095-1096.
136. Krokhin, O. V.; Adamov, A. V.; Hoshino, H.; Shpigun, O. A.; Yotsuyanagi, T. *J. Chromatogr. A* **1999,** *850,* 269-276.
137. Kvasnicka, F.; Míková, K. *J. Food Comp. Anal.* **1996,** *9,* 231-242.
138. Owens, G.; Ferguson, V. K.; McLaughlin, J.; Singleton, I.; Reid, R. J.; Smith, F. A. *Environ. Sci. Technol.* **2000,** *34,* 885-891.
139. Pozdniakova, S.; Ragauskas, R.; Dikcius, A.; Padarauskas, A. *Fresenius J. Anal. Chem.* **1999,** *363,* 124-125.
140. Sheppard, R. L.; Henion, J. *Anal. Chem. News Feat.* **1997,** 477A-480A.
141. Sheppard, R. L.; Henion, J. *Anal. Chem.* **1997,** *69,* 2901-2907.
142. Soga, T.; Imaizumi, M. *Electrophoresis* **2001,** *22,* 3418-3425.
143. Sheppard, R. L.; Henion, J. *Electrophoresis* **1997,** *18,* 287-291.
144. Schäffer, S.; Gareil, P.; Carpot, L.; Dezael, C. *J. Chromatogr. A* **1995,** *717,* 351-362.
145. Okemgbo, A. A.; Hill, H. H.; Metcalf, S. G.; Bachelor, M. *Anal. Chim. Acta* **1999,** *396,* 105-116.
146. Bullock, J. *J. Chromatogr. B* **1995,** *669,* 149-155.

147. Wiley, J. P. *J. Chromatogr. A* **1995,** *692,* 267-274.
148. Ballou, N. R.; Ducatte, G. R.; Quang, C.; Remcho, V. T. *J. High Res. Chromatogr.* **1996,** *19,* 183-188.
149. Ye, L.; Wong, J. E.; Lucy, C. A. *Anal. Chem.* **1997,** *69,* 1837-1843.
150. Zhu, Z. W.; Zhang, L. F.; Marintuthu, A.; Yang, Z. G. *Am. Lab.* **2003,** *35*(2), 18-19.
151. Zhu, Z. W.; Zhang, L. F.; Marimuthu, A.; Yang, Z. G. *Electrophoresis* **2002,** *23*(17), 2880-2887.
152. Kaiser, K. L. E. *Water Res.* **1973,** *7,* 1465-1473.
153. Schramm, C.; Maier, S.; Bobleter, O.; Blankenhorn, P. *GIT Fachz. Lab.* **1994,** *3,* 184-189.
154. DIN 38413-5. 1990.
155. Metrohm *Application Bulletin* **2002** No. 76/3 d.
156. Belal, F.; Aly, F. A.; Walash, M. I.; Mesbah, A. O. *J. Pharm. Biomed. Anal.* **1998,** *17,* 1249-1256.
157. Metrohm *Application Bulletin* **2002** No. 143/2 d.
158. Voulgaropoulos, A.; Tzivanakis, N. *Electroanalysis* **1992,** *4,* 647-651.
159. Schön, P.; Bauer, K.-H.; Wiskamp, V. *Fresenius J. Anal. Chem.* **1997,** *358,* 699-702.
160. Voulgaropoulos, A.; Valenta, P.; Nürnberg, H. W. *Fresenius Z. Anal. Chem.* **1984,** *317,* 367-371.
161. Esser, S.; Wenclawiak, B. W.; Gabelmann, H. *Fresenius J. Anal. Chem.* **2000,** *368,* 250-255.
162. Esser, S.; Wenclawiak, B. W.; Gabelmann, H. L. *Vom Wasser* **1999,** *92,* 61-69.
163. Kratochvil, B.; White, M. C. *Anal. Chem.* **1965,** *37*(1), 111-113.
164. Pilipenko, A. T.; Tulyupa, M. F. *J. Anal. Chem. USSR* **1990,** *45*(5), 698-702.
165. Fujita, Y.; Mori, I.; Matsuo, T. *Anal. Sci.* **1998,** *14*(6), 1157-1159.
166. Cherny, P. J.; Crafts, B.; Hagermoser, H. H.; Boule, A. J.; Harbin, R.; Zak, B. *Anal. Chem.* **1954,** *26*(11), 1806-1809.
167. Mosher, R. E.; Burcar, P. J.; Boyle, A. J. *Anal. Chem.* **1963,** *35*(3), 403.
168. Saito, K.; Hasuo, T.; Nakano, H. *Nippon Jozo Kyokai Zasshi* **1968,** *63,* 1193-1197.
169. Bruno, E.; Calapaj, R.; Sergi, G. *Ann. Fac. Econ. Commerc. Univ. Stud. Messina* **1969,** *7,* 3-12.
170. Hill-Cottingham, D. G. *Soil Sci.* **1957,** *84,* 43-49.
171. Bhattacharyya, S. N.; Kundu, K. P. *Talanta* **1971,** *18*(4), 446-449.
172. Clinckemaille, G. G.; Vanwelssenaers, N. *Anal. Chim. Acta* **1972,** *58,* 243-245.
173. Menis, O.; House, H. H.; Rubin, I. B. *Anal. Chem.* **1956,** *28*(9), 1439-1441.

174. Hamano, T.; Mitsuhashi, Y.; Tanaka, L.; Matsuki, Y.; Oji, Y.; Okamoto, S. *Z. Lebensm. Unters. Forsch.* **1985,** *181*(1), 35-39.
175. Parkash, R.; Bansal, R.; Rehani, S. K.; Dixit, S. *Talanta* **1998,** *46*(6), 1573-1576.
176. Parkash, R.; Bansal, R. *Anal. Lett.* **1990,** *23*(7), 1159-1166.
177. Bersin, T.; Schwarz, H. *Schweiz. Med..Wochenschr.* **1953,** *83*(33), 756-766.
178. Darbey, A. *Anal. Chem.* **1952,** *24*(2), 373-378.
179. Honová, D.; Nemcová, I.; Suk, V. *Talanta* **1988,** *35*(10), 803-804.
180. Wanke, T. Ph.D. thesis, , Kernforschungszentrum, Karlsruhe, 1993.
181. Kawasaki, N.; Lee, Y. C.; Hashimoto, O.; Yamamoto, M.; Kawanishi, T.; Hayakawa, T. *Anal. Biochem.* **1999,** *270,* 329-331.
182. Parker, C. J. *Anal. Chem.* **1964,** *36*(1), 236.
183. Shenker, M.; Hadar, Y.; Chen, Y. *Soil Sci. Soc. Am. J.* **1995,** *59,* 1612-1618.
184. Kunkel, R.; Manahan, S. E. *Anal. Chem.* **1973,** *45*(8), 1465-1468.
185. Itabashi, H.; Umetsu, K.; Teshima, N.; Satoh, K.; Kawashima, T. *Anal. Chim. Acta* **1992,** *261,* 213-218.
186. Nemcova, I.; Pesinova, H.; Suk, V. *Microchem. J.* **1984,** *30*(1), 27-32.
187. Qureshi, S. Z.; Bansal, R. *Analusis* **1984,** *12*(1), 42-44.
188. Haddad, P. *Talanta* **1977,** *24*(1), 1-13.
189. Hamano, T.; Mitsuhashi, Y.; Kojima, N.; Aoki, N.; Shibata, M.; Ito, Y.; Yoshikiyo, O. *Analyst* **1993,** *118,* 909-912.
190. Campana, G. A. M.; Barrero, F. A.; Ceba, M. R. *Anal. Chim. Acta* **1996,** *329,* 319-325.
191. Lee, Y. C. *Anal. Biochem.* **2001,** *293,* 120-123.
192. Jones, D. R.; Manahan, S. E. *Anal. Lett.* **1975,** *8*(7), 421-434.
193. Milosavljevic, E. B.; Solujic, L.; Hendrix, J. L.; Nelson, J. H. *Analyst* **1989,** *114,* 805-808.
194. Güclü, K.; Hugül, M.; Demirci-Cekic, S.; Apak, R. *Talanta* **2000,** *53,* 213-222.
195. Belal, F.; Aly, F. A.; Walash, M. I.; Kenawy, I. M.; Osman, A. M. *FARMACO* **1998,** *53*(5), 365-368.
196. Clinckemaille, G. G. *Anal. Chim. Acta* **1968,** *43,* 520-522.
197. Blijenberg, B. G.; Leijnse, B. *Cin. Chim. Acta* **1969,** *26* 577-579.
198. van der Deelen, J.; van der Hende, A. *Chemie Analytique* **1968,** *50,* 237-241.
199. Milwidsky, B. M. *Soap Cosmetics Chem. Spec.* **1971,** *47,* 46-49.
200. Ternero, M.; Pino, F.; Perezbendito, D.; Valcarel, M. *Anal. Chim. Acta* **1979,** *109*(2), 401-409.
201. Rayasaro, T.; Perezbendito, D. *Analyst* **1983,** *108*(1288), 857-863.
202. Fuhrman, D. L.; Latimer, G. W.; Bishop, J. *Talanta* **1965,** *13*(1), 103-108.

203. Hulden, S. G.; Harju, L. *Talanta* **1980,** *27*(10), 815-817.
204. Ratina, M. A.; Zolotova, G. A.; Dolmanova, I. F. *J. Anal. Chem. USSR* **1987,** *42*(6), 774-785.
205. BUA *Ethylendiamintetraessigsäure/Tetranatriumethylendiamintetraacetat (H_4EDTA/Na_4EDTA); BUA-Stoffbericht 168;* S. Hirzel Wissenschaftliche Verlagsgesellschaft: Stuttgart, 1996.
206. Klinger, J.; Sacher, F.; Brauch, H.-J.; Maier, D. *Acta hydrochim. hydrobiol.* **1997,** *25*(2), 79-86.
207. Klinger, J. Ph.D. thesis, Technische Universität, Dresden, 1997.
208. Ismail, Z.; Aldous, S.; Triggs, E. J.; Smithurst, B. A.; Barry, H. D. *J. Chromatogr.* **1987,** *404*(2), 372-377.
209. Thompson, R.; Grinberg, N.; Perpall, H.; Bicker, G.; Tway, P. *J. Liquid Chromatogr.* **1994,** *17*(11), 2511-2531.
210. Tsai, E. W.; Chamberlin, S. D.; Forsyth, R. J.; Bell, C.; Ip, D. P.; Brooks, M. A. *J. Pharm. Biomed. Anal.* **1994,** *12*(8), 983-991.
211. Lovdahl, M. J.; Pietrzyk, D. J. *J. Chromatogr. A* **1999,** *850*(1-2), 143-152.
212. Pacholec, F.; Rossi, D. T.; Ray, L. D.; Vazopolos, S. *Liquid Chromatogr. HPLC Mag.* **1985,** *3,* 1068-1069.
213. Daley-Yates, P. T.; Gifford, L. A.; Hoggarth, C. R. *J. Chromatogr.* **1989,** *490*(2), 329-338.
214. Meek, S. E.; Pietrzyk, D. J. *Anal. Chem.* **1988,** *60*(14), 1397-1400.
215. Fitchett, A. W.; Woodruff, A. *Liquid Chromatogr. HPLC Mag.* **1983,** *1,* 48-49.
216. Vaeth, E.; Sladek, P.; Kenar, K. *Fresenius Z. Anal. Chem.* **1987,** *329,* 584-589.
217. Frigge, E.; Jackwerth, E. *Anal. Chim.Acta* **1991,** *254*(1-2), 65-73.
218. Tschabunin, G.; Fischer, P.; Schwedt, G. *Fresenius Z. Anal. Chem.* **1989,** *333*(2), 111-116.
219. Tschabunin, G.; Fischer, P.; Schwedt, G. *Fresenius Z. Anal. Chem.* **1989,** *333*(2), 117-122.
220. Tschabunin, G.; Schwedt, G.; Fischer, P. *Fresenius Z. Anal. Chem.* **1989,** *333*(2), 123-128.
221. Wong, D.; Jandik, P.; Jones, W. R.; Hagenaars, A. *J. Chromatogr.* **1987,** *389*(1), 279-285.
222. Tewari, K. M. J.; vanStroeBieze, S. A. M. *J. Chromatogr. A* **1997,** *771*(1-2), 155-161.
223. Felber, H.; Hegetschweiler, K.; Muller, M.; Odermatt, R.; Wampfler, B. *Chimia* **1995,** *49*(6), 179-181.
224. Waldhoff, H.; Sladek, P. *Fresenius Z. Anal. Chem.* **1985,** *320*(2), 163-168.
225. Schulze, B. Diplomarbeit, Fachhochschule, Mannheim, 1997.

226. Reichert, J. K. *Organophosphonsäuren - Bewertung der Leistungsfähigkeit biologischer Abwasserbehandlungsanlagen zu ihrer Eliminierung und Beurteilung ihres Umweltgefährdungspotentials,* Status report Deutsche Bundesstiftung Umwelt, 1995.
227. Knepper, T. P.; Driemler, J.; Maes, A.; Müller, J.; Soßdorf, D. Einträge synthetischer Komplexbildner in die Gewässer, Final report FKZ 299 24 284, 2001.
228. Forbes, K. A.; Vecchiarelli, J.; Uden, P. C.; Barnes, R. M. In *Advance in Ion Chromatography;* Jandik, P.; Cassidy, R. M., Ed.; Century International Inc.: Franklin, MA, 1989, pp 487-502.
229. Chester, T. L. *Anal. Chem.* **1980,** *52*(11), 1621-1624.
230. Li, C. Y.; Gao, L. Z.; Zhao, G. H.; Kang, J. W.; He, H. H. *Chromatographia* **1997,** *46*(9-10), 489-494.
231. Li, C. Y.; Gao, J. Z.; Yu, S. Y.; Han, X. Q.; Li, B. Y.; Liu, H. T. *Chromatographia* **2001,** *54*(1-2), 114-116.
232. Nowack, B. *J. Chromatogr. A* **1997,** *773* 139-146.
233. Nowack, B.; Stone, A. T. *Environ. Sci. Technol.* **2000,** *34*(22), 4759-4765.
234. Chester, T. L.; Lewis, E. C.; Benedict, J. J.; Sunberg, R. J.; Tettenhorst, W. C. *J. Chromatogr.* **1981,** *225*(1), 17-25.
235. Julin, B. G.; Vandenborn, H. W.; Kirkland, J. J. *J. Chromatogr.* **1975,** *112,* 443-453.
236. Knepper, T. P.; Weil, H. *Vom Wasser* **2001,** *97* 193-232.
237. Liggett, S. J. *Biochem. Med.* **1973,** *7*(1), 68-77.
238. Flesch, G.; Hauffe, S. A. *J. Chromatogr.* **1989,** *489*(2), 446-451.
239. Flesch, G.; Tominaga, N.; Degen, P. *J. Chromatogr.* B **1991,** *568*(1), 261-266.
240. Liggett, S. J.; Libby, R. A. *Talanta* **1970,** *17*(11), 1135-1140.
241. Bisaz, S.; Felix, R.; Fleisch, H. *Clin. Chim. Acta* **1975,** *65*(3), 299-307.
242. Benyoseph, O.; Sparkes, M. J.; Dixon, H. B. F. *Anal. Biochem.* **1993,** *210*(1), 195-198.
243. Sloat, S. S.; Buck, M. *Combustion* **1979,** *51*(2), 10-13.
244. Gonzalez, O. F. I.; Statham, P. J. *Water Res.* **1996,** *30*(11), 2739-2747.
245. Priebe, S. R.; Howell, J. A. *J. Chromatogr.* **1985,** *324*(1), 53-63.
246. Woo, L.; Maher, W. *Anal. Chim. Acta* **1995,** *315*(1-2), 123-135.
247. Cembella, A. D.; Antia, N. J. *Mar. Chem.* **1986,** *19*(3), 205-210.
248. Ratina, M. A.; Zolotova, G. A.; Dolmanova, I. F. *Zurnal analitizescoj chimii* **1980,** *35,* 1366-1371.
249. Taulli, T. A. *Anal. Chem.* **1967,** *39*(14), 1901.
250. Sawada, K.; Araki, T.; Suzuki, T. *Inorg. Chem.* **1987,** *26*(8), 1199-1204.
251. Wittmann, Z.; Szebenyi, N. *Acta Chim. Hung.* **1983,** *114*(1), 89-93.
252. Glonek, T.; Henderson, T. O.; Hilderbrand, R. L.; Myers, T. C. *Science* **1970,** *169*(941), 192-194.

253. Perosa, D.; Zanette, M. L.; Magno, F.; Bontempelli, G. *Analyst* **1986,** *111*(3), 365-369.
254. Shamsi, S. A.; Danielson, N. D. *Anal. Chem.* **1995,** *67*(11), 1845-1852.
255. Huikko, K.; Kostiainen, R. *J. Chromatogr. A* **2000,** *893*(2), 411-420.
256. Stover, F. S.; Wagenknecht, J. H. *Anal. Chim. Acta* **1982,** *135*(2), 347-350.
257. Koudelkova, M.; Jedinakova-Krizova, V. *J. Chromatogr. A* **2003,** *990*(1-2), 317-323.
258. Fitzgerald, E. A. *J. Chromatogr. Sci.* **1983,** *21*(4), 188-189.

Chapter 5

Speciation of Aminopolycarboxylate and Aminophosphonate Metal Complexes by AEX ICP–MS in Environmental Water Samples

Adrian A. Ammann

Analytical Chemistry of the Aquatic Environment, EAWAG, Swiss Federal Institute for Environmental Science and Technology, P.O. Box 611, CH–8600 Dübendorf, Switzerland

The widespread diverse applications, the persistency and fate of chelating agents in the environment are inevitably associated with metal binding. Depending on the metal bound to a chelating agent, such a species may be well or not at all bio-degradable. So biodegradability of free chelator is not a criterion to assess whether chelating agents are dangerous pollutants or not. Direct and reliable observation of single metal chelates is required in order to gain adequate knowledge on persistency, mobility, chemical transformation, uptake and effects of metal chelates. This is provided by sensitive species specific analytical methods as the one described here. Determination of so called total chelator concentrations is not intended to and cannot provide such a realistic picture since free chelators do not exist in the environment and speciation cannot be calculated from the total amount of chelators.

A low hydrophobic narrow bore anion exchange column coupled to an ICP-MS provided new exceptional selectivity and sensitivity in metal speciation analysis of anionic synthetic and biogenic chelates and oxo-species in environmental waters. Moreover the selectivity produced individual metal separation patterns unique for many chelators which make it possible to identify them independent of retention times.

Introduction

Chelating agents form stable metal complexes that keep metals in solution and prevent the formation of undesired aggregates and precipitates. This sequestering effect leads to their widespread application in consumer goods, fertilizers and production processes. In cases of insufficient degradation or elimination chelators enter environmental water bodies through waste water treatment plants (*1,2,3*) and through soil water (*4*) where they efficiently increase the mobility of heavy metals and dominate its speciation in the final receiving waters. Investigations in to the biodegradation of free chelators have revealed structural aspects (secondary or primary amine) that are more susceptible to degradation but these also give rise to less stable metal chelates (*5*). Once released to the environment, uncontrolled and innumerable conditions for species formation and partial degradation occur, which make metal chelator species and chelator degradation unpredictable. For this reason research on environmental behavior of metals and of chelating agents are closely interrelated. Speciation of chelating agents in the environment cannot be carried out without considering complexing metals (*6*) and environmental metal speciation cannot be done without considering strongly binding biogenic or anthropogenic chelators (*1*). Accordingly, this requires speciation methods that can determine individual chelating agents including the metals bound to it (species). One example that illustrates how single metal chelates are related to real world problems is the non biodegradability of metal-EDDS (metal = Co, Cu, Ni, Hg, Zn) species (*7*) whereas EDDS itself is a biogenic and biodegradable chelator recommended to replace EDTA in many applications. So, unfortunately, biodegradability of free or weakly bound chelator is not a criterion to assess whether chelating agents are dangerous pollutants or not. Whatever the source of EDDS, if non-biodegradable metal complexes are formed before degradation, EDDS certainly will contribute to the mobility of Cu, Ni, Zn and Hg. Research in this area strongly depend on analytical speciation methods (*6*). In order to meet the research requirements for multiple aspects of environmental chelation chemistry (*6*) a selective anion exchange (AEX) procedure was developed to separate metal chelates of several of the most often used chelators.

With respect to above problems it makes sense to classify published analytical methods according to the ability to determine individual metal chelates, that is species specific (speciation) and non species specific (non speciation) methods. They were developed in different times and nowadays fulfill different tasks. Both methods use diverse separation techniques such as gas chromatography (GC), capillary electrophoresis (CE), and several liquid chromatographic (LC) methods such as ion exchange chromatography (IEC) and ion pair chromatography (IPC) on reversed phase HPLC columns. These techniques are linked to diverse detectors according to the analytical task the system has to provide. Traditionally, non speciation analyses are conveniently performed by non specific but universal detectors like flame ionization, UV, and conductivity etc. However, research and validation requires more species specific information which can be obtained from molecule or element specific detectors e.g. mass spectrometers.

Non species specific analytical methods

Non speciation methods are used to determine so called "total" concentration of a chelator. Total implies here that a chelating agent can be isolated and determined quantitatively irrespective of its binding to metals, which is at least a delicate task and in some cases even impossible (see below).

Weiss (*8*) and Schwedt (*9*) were among the first who separated free polyaminocarboxylates (NTA, EDTA, DTPA), polyaminophosphonates (ATMP; EDTMP, DTPMP), polyphosphonic and polyphosphinic acids. These non UV-absorbing chelators were transformed into UV-absorbing iron chelates which were successfully detected by these authors. Later this derivatisation step became the most often used for UV detection frequently in combination with ion pair separation procedures (a detailed review is given by Schmidt & Brauch in this book). However there are several crucial points which need careful attention and optimization (*10*). Environmental samples especially, usually contain the most refractory (non biodegradable) metal bound chelators. Among them are metals that are not readily exchangeable. But all the metals bound to a particular chelator including the slow reacting have to be quantitatively exchanged (*11*) by iron since free chelators are exclusively used for calibration. Beside different reaction velocities, the stability of the Fe-chelates has to be considered, e.g. Fe(III)EDTA is photo-labile (*12*, *13*) and is destroyed by day light or intense lab light. In fact differences in reaction speed (*14*) and reaction conditions (*15*) have been used to differentiate metal chelates. The derivatisation step was also performed after separation (*16*) as post column in line reaction with higher detection limits (DL) (> 60 $nmolL^{-1}$) because of dilution and shorter reaction times available to slowly reacting metal chelates. The most severe limitation of this derivatisation arises from the fact that some chelates do not form complexes with iron; either because they are not stabile enough or they undergo side reactions (*17*). Additionally, the method cannot tell the analyst if there are metal (e.g. CrIII, CoIII) complexes in the sample that will not exchange with iron (*11*) except under condition that destroys the chelator. Usually not all these conditions are checked for in "total" chelator determinations and it can be concluded that not in all cases a "total" value corresponds to the real total amount of a chelator.

Species specific analytical methods

Gains in sensitivity and information on composition and structure are the main reasons for the use of mass spectrometers as detectors in chelator separation. Some information on elemental and isotopic composition of chelates especially those containing transition or heavy metals are most conveniently obtained from inductively coupled plasma mass spectroscopy (ICP-MS) (*18,19*). The high temperature in the plasma (5000-7000°) produces ultra fast and efficiently elemental ions almost independent of any molecule binding. As the elements are atomized information on molecule structure is lost except in cases where a structure specific separation is coupled to the ICP-MS (*20*). Such hyphenated techniques (*21*) can identify and quantitate known metal chelates based on reference materials. Structural details of unknown chelates might be obtained from low energy ion source MS like e.g. electro spray ionization MS (ESI-MS) (*22*). But ESI mass spectra of metal chelates can be extremely complicated caused by metal redox reaction and/or many additional fragmentations of the organic molecule catalyzed by the metal.

Among the separation techniques used CE is a newer method providing superior separation efficiency. Application of CE connected to non specific detectors is reviewed extensively in the following two papers. However, the inherently higher DL in CE is due to the low flow (~10 $nLmin^{-1}$) available. But such low flow rates are also a problem for sample introduction systems of MS. There are no nebulizers that can directly convert such a low flow into an aerosol. An additional flow (make-up or sheath flow ~ 10 -20 $\mu L/min^{-1}$) is required which has caused additional problems in the past (*24*). Recent developments (*25*) in CE coupling to ICP-MS, however, seem to omit these problems since DL in the range 10–100 $nmolL^{-1}$ for determination of metallothioneins (*26,27*) and Se- and As-species (*28*) have been reported. No investigation of metal chelates by CE-ICP-MS has been published so far but EDTA was used as a competitive ligand in determining rare earth metals fulvic acid binding constants by this speciation method (*29*). Little is known (*30,31*) on the applicability of CE-ESI-MS to metal chelate speciation.

IC is an older and established separation technique that was more often applied in speciation analysis. Coupling IC to ESI-MS with a carbonate eluent as commonly used in IC was preferred over other inorganic eluents since carbonate can be chemically suppressed providing much better signal to noise ratios (*32,33*). However carbonate eluents have two main drawbacks. Firstly, the anionic MeEDTA species eluted at similar retention times like the high concentrated anions Cl^-, NO_3^-, HCO_3^- and SO_4^{2-} and secondly, a carbonate eluent needs a high pH (>9) to reduce the amount of HCO_3^- in order to increase CO_3^{2-} and the elution strength. But at such high pH's some metal (Fe, Al, Zn) complexes are no longer stable.

Other eluents can be used in IC coupled to ICP-MS. NH_4NO_3 was preferably applied because it exhibits the best plasma compatibility and has the lowest interferences with elements to be detected. All other eluents usually used in IEC and IPC would produce deposits in the MS and interferences that prevent sensitive determination of some metals. NH_4NO_3 was used to separate CrIIIEDTA in a CrIII/CrVI speciation (*34*). As metal chelates exhibit a high affinity for polymeric column materials, high eluent salt concentrations have to be applied. This generates a high salt load in the plasma that can be lowered by chemical suppression of the eluent. On the other hand, the plasma load was also considerably reduced by utilizing a micro bore system that requires lower eluent concentrations and lower flow rates which introduce lower amounts of salt into the plasma. This setup was specifically developed (*35*) to determine low concentrated metal chelates of commonly used chelators.

Materials and Methods

The chromatography and the coupling to ICP-MS have been described (*35*) in details. In brief, an all PEEK micro bore system and a low capacity anion exchange column (AS11, 250 x 2 mm, Dionex) was used. The high selectivity observed was due to lowest hydrophobicity and high efficiency in mass transfer of the column material. As an eluent NH_4NO_3 was chosen because of its ideal plasma compatibility and eluent properties (*35*). The pH of the eluent was adjusted (HNO_3/NH_4OH) and controlled at the column exit. In a direct coupling without splitting the eluent flow (440 $\mu Lmin^{-1}$) was optimized for fast separation and to the highest nebulizer to plasma mass transfer. On column sample preconcentration was achieved either by a sample loop or, for larger volumes (1-5 mL), by injection on a preconcentrator column (AG11, 50x4 mm, Dionex). All given stability constants (I=0.1, 25°) and pK_a values were taken from (*36*) if not otherwise indicated.

Results and Discussion

The investigations (*35*) of free chelators like EDA_3, $HEDA_3$, NTA, EDDS, EDTA, CDTA, DTPA, NTMP, EDTMP and their metal (Fe, Mn, Co, Ni, Cu,

Zn, Cd, Pb) complexes comprised of several aspects of complexation chemistry (charge density, equilibrium stability constants, reaction kinetics) and environmental concern. Application of a particular isocratic eluent concentration provided the high selectivity for the separation of several metal chelates that do not differ by more than one charge unit. A pure anion exchange mechanism was shown to be active in separation (*35*). Multiply charged species required increasing eluent concentrations of about 32 $mmolL^{-1}$ NO_3^- per formal charge unit. It allowed separation of several metal species of the same chelator which was used for chelator identification (see below). However, in environmental waters chelators were found that differ by more than one charge unit. In order to separate as many as possible of these higher charged chelates in the same run, gradient elution had to be applied. It was found that gradients (20-200 $mmolL^{-1}$ NH_4NO_3, 0-8 min) up to such high concentrations were well tolerated by the plasma and they separated in the same run anionic species that differ approximately by five charge units. The selectivity was still excellent because the gradient narrowed the peak width considerably so that small structural changes which affected charge density resulted in retention time (t_R) shifts.
The selectivity by gradient elution was further investigated using standard solutions of MeEDTA, MeCDTA and MeDTPA. The separation achieved for the metal chelates (Me= Cu, Ni, Zn, Mn, 0.1 $\mu molL^{-1}$ each) of these three most often applied polyaminocarboxylates is shown in Figure 1. The chelates of each ligand are well separated from each other. This is even the case for $MeEDTA^{2-}$ and $MeCDTA^{2-}$ which both bear the same negative formal charge. An explanation for the elution order is given in (*35*). Moreover, the four metals resulted in individual metal separation patterns which are unique for each chelator. This allows identification of each ligand structure independently of the actual t_R by simply measuring only three commonly occurring metals (e.g. Cu, Zn and Mn). It was also noticed that a loss of a carboxylic or a phosphonate group altered the metal chelate separation pattern as well, beside a large shift in retention time.

Influence of pH

As a master variable the pH strongly influences the stability and other properties of metal species. However, the pH of the so far most often used eluents in IC (HNO_3, CO_3^{2-} and OH^-) cannot be altered without changing the eluent strength. However a change in eluent strength vastly changes ion chromatography. This is not the case for the eluent chosen. Here the eluent pH can be adjusted in the range of highest metal complex stabilities (*37,7*) without changing the eluent strength. This means that the pH of the NH_4NO_3 eluent can be used as an independent variable to control the separation selectivity.

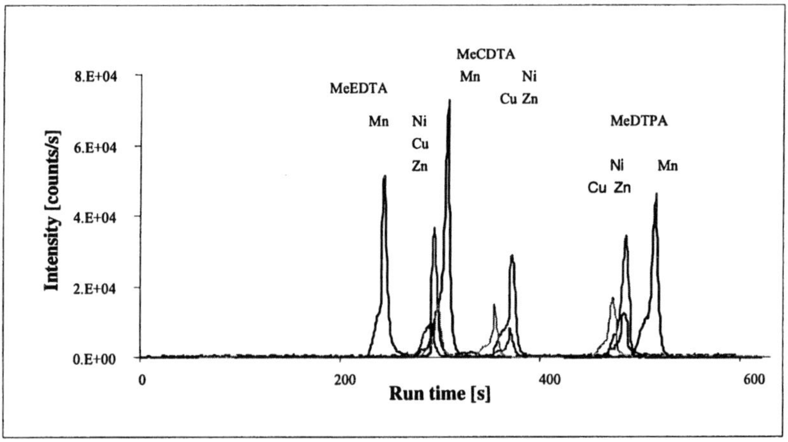

Figure 1. Separation and ICP MS detection of polyaminocarboxylate metal chelates. The sample was injected by a loop (250 μL) and a NH_4NO_3 gradient (20-150 mmol/L^{-1}) was applied.

Tunable Selectivity

Since the pH of this eluent can be varied in the most relevant range (pH 6-8) conditions were established that separated free $DTPAH_2^{3-}$ (pK_{a4}=8.6) from $EDTAH^{3-}$ and $CDTAH^{3-}$ at pH 8.0. Such a partial deprotonation was also helpful in separation of $MeEDTA^{2-}$ from $MeEDDS^{2-}$. Since EDTA and EDDS are built from the same number and type of atoms (constitution isomers) similar size and charge density can be expected. However, the deprotonation of a coordinated H_2O molecule differs by one pK-unit which was used to separate CuEDTA from CuEDDS at pH = 8.7 (see Figure 2). Since the pK_a of their Ni-analogues is higher their separation is only slightly improved by the applied increase in pH.
The protonation degree is not the only reason for different charge densities of metal complexes. Oxidation state, composition and stereo electronic effects cause also differences in charge densities which result in different t_R (*35*). This is illustrated in Figure 3 for $MeEDTA^{2-}$ (Me = Fe, Co, Ni, Cu, Cd, Pb). Retention factors (k') were lower (=shorter t_R) for the larger metals (Cd, Pb) complexes having lower charge densities, and there was almost no pH

dependency of k' except for Fe(III)EDTA which exhibited a large increase in k' between pH 7 and 8.

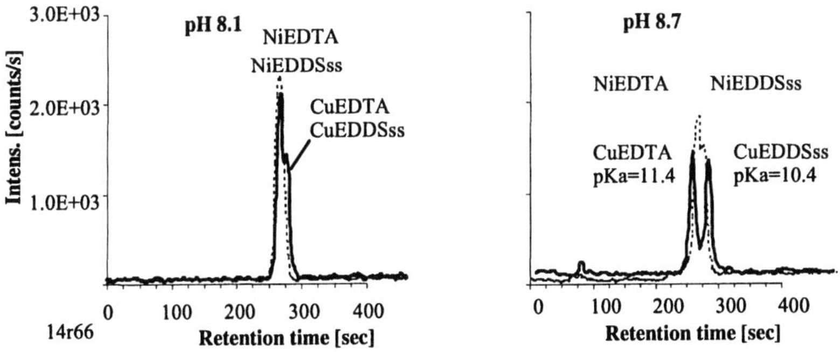

Figure 2. Separation of CuEDTA from the natural enantiomer $CuEDDS_{SS}$ (both 0.08 $\mu molL^{-1}$, 100 μL injected) occurs by a gradient run with an eluent-pH closer to the pK_a of the coordinated H_2O in the molecules.

This is in agreement with spectroscopic investigations (*38*) which showed that the coordinated H_2O molecule in these complexes can be deprotonated below pH=8 only in case of $Fe(H_2O)EDTA$ (pK_a=7.5). The pK_a of all other species are above 10 so they do not change charge density by varying the pH below 8.

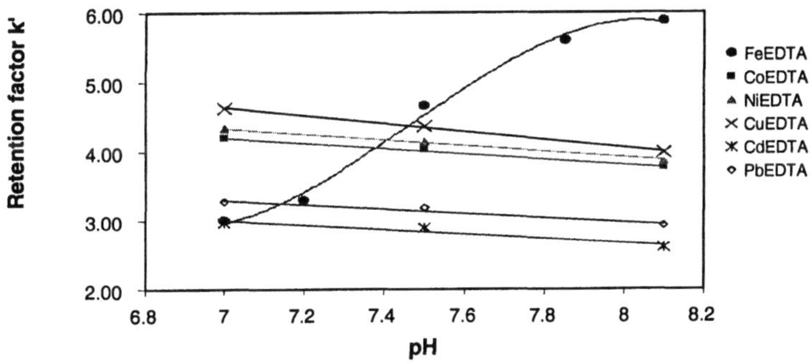

Figure 3. Variation of retention factors k' ($=(t-t_o)/t_o$) at different eluent pH.

DL and Sensitivity

The sensitivity depends on a variety of factors, mainly on species stability, preconcentration capability of the procedure, mass transfer efficiency of the sample introduction system, salt load in the plasma and the isotope measured.
The chelate stability is an important prerequisite since there is competition in metal binding between the chelator and the column. With the column used chelate stabilities pK>10 are required for concentrations below micro molar. The greater the stability, the lower the amount of the metal taken up by the column and the lower the concentration that can be determined. The ion content is another important factor in species preconcentration. The higher the concentration e.g. of doubly charged anions CO_3^{2-} and SO_4^{2-} in a sample, the lower the preconcentration capability of a column. Ion exchange is especially suited to preconcentrate selectively multiply charged ionic species by a factor 10-100. As in this case, the matrix is separated and at the same time exchanged by the eluent with approximately similar ionic strength and a minimal shift in pH ($\Delta pH<0.5$). It is very unlikely that a shift in speciation occurs due to sample preconcentration. Moreover, if such a shift should occur, it is accounted for in calibration by standard addition to the sample before preconcentration. The sample inlet system (*39*) to the ICP-MS has a large impact on DL as well. Compared to a cross flow nebulizer, a micro concentric nebulizer (MCN) and an ultrasonic nebulizer (USN) achieved lower DL by a factor of 2 and 10, respectively. The USN reduces mainly oxygen background and interference levels since most of the water (oxygen source) is dried out from the heated aerosol. On the other hand eluent salts (e.g. CO_3^{2-}, AcO^-, SO_4^{2-}) in the plasma produce other background levels that can interfere with masses of important elements like ^{52}Cr ($^{40}Ar^{12}C$), ^{53}Cr ($^{40}Ar^{13}C$), ^{64}Zn ($^{32}S_2$), ^{65}Cu ($^{32}S^{33}S$), ^{66}Zn ($^{33}S_2$) etc. Finally, DL depends on the element and isotope measured. Some elements are mono-isotopic (^{31}P, ^{55}Mn, ^{75}As, ^{127}I, ^{197}Au) while others are composed of several isotopes each of low abundance (Cd, Sn). Not all the elements form the same amount of ions in the plasma. Elements with a higher ionization potential (N, O, Cl, C, and Br) are ionized in the plasma by less than 10% (*40*).
For several stable (e.g. EDTA) metal complexes (Me=Mn, Co, Ni, Cu, Zn, Cd, Pb) different DL were achieved (*35*) depending on the isotope and with a total sample ion content of ~10 mequiv (surface water) and a MCN sample inlet. Sample introduction by USN lowered DL by a factor of ten.

Application

The gradient separation procedure was applied in speciation analysis of diverse sample matrices and concentrations. In ecotoxicological studies heavy metals stabilized by EDTA in algal nutrition broth were analyzed (*41*). The MeEDTA species (Me = Mn^{2+}, Fe^{3+}, Cu^{2+}, Co^{2+}, Zn^{2+}) identity have been monitored before and during the contact with algae. Nutrition salts and 10 mmol L^{-1} organic buffer salt (MOPS) did not hamper the separation and quantification of MeEDTA (0.05 – 1 $\mu molL^{-1}$) after direct injection of the nutrition medium.

In environmental water metal chelates occur in low concentrations ($nmolL^{-1}$), so the applicability depends on sample preconcentration. On a concentrator column 2-3 mL of river water were preconcentrated which allowed determining a variety of anionic metal chelates (see Figure 4).

In polluted river water metal complexes of polyaminocarboxylates (EDTA, DTPA), phosphonates (NTMP, EDTMP) and not jet identified species were found. Interestingly, beside these coordinated metal compounds, covalent bound metal oxoanions can be separated and quantified, here e.g. the highly toxic chromate (5 $nmolL^{-1}$).

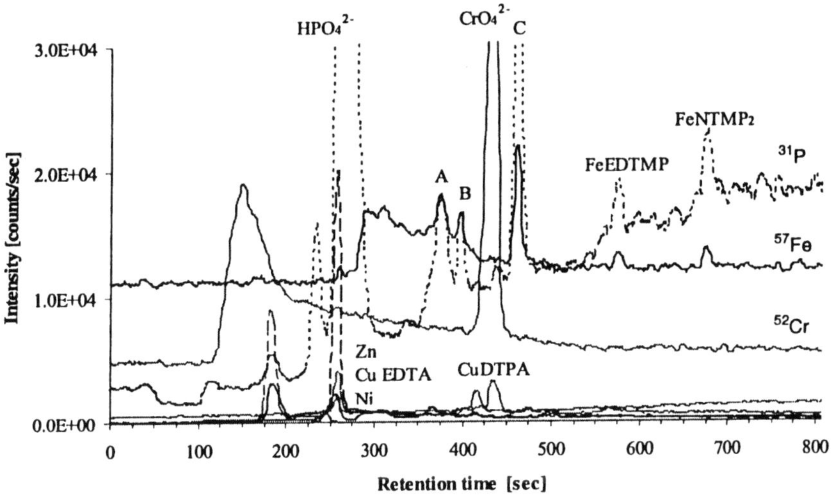

Figure 4. Speciation analysis of river water 500m downstream from a sewage treatment plant discharge. Beside identified metal polyamino-carboxylates (0.5-4 $nmolL^{-1}$) unidentified Fe-phosphonates (A, B, C) were detected, most probably degradation products.

Conclusions

The combination of all the features presented renders the method exceptionally well suited for application to environmental waters. For the first time direct observation of all the metal species, chelated and covalent bund oxo-anions is available. It certainly will be used to fill the gap left by non species based total chelator measurements which cannot answer the question about real metal chelate species in a sample. As the method allows gathering of much greater and more detailed information it can contribute in research on persistency, mobility, chemical transformation, uptake and effects of metal chelates which are other compounds that have different properties compared to free chelators. This provides a much more realistic picture than total chelator concentrations since free chelators do not exist in the environment and speciation cannot be calculated from the total amount of chelator.

Acknowledgement

The help by S. Tandy to improve the language is gratefully acknowledged.

References

1. Bucheli-Witschel, M.; Egli, T. *FEMS Microbiol. Rev.* **2001**, *25*, 69-106.
2. Nowack, B. *Water Res.* **2002**, *32*, 4636-4642.
3. Nowack, B. *Water Res.* **1998**, *36*, 1271-1279.
4. Stumpf, M.; Ternes, T.A.; Schuppert, B.; Haberer, K.; Hoffmann, P.; Ortner, H.M., *Vom Wasser* **1996**, *86*, 157-171.
5. Sykora, V.; Pitter, P.; Bittnerova, I.; Lederer, T. *Water Res.* **2001** *35*, 2010-2016.
6. Nowack, B. *Environ. Sci. & Technol.* **2002**, *36*, 4009-4016.
7. Vandevivere, P.C.; Hammes, F.; Verstraete, W.; Feijtel, T.C.J.; Schowanek, D.R. *Environ. Sci. Technol.* **2001**, *35*, 1765-1770.
8. Weiss J.; Hägele G. Fresenius J. *Anal. Chem.* **1987** *328,* 46-50.
9. Tschäbunin, G.; Fischer, P.; Schwedt, G., *Fresenius Z. Anal. Chem.* **1989**, 333, 111-128.
10. Loyaux-Lawniczak, S.; Douch, J.; Behra, Ph. *Fresenius J. Anal. Chem.* **1999**, *364*, 727.

11. Nowack, B.; Kari, F.G.; Hilger, S.U.; Sigg, L. *Anal. Chem.* **1996** *68,* 561-566.
12. Karametaxas, G.; Hug, S.J.; Sulzberger, B. *Environ. Sci. Technol.* **1995,** *29,* 2992-3000.
13. Kari, F.G.; Hilger, S.; Canonica, S. *Environ. Sci. Technol.* **1995,** *29,* 1008-1017.
14. Nirel, P.M.; Pardo, P.M.; Landry, J.C.; Revaclier, R. *Water Res.* **1998** *32,* 3615-3620.
15. Nowak, B.; Xue, H.; Sigg, L. *Environ. Sci. Technol.* **1997,** *31,* 866-872.
16. Bedsworth, W.W.; Sedlak, D.L. *J. Chromatogr. A* **2001,** *905,* 157-162.
17. *Ionenchromatographie* 3rd Ed., Weiss, J. Wiley-VCH, Weinheim, 2001, pp 456.
18. *Inductively Coupled Plasma Mass Spectrometry*; Montaser, A. Wiley-VCH: New York, 1998.
19 *Handbook of Inductively Coupled Plasma Mass Spectrometry;* Jarvis, K.E.; Gray, A.L.; Houk, R.S., Eds.; Blackie: London, 1992.
20. *Handbook of Elemental Speciation Techniques and Methodology*; Cornelis, R.; Caruso, J.; Crews, H.; Heumann, K., Eds.; Wiley: Chichester, 2003.
21. *Elemental Speciation - New Approaches for Trace Elemental Analysis*; Caruso, J.A.; Sutto, K.L.; Ackley, K.L., Eds.; Comprehensive Analytical Chemistry 33; Barcelo D., Ed.; Elsevier: Amsterdam, 2000.
22. Stewart, I.I. *Sectrochim. Acta B* **1999,** *54,* 1649-1695.
23. Dabek-Zlotorzynska, E.; Lai, E.P.C.; Timerbaev, A.R. *Anal. Chim. Acta* **1998,** *359,* 1-26.
24. Kannamkumarath, S.S.; Wrobel, K.; B'Hymer, C.; Caruso, J.A. *J. Chromatogr. A* **2002,** *975,* 245-266.
25. Schaumlöffel, D.; Prange, A. *Fresenius J. Anal. Chem.* **1999,** *364,* 452-456.
26. Schaumlöffel, D.; Prange, A.; Marx, G.; Heumann, K.G.; Brätter, P. *Anal. Bioanal. Chem.* **2002,** *372* 155-163.
27. Połeć, K.; Szpunar, J.; Palacios, O.; Gonzalez-Duarte, P.; Atrian, S.; Łobinski, R. *J. Anal. Atom. Spectrom.* **2001,** *16,* 567-574.
28. Schaumlöffel, D.; Prange, A.; *J. Anal. Atom. Spectrom.* **1999,** *14,* 1329-1332.
29. Sonke, J.E.; Salters, V.J.M. *J. Anal. Atom. Spectrom.* **2004,** *19,* 235-240.
30. Schramel, O.; Michalke, B.; Kettrup, A. *J. Chromatogr. A* **1998** *819,* 117-128.
31. Baron, D.; Hering, J.G. J. Environ. Quality **1998,** *27,* 844-850.
32. Bauer, K.H.; Kepper, T.P.; Maes, A.; Schatz, V.; Voihsel, M.J. *J. Chromatogr. A* **1999,** *837,* 117-128.
33. Collins, R.N.; Onisko, B.C.; McLaughlin, M.J.; Merrington, G. *Environ. Sci. Technol.* **2001,** *35,* 2589-2593.

34. Gürleyük, H.; Wallschläger, D. *J. Anal. Atom. Spectrom.* **2001**, *16*, 926-930.
35. Ammann, A.A. *J. Chromatogr. A* **2002**, *947,* 205-216.
36. Smith R.M.; Martell A.E. NIST Critically Selected Stability Constants of Metal Complexes. NIST Standard Database 46. 2001.
37 Buchberger, W.; Mülleder, S. *Mikrochim. Acta* **1995**, *119,* 103-111.
38. Oakes, J. Smith, E.G. J. Chem. Soc. Dalton Trans. **1983**, 601-605.
39. *Discrete Sample Introduction Techniques for Inductively Coupled Plasma Mass Spectrometry;* Beauchemin, D.; Grégoire, D.C.; Günther, D.; Karanassios, V.; Mermet, J.-M.; Wood, T. J., Eds.; Comprehensive Analytical Chemistry 34; Elsevier: Amsterdam, 2000.
40. *Handbook of Inductively Coupled Plasma Mass Spectrometry;* Jarvis, K.E.; Gray, A.L.; Houk, R.S., Eds.; Blackie: London, 1992; pp 15-18.
41. Ammann, A.A. *Anal. Bioanal. Chem.* **2002,** *372*, 448-452.

Chapter 6

Analysis of Metal-Chelating Agent Complexes by Capillary Electrophoresis

Scott C. Brooks

Environmental Sciences Division, Oak Ridge National Laboratory, P.O. Box 2008, MS 6038, Oak Ridge, TN 37831–6038 (brookssc@ornl.gov)

Direct measurement of metal-ligand (ML) complexes is essential for understanding metal speciation and ML reactivity in natural and experimental settings. Capillary electrophoresis (CE) is highly efficient separation technique that can provide direct measurement of ML species revealing slow reaction kinetics, subtle changes in coordination, and oxidation-reduction reactions involving the metal and ligand. Applications of this technique to understanding ML reactivity in multicomponent systems are discussed.

Introduction

The land based codisposal of radionuclides with synthetic organic chelating agents has created a legacy of vast amounts of contaminated soils and groundwater within the U.S. Department of Energy complex *(1)*. A survey of EPA Superfund and other agency sites suggests that similar mixed waste contamination problems exist across the United States *(2)*. Shallow land burial of mixed wastes was considered acceptable, in part, because of the large sorption capacity of subsoils and aquifer solids for many metals and radionuclides. Nevertheless, the soluble metal-chelate (ML) complexes often exhibit sorption behavior that may promote the undesirable movement of contaminant metals and radionuclides away from primary disposal areas.

During transport through subsoils and aquifers, the fate of metals/ radionuclides and chelating agents is inescapably linked. As demonstrated through prior research on ML complexes in contact with synthetic and natural geologic materials, aqueous complexation, geochemical dissociation (competition among metals for ligand, competition between surfaces and aqueous ligands for metals), and geochemical oxidation of the metal, organic ligand, or both are all important reactions to consider when evaluating the coupled fate of chelated metals.

To improve our understanding of the coupled fate and transport of metal-chelate complexes in the environment, the development and use of analytical techniques that allow direct measurement of chelate speciation is essential. The influence of chelating agents on the fate and transport of metals and radionuclides has been studied for years. Due to a lack of adequate techniques, most of these studies have relied on the separate analysis of total metal (e.g., AAS, ICP-MS) and chelate (e.g., GC, HPLC) concentrations followed by equilibrium speciation calculations to predict metal speciation. As good as these analytical methods are they do not provide direct analysis of the ML species and therefore may lead to an incomplete or incorrect understanding of the underlying mechanisms governing ML behavior. For example, slow exchange kinetics may invalidate the equilibrium assumption implicit in the use of equilibrium speciation models. Measurement of total metal concentration will miss changes in oxidation state of the metal (e.g., Co(II) versus Co(III)). Radiometric assays using ^{14}C labeled organic compounds can miss changes to the organic chelate (e.g., oxidation of nitrilotriacetate to iminodiacetate).

Over the past ten years analytical techniques using standard chromatography equipment have been reported that allow the direct determination of some metal-EDTA (M-EDTA) complexes. Taylor and Jardine *(3)* described an ion chromatography (IC) method for the direct determination of $Co^{II}EDTA$ and $Co^{III}EDTA$. The method was later extended to include $Cd^{II}EDTA$ *(4)*. Nowack et al. *(5)* have described a high performance liquid chromatography (HPLC) method for quantifying EDTA following precolumn conversion to its Fe(III) complex. The method has also been used to quantify Co(III), Cr(III), and Bi(III) complexes of EDTA without conversion to the Fe(III) form *(6)*. The purpose of this chapter is to introduce the reader to capillary electrophoresis (CE), an analytical technique that has proved useful for the direct determination of organic chelates and their metal complexes. A brief overview of CE is followed by a summary of CE methods that have been used for ML analysis and finally examples of applications of those methods for understanding ML reactions. Results of CE analyses conducted by this author (viz. Figures 3, 4, 5, 6, and 7) were collected using a Waters Quanta 4000E instrument (Waters Corp, Milford, MA).

Fundamentals of Capillary Electrophoresis

Capillary electrophoresis is a highly efficient separations technique that has been used for the analysis of inorganic and organic ions. The separation mechanism is based on the differential mobility of charged solutes under an applied electric field. Several texts and reviews are available for details on the different modes capillary electrophoresis and their applications *(7-9)*. The following serves as a cursory introduction to some of the basic fundamentals.

The CE System

A CE system consists of a run buffer (electrolyte), high voltage power supply, capillary, and detector (Figure 1). Two electrolyte reservoir vials are connected by the capillary, typically 50 µm to75 µm i.d., that is also filled with the electrolyte. Each reservoir also contains an electrode connected to the power supply. The electrolyte has some pH buffering capacity to minimize variability between runs and to offset H^+ production at the anode and OH^- at the cathode. Samples are usually introduced into the capillary using either hydrodynamic injection, in which the capillary is placed in the sample vial and a positive pressure is applied to the sample vial, or hydrostatic injection in which the capillary is placed in the sample vial and the vial is raised a preset distance above the collection reservoir causing sample to siphon into the capillary.

Solute Separation

Following sample introduction, the capillary end is returned to the electrolyte reservoir and when an electric field is applied across the liquid, ions in the fluid will migrate towards the electrode of opposite polarity. The observed solute mobility results from the vector sum of the electroosmotic flow (EOF) and the electrophoretic velocity. EOF is the bulk movement of electrolyte as a result of a charged inner capillary wall and the applied potential. At pH > ~3 silanol groups of fused silica dissociate imparting a net negative charge at the surface. Cations in the buffer are attracted to the silanoate ($Si\text{-}O^-$) groups forming a diffuse double layer: an inner layer held tightly by the $Si\text{-}O^-$ groups and an outer mobile layer. When an electric field is applied the outer later of cations, along with their hydration shells, move toward the cathode resulting in the bulk movement of buffer creating the EOF.

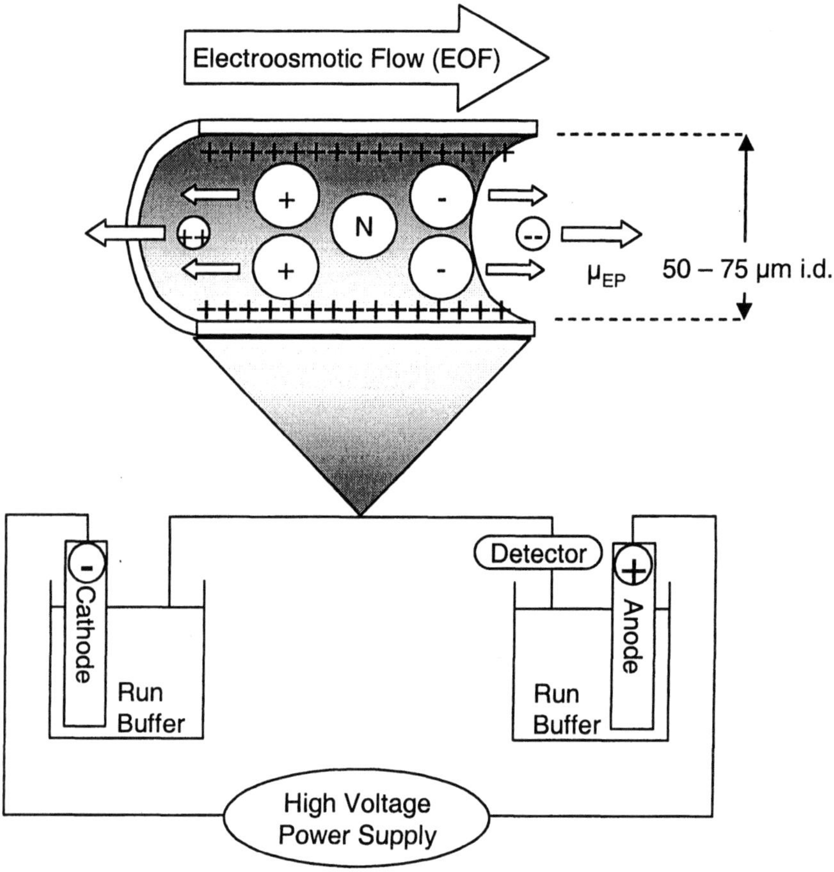

Figure 1. Schematic diagram of capillary electrophoresis. As depicted, the instrument is set up for routine anion analysis with injection at the cathodic end of the capillary and detection at the anodic end. The inner surface of the capillary has been treated with electroosmotic flow (EOF) modifier reversing the surface charge of the fused silica capillary resulting in EOF towards the anode. Smaller, more highly charged ions have higher electrophoretic mobility (μ_{EP}) and migrate more rapidly toward their respective electrodes than larger less highly charged ions. Neutral solutes are not separated and move with the bulk EOF. Depending on the relative magnitude of the EOF to the μ_{EP} of the cations, the cations may also reach the detector after neutrals and water.

The magnitude and direction of the EOF can be manipulated by changing the run voltage and polarity or by manipulating the electrolyte chemistry such as pH, ionic strength, viscosity, or the addition of surfactants to modify the charge of the capillary wall (EOF modifier). Reversal of the EOF toward the anode is commonly achieved by using the cationic surfactant tetradecyltrimethylammonium bromide (TTAB) or its hydroxide form (TTAOH). In general, if the charge on the inner capillary wall is negative, EOF is towards the cathode, if it is positive, EOF is towards the anode. There is no EOF if the inner wall of the capillary is neutral.

The electrophoretic velocity for a solute depends on the applied electric field and the electrophoretic mobility. In general, ions are separated based on their charge to size ratio. Smaller more highly charged ions have a higher electrophoretic mobility than larger less highly charged ions.

The Capillary

Typically fused silica capillaries are used in CE. Before their first use and routinely during analytical runs, the capillaries are conditioned by rinsing with some combination of dilute acid, base, deionized water, and electrolyte. Periodic conditioning of the capillary surface minimizes variability in performance. The inner surface of the capillary can be dynamically coated (adding surfactant to run buffer) or permanently coated by a variety of methods of varying complexity *(10-14)*. Capillary coatings are used to modify the EOF (see above) or to minimize ion interaction with the inner surface of the capillary.

Detection

Several on-column detectors with zero dead volume have been developed for CE analysis including fluorescence and conductivity detectors. For environmental analyses UV/Vis detectors are most commonly used in either the direct or indirect detection mode. Direct photometric detection relies on the light absorbing characteristics of the solutes of interest. Indirect photometric detection is used when the solutes of interest do not absorb well at the available wavelengths. A light absorbing compound (a chromophore, e.g. chromate or *p*-aminobenzoic acid) which absorbs strongly at a specific wavelength is added to the run buffer creating a high background absorbance. When nonabsorbing solutes pass through the detector a decrease in absorbance is recorded. In this case the detector polarity can be reversed so the peaks take on their conventional appearance.

In CE the applied electric field provides the driving force generating a flat electroosmotic flow profile across the capillary. The solutes being separated experience the same velocity regardless of their cross-sectional position in the capillary and migrate in narrow zones generating narrow peaks, and thus the high efficiency of CE. The parabolic velocity profile generated in liquid chromatography tends to generate broader peaks (Figure 2). Images of solute migration under electrokinetically and pressure driven flow regimes verify this phenomenon *(15,16)*.

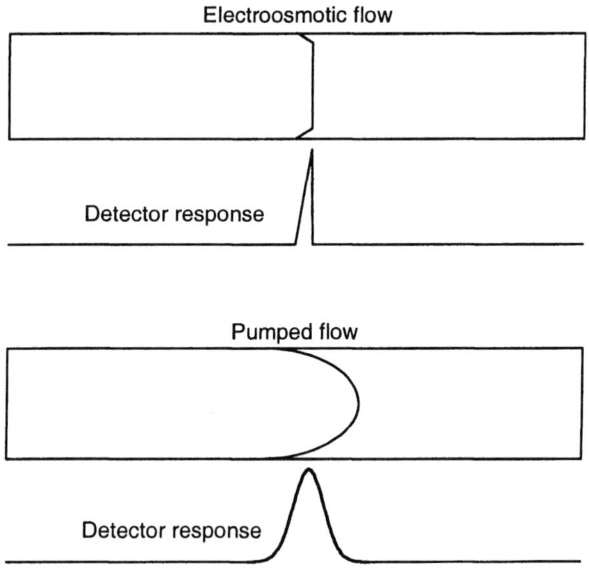

Figure 2. Electroosmotic flow has a relatively flat velocity profile resulting in narrow peaks and high efficiency of capillary electrophoresis.

CE offers a number of advantages relative to more traditional chromatographic techniques. The most widely recognized advantage is the high efficiency of CE with greater than 10^5 theoretical plates routinely achieved compared to ~10^5 for gas chromatography (GC) or 10^4 for HPLC. CE is compatible with a wide range of sample and separation chemistries. It can be applied over a wide range of pH with or without organic modifiers (e.g., methanol, acetonitrile) for the determination of solutes with a range of hydrophobicities. Solutes ranging in size from simple ions (Cl^-, Br^-, NO_3^-) to high molecular weight compounds such as proteins can be determined by CE. In addition, much smaller sample and reagent volume is required. Usually, only a few nanoliters of sample are injected and an entire day's analyses use less than 100 mL of electrolyte.

Perhaps the primary disadvantage to CE is the lower sensitivity due to the shorter optical path length. This limitation can be partially compensated by the use of "Z-cells" or bubble cells but these can also lead to peak broadening compromising resolution. In some cases the lower sensitivity can be overcome by using electrokinetic injection in which sample is introduced by applying a voltage to the sample causing sample ions to migrate into the capillary due to electroosmosis and electrophoretic mobility. Dahlen et al. *(17)* improved detection limits for low molecular weight organic acids by two orders of magnitude using electrokinetic injection (from ~0.1 mg/L to ~0.001 mg/L). Finally, the community of CE users for environmental applications is relatively small compared to other forms of chromatography and fewer "standard" methods for these applications are available.

Metal-Chelate Speciation by Capillary Electrophoresis

Successful metal-chelate separation and quantification by CE depends on the kinetic and thermodynamic stability of the complex under the separation conditions employed. Complexes that undergo rapid ligand exchange dissociate and reassociate more often than complexes with slow exhange rates. Uncomplexed metal and ligand migrate towards opposite electrodes during separation hampering reliable analysis. ML complexes with large stability constants are suitable for CE analysis, regardless of their rate of ligand exchange. Finally, the interaction of the complex with the silica capillary or with constituents of the electrolyte will influence ML separation and quantification. For example, Burgisser and Stone *(18)* obtained linear calibration curves for $Co^{II}EDTA^{2-}$ (log $K_{formation}$ = 18.16) with a reported detection limit of 2×10^{-6} M, but $Co^{II}NTA$ (log K = 11.66) gave nonlinear calibration curves with an approximate detection limit of 1×10^{-4} M. The authors attributed this to ligand exchange reactions in which phosphate in the electrolyte exchanged with NTA. The fractional amount of exchange increased as the total $Co^{II}NTA$ in the sample decreased. This problem can be resolved using a different buffer composition. For example, using a run buffer of 25 mM $Na_2B_4O_7$, 0.5 mM TTAOH, pH 9.3, we are able to achieve linear calibration curves with $Co^{II}NTA$ detection limits of 8.7×10^{-6} M (Figure 3)

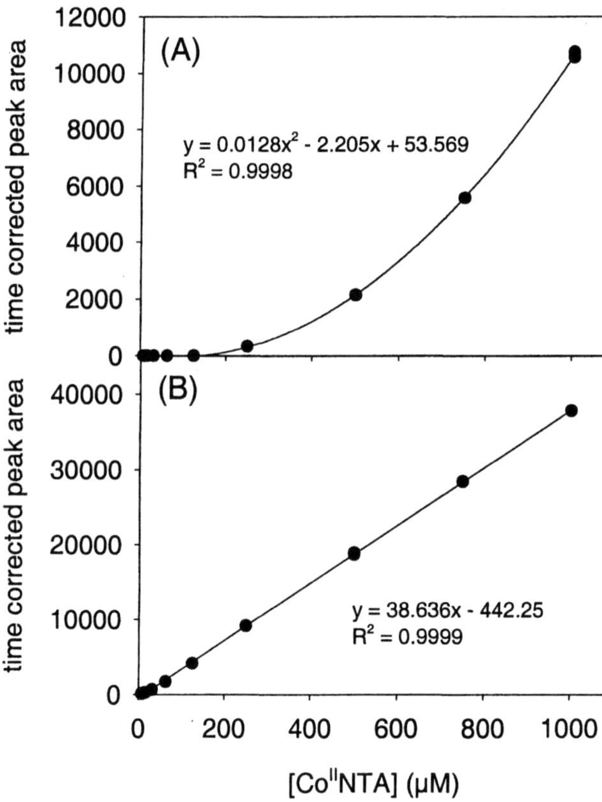

Figure 3. $Co^{II}NTA$ calibration plot for standards analyzed using phosphate (A) or borate (B) buffer. Separation conditions: for both panels 75 μm × 60 cm (52 cm to detector) capillary, 30s hydrostatic sampling, direct detection at 185 nm. Buffer composition (A) 25 mM phosphate, 0.5 mM TTAOH, pH 7.2; (B) 25 mM $Na_2B_4O_7$, 0.5 mM TTAOH, pH 9.3.

Given that the first commercially available CE apparatus was introduced in the early 1980's the analysis of metal-chelate complexes by capillary electrophoresis has a fairly rich history. However, many of the reported methods have the goal of separating and analyzing mixtures of alkali, alkaline earth, and transition metals *(19-28)*. Optimum methods reported generally require the addition of excess chelate to the sample and/or the electrolyte making these methods not directly applicable to ML speciation studies in water samples. Schaffer et al. *(29)* analyzed NTA and its breakdown products in water from a desulfuration process. The mixture components were separated but excess EDTA was added to samples to remove interference from Fe(III). All components could not be analyzed under a single set of run conditions. Buchberger and Mulleder *(30)* developed a method for analysis of DTPA and its Fe(III), Bi(III), Cu(II), Pb(II), and Ni(II) complexes using 10 mM borate buffer, pH 9.3, with 0.2 mM TTAB as EOF modifier and direct detection at 254 nm. Reported detection limits ranged from 5×10^{-7} M for $Fe^{III}DTPA$ to 3.5×10^{-6} M for $Bi^{III}EDTA$. The method was successfully applied to the analysis of waste water from a paper mill. Harvey *(31)* described a method for the separation of a mixture of aminopolycarboxylates (EDTA, NTA, ED3A, DTPA, HEDTA) by on-column formation of their Cu(II) complexes. Using a variation of CE, micellar electrokinetic chromatography (MEKC), the method uses differential affinity for surfactant micelles in the run buffer to assist in the separation of solutes. The performance of several buffer compositions and separation conditions is reported. While the method is not directly applicable to the study of metal-chelate complexes in water it does illustrate the rapid (< 7.5 minutes) separation of complexes with very similar electrophoretic mobility. Ballou et al. *(32)* also described a method for separation of a mixture of aminopolycarboxylates (EDDA, HEDTA, EDTA, NTA, DTPA) as their Cu(II) complexes using 10 mM phosphate buffer, pH 11.5, and direct detection at 254 nm. The authors demonstrated the successful application of the method to the analysis of a high level radioactive waste tank simulant. Method detection limits for standards prepared in the waste simulant were on the order of 10^{-5} M.

Few studies have been published that focus on the application of CE for the purpose of quantifying free chelates and their metal complexes in aqueous samples. Burgisser and Stone *(18)* described a method for analyzing the Co(II) and Co(III) complexes of EDTA, NTA, and 13 other aminocarboxylate ligands using 25 mM phosphate buffer, pH 7, as an electrolyte with 0.5 mM TTAB as EOF modifier and direct detection at 185 nm. The fifteen chelates were separated without pre-column or on-column complexation. Changes in metal oxidation state, subtle changes in coordination and stereochemistry were readily determined. The Co(II) and Co(III) complexes with EDTA (as $Co^{II}EDTA^{2-}$ and $Co^{III}EDTA^{-}$, respectively) were separated and quantified. Co(III) complexes with NTA were synthesized and analyzed. Acidification of solutions of α-$Co^{III}NTA$ yielded a new peak with slower migration time confirming the formation of a new Co(III) complex. The CE results were consistent with a dimer-monomer transition upon acidification that was first proposed by Smith and Sawyer *(33)* based on proton NMR spectra. Hexadentate $Co^{III}EDTA^{-}$ can be separated from the pentadentate $Co^{III}EDTA^{-}$ that forms under acidic conditions (Stone, personal communication). Diastereomers of $Co^{III}IDA_2$ (IDA = iminodiacetic acid) were also separated by CE despite having the same molecular weight and ionic charge. The authors hypothesized that the greater degree of symmetry for *s-fac*-$Co^{III}IDA_2^{-}$ may result in a slightly smaller size and consequently higher charge to size ratio than the *u-fac*-$Co^{III}IDA_2^-$ diastereomer enabling their separation by CE (Figure 4).

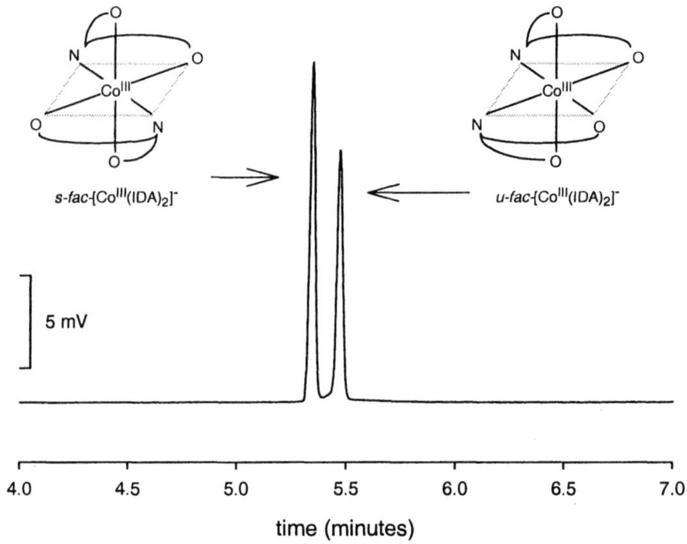

Figure 4. Separation of $Co^{III}(IDA)_2$ diastereomers by capillary electrophoresis. Despite having the same molecular weight and charge, the two compounds are separated by CE. Separation conditions: 25 mM $Na_2B_4O_7$, 0.5 mM TTAOH, pH 9.3; -15 kV, 30s hydrostatic sampling. Capillary 75 μm i.d. × 60 cm (52 cm to detector); direct detection at 254 nm. 100 μM each compound.

The CE method developed by Burgisser and Stone *(18)* was applied to study the reactions of aminocarboxylate chelates with the minerals heterogenite (CoOOH) and manganite (γ-MnOOH) *(34)*. Suspensions of CoOOH in EDTA containing solution generated products resulting from ligand assisted dissolution (Co^{III}EDTA) as well as reductive dissolution of the solid (e.g., Co^{II}EDTA, ED3A, Co^{III}ED3A) reflecting the reduction of Co(III) and the oxidation of the organic ligand. Similarly, when NTA was substituted for EDTA, Co^{III}NTA, $(Co^{III}NTA)_2$, Co^{II}NTA, and IDA were identified in solution. In contrast, dissolution of CoOOH by IDA proceeded via ligand promoted dissolution only with the formation of isomers of $Co^{III}IDA_2$; there was no evidence for a reductive dissolution mechanism in the presence of IDA. Reaction of EDTA with manganite resulted in the production of Mn^{II}EDTA, ED3A, and EDDA. It was unclear whether the failure to detect $Mn^{III}EDTA$ was due to the predominance of the reductive dissolution pathway or the rapid reduction of aqueous $Mn^{III}EDTA$ once formed via ligand promoted dissolution.

Owens et al. *(35)* analyzed EDTA and NTA speciation in the presence of a mixture of metals (Ca, Co(II), Cu(II), Cd(II), Zn(II), Pb(II), and Fe(III)) using 50 mM phosphate buffer, 0.5 mM TTAB, pH 6.86, and direct detection at 185 nm. Excess chelating agent was not needed in either the samples or the electrolyte for the analysis. The measured ML concentrations agreed well with concentrations predicted using an equilibrium speciation model. Detection limits for the M-EDTA and M-NTA complexes were on the order of 10^{-6} M and 10^{-5} M, respectively. However, not all the M-EDTA complexes could be resolved under the same run conditions due to similarities in their electrophoretic mobilities. The authors applied the method to the analysis of a hydroponic solution for NTA, ZnNTA, CaEDTA, and $Fe^{III}EDTA$. As observed by Burgisser and Stone *(18)* $Co^{II}NTA$ exhibited very high detection limits due to ligand exchange reactions with phosphate (cf. Figure 3); the problem was exacerbated due to the higher phosphate concentration. Similarly, Pb- and Cd-NTA were not stable under the separation conditions and injection of pure standards produced no peaks in the electropherograms. Although the stability constants for Pb- and Cd-NTA are similar to that for $Co^{II}NTA$, the larger ionic radius of Pb and Cd relative to Co may make their complexes with NTA more kinetically labile to ligand exchange reactions under the separation conditions.

Kinetics of ML Speciation

The slow kinetics of ML reactions can make metal-ligand mixtures slow to attain their predicted equilibrium state, making direct analysis of ML complexes a valuable asset in understanding metal speciation. A 100 μM solution of $Fe^{III}EDTA$ in 10 mM $NaNO_3$, 10 mM 3-[N-morpholino]propanesulfonic acid (MOPS), pH 7 was spiked with Ni(II) at a final concentration of 100 μM. Equilibrium speciation of the M-EDTA system was predicted using the measured total concentrations and the equilibrium speciation code PHREEQC with the included MINTEQ database *(36)*. PHREEQC is a geochemical model with a broad range of functions including aqueous speciation and saturation index calculations. The model is a publicly available freeware program (wwwbrr.cr.usgs.gov/projects/GWC_coupled/phreeqc/index.html). The initial solution is undersaturated with respect to Ni(II) solids but oversaturated with respect to $Fe(OH)_{3,am}$. If precipitation is not allowed in the model, predicted equilibrium concentrations for $Fe^{III}EDTA$ and $Ni^{II}EDTA$ are 28 μM and 72 μM, respectively. If $Fe(OH)_{3,am}$ precipitation is allowed, virtually all the Fe(III) precipitates and the predicted $Ni^{II}EDTA$ concentration is 100 μM (ignoring sorption onto precipitated Fe(III) solids for simplicity). These model predictions are compared to direct measurement of ML speciation over time.

Dynamics of the ML system were monitored over time with direct measurement of M-EDTA species using CE for comparison to the predicted equilibrium speciation (Brooks, unpublished results). Total aqueous Ni(II) remained constant over 864 hours (36 days; Figure 5). The concentration of $Fe^{III}EDTA$ decreased slowly over the first 264 hours with corresponding increases in the concentration of $Ni^{II}EDTA$. Total EDTA (sum of $Fe^{III}EDTA$ and $Ni^{II}EDTA$) remained constant over the same interval. Between 264 and 456 hours an orange tint developed in the solution and CE analysis indicated that the loss of $Fe^{III}EDTA$ exceeded the increase in $Ni^{II}EDTA$, consistent with precipitation of Fe(III) solids. Total EDTA in solution decreased, presumably by sorption onto the Fe(III) precipitate. The measured concentration of $Ni^{II}EDTA$ remained well below the predicted equilibrium concentration (adjusted for measured concentrations of Fe(III), Ni(II), and EDTA) over 36 days. Constant concentrations of total Ni(II) demonstrated that neither Ni(II) nor $Ni^{II}EDTA$ sorbed onto the Fe(III) precipitate. The results illustrate slow Fe(III) precipitation in the presence of EDTA and how slow exchange kinetics can lead to an incorrect interpretation of metal speciation in the absence of direct ML measurement.

Reactions of $Co^{II}NTA$ with Natural Mineral Assemblages

Measurement of total concentrations can mask oxidation/ reduction reactions. Solids from the vadose zone of Melton Branch Watershed within the proposed solid waste storage area 7 on the Oak Ridge Reservation in eastern Tennessee were mixed with a 1 mM solution of $Co^{II}NTA$ in 5 mM $CaCl_2$, pH 7.6 (natural pH of the solids). Total concentrations of Co (atomic absorption spectrometry), NTA (total organic carbon (TOC) and CE), and aqueous Co-NTA species were monitored over time (Figure 6). Total NTA in solution remained constant reflecting lack of sorption of this constituent; the results from TOC and CE analyses agreed within 5%. By contrast, total Co in solution decreased over 21 days suggesting that mineral surfaces in the solids had a higher affinity for Co than did the aqueous NTA. Assuming all the Co to be Co(II), $Co^{II}NTA$ is predicted to account for all the aqueous Co using equilibrium

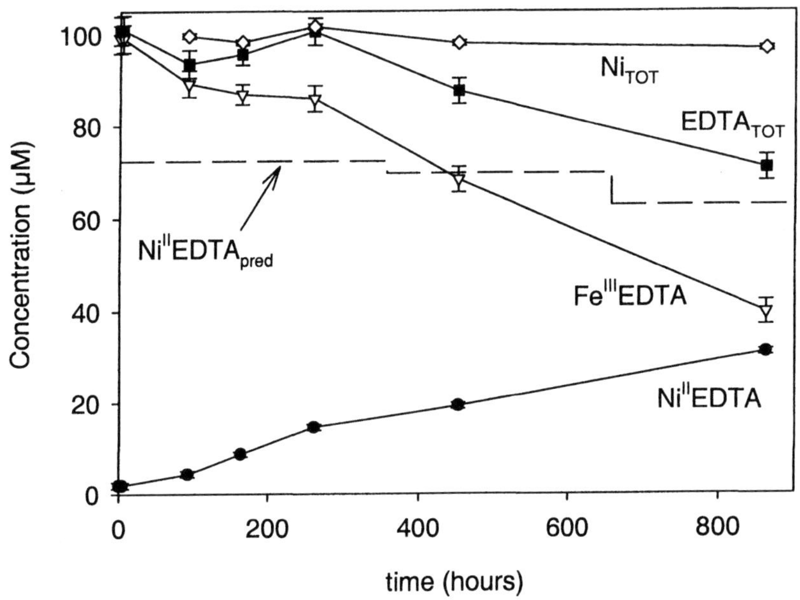

Figure 5. Dynamics of Ni(II) – Fe^{III}EDTA exchange in aqueous solution determined by capillary electrophoresis. 100 μM Ni(II) added to 100 μM Fe^{III}EDTA. Predicted concentration Ni^{II}EDTA calculated using PHREEQC and the measured total concentrations. Separation conditions: 25 mM phosphate, 0.5 mM TTAOH, pH 7.1, -20 kV constant voltage, 30s hydrostatic sampling. Capillary 75 μm i.d. × 60 cm (52 cm to detector); direct detection at 185 nm.

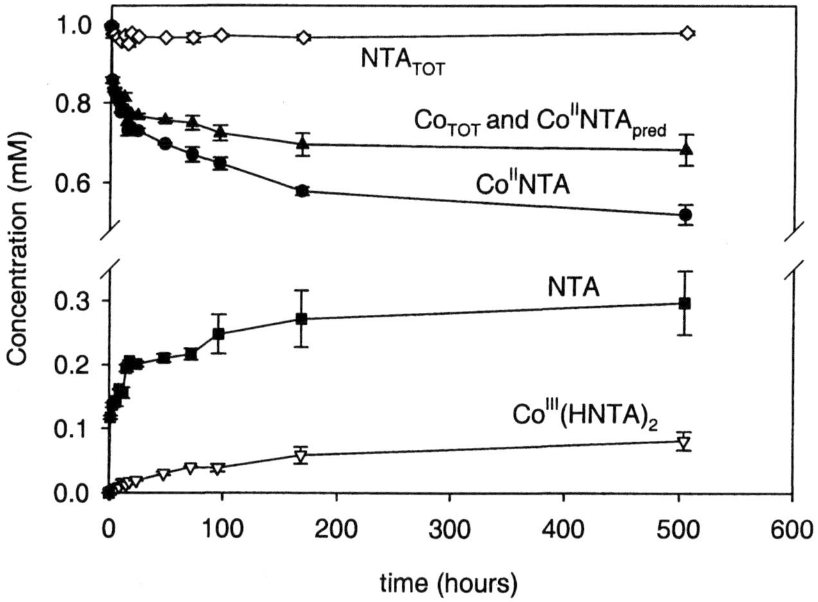

Figure 6. Reactions of $Co^{II}NTA$ in contact with natural subsurface media. Measurement of total NTA and Co only mask important changes in the oxidation state of Co. Separation conditions: 25 mM phosphate, 1 mM TTAB, pH 7.8, -55 μA constant current, 30s hydrostatic sampling. Capillary 75 μm i.d. × 60 cm (52 cm to detector); direct detection at 185 nm plus 254 nm for Co(III) species.

thermodynamic calculations. Co(II)NTA determined by CE revealed lower than expected Co(II)NTA concentrations. Sample analysis by CE, coupled with comparison to a reference standard, revealed the formation of the Co(III) species $Co^{III}(HNTA)_2$. Formation of this oxidation product likely is the result of interaction with mineral surfaces as this species was not seen in controls without solids. Total Co determined by AAS was in excellent agreement with total Co determined from summing the identified species from the CE analysis. After 21 days, $Co^{III}(HNTA)_2$ accounted for 12% of the aqueous Co and 16% of the aqueous NTA.

The formation of Co(III) complexes with NTA in contact with natural media has important implications for the fate and transport of this metal. Co(III) complexes are kinetically and thermodynamically stable, exhibit lower sorption, and are more resistant to biodegradation than their analogous Co(II) complexes *(37,38)*. These characteristics may combine to facilitate the undesirable movement of Co in the subsurface. Laboratory experiments have investigated the fate and transport of $Co^{II}NTA$ through intact cores of the same material used in Figure 6. The concentration history (breakthrough curve) for total Co exhibited two plateaus which is indicative of multispecies transport (Figure 7) *(39-41)*. CE analysis of samples revealed formation of the Co(III) species $Co^{III}(HNTA)_2$, $Co^{III}NTA$, and *s-fac*-$Co^{III}(IDA)_2$ reflecting the oxidation of both Co and NTA. The Co(III) species were transported through the column more rapidly than $Co^{II}NTA$ and accounted for 44% of the recovered Co.

Capillary electrophoresis is a useful and relatively easy technique for the direct determination of ML species. Application of this analytical method can reveal slow exchange kinetics, subtle changes in coordination, and oxidation/ reduction reactions involving both the metal and ligand.

Acknowledgement

This research was funded by the U.S. Department of Energy's Office of Science Biological and Environmental Research, Natural and Accelerated Bioremediation Research (NABIR) program. This work benefited greatly form conversations with Alan Stone, Johns Hopkins University. Oak Ridge National Laboratory is managed by UT-Battelle, LLC, for the U.S. Dept. of Energy under contract DE-AC05-00OR22725.

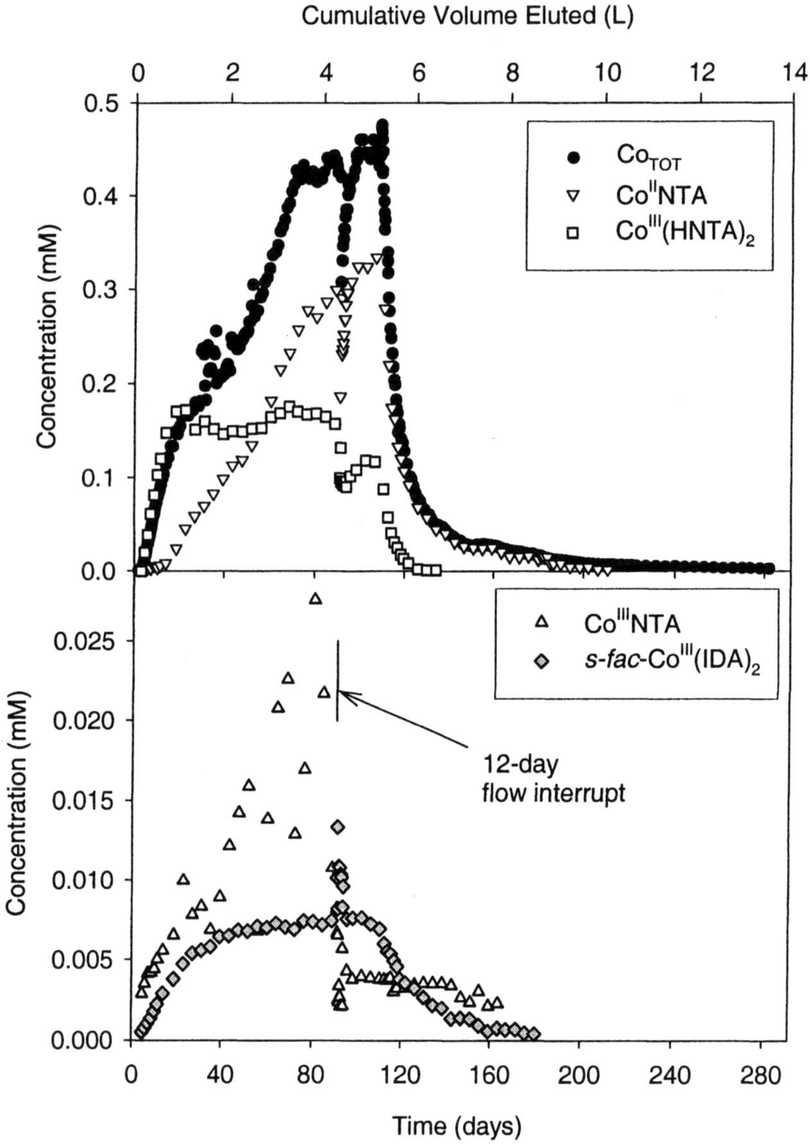

Figure 7. Transport of Co^{II}NTA through intact cores of weathered saprolite. Formation of Co(III) species resulted in more rapid transport of Co through the column and accounted for 44% of the total recovered Co (adapted from (37)). Separation conditions same as for Figure 6.

References

(1) Riley, R. G.; Zachara, J. M.; Wobber, F. J. "Chemical contaminants on DOE lands and selection of contaminant mixtures for subsurface science research.," U S Department of Energy, 1992.

(2) USEPA; USDOE; USNRC "Environmental Characteristics of EPA, NRC, and DOE Sites Contaminated with Radioactive Substances," Office of Radiation and Indoor Air (USEPA), Office of Environmental Restoration (USDOE), Office of Nuclear Material Safety and Safeguards (USNRC), 1993.

(3) Taylor, D. L.; Jardine, P. M. *J. Environ. Qual.* **1995**, *24*, 789-792.

(4) Mayes, M. A.; Jardine, P. M.; Larsen, I. L.; Brooks, S. C.; Fendorf, S. E. *J. Contam. Hydrol.* **2000**, *45*, 243-265.

(5) Nowack, B.; Kari, F. G.; Hilger, S. U.; Sigg, L. *Anal. Chem* **1996**, *68*, 561-566.

(6) Nowack, B.; Sigg, L. *J Coll Interfac Sci* **1996**, *177*, 106-121.

(7) Jandik, P.; Bonn, G. *Capillary Electrophoresis of Small Molecules and Ions*; VCH, 1993.

(8) Baker, D. R. *Capillary Electrophoresis*; John Wiley & Sons: New York, 1995.

(9) Weinberger, R. *Practical Capillary Electrophoresis*; 2nd ed.; Academic Press, 2000.

(10) Terabe, S.; Isemura, T. *Anal. Chem* **1990**, *62*, 650-652.

(11) Krokhin, O. V.; Hoshino, H.; Shpigun, O. A.; Yotsuyanagi, T. *J Chrom A* **1997**, *776*, 329-336.

(12) Fritz, J. S.; Breadmore, M.; Hilder, E. F.; Haddad, P. R. *J Chrom A* **2001**, *942*, 11-32.

(13) Fritz, J. S.; Steiner, S. A. *J Chrom A* **2001**, *934*, 87-93.

(14) Baryla, N. E.; Lucy, C. A. *J Chrom A* **2002**, *956*, 271-277.

(15) Paul, P. H.; Garguilo, M. G.; Rakestraw, D. J. *Anal. Chem* **1998**, *70*, 2459-2467.

(16) Ross, D.; Johnson, T. J.; Locascio, L. E. *Anal. Chem* **2001**, *73*, 2509-2515.

(17) Dahlen, J.; Hagberg, J.; Karlsson, S. *Fresenius J Anal. Chem* **2000**, *366*, 488-493.

(18) Burgisser, C. S.; Stone, A. T. *Environ. Sci. Technol.* **1997**, *31*, 2656-2664.

(19) Motomizu, S.; Oshima, M.; Matsuda, S.-y.; Obata, Y.; Tanaka, H. *Anal. Sci.* **1992**, *8*, 619-625.

(20) Baraj, B.; Martines, M.; Sastre, A.; Aguilar, M. *J Chrom A* **1995**, *695*, 103-111.

(21) Krokhin, O. V.; Xu, W.-z.; Hoshino, H.; Shpigun, O. A.; Yotsuyanagi, T. *Chem. Lett.* **1996**, 1095-1096.

(22) Hilder, E. F.; Macka, M.; Bogan, D. P.; Haddad, P. R. *Anal. Comm.* **1997**, *34*, 63-65.

(23) Padarauskas, A.; Schwedt, G. *J Chrom A* **1997**, *773*, 351-360.
(24) Hilder, E. F.; Macka, M.; Haddad, P. R. *Analyst* **1998**, *123*, 2865-2870.
(25) Pozdniakova, S.; Padarauskas, A. *Analyst* **1998**, *123*, 1497-1500.
(26) Krokhin, O. V.; Adamov, A. V.; Hoshino, H.; Shpigun, O. A.; Yotsuyanagi, T. *J Chrom A* **1999**, *850*, 269-276.
(27) Krokhin, O. V.; Hoshino, H.; Shpigun, O. A.; Yotsuyanagi, T. *J Chrom A* **2000**, *895*, 255-261.
(28) Naujalis, E.; Cepyte, J.; Padarauskas, A. *Anal. Bioanal. Chem* **2003**, *376*, 759-762.
(29) Schaffer, S.; Gareil, P.; Carpot, L.; Dezael, C. *J Chrom A* **1995**, *717*, 351-362.
(30) Buchberger, W.; Mulleder, S. *Mikrochim Acta* **1995**, *119*, 103-111.
(31) Harvey, S. D. *J Chrom A* **1996**, *736*, 333-340.
(32) Ballou, N. E.; Ducatte, G. R.; Quang, C.; Remcho, V. T. *J High Resol. Chromatogr* **1996**, *19*, 183-188.
(33) Smith, B. B.; Sawyer, D. T. *Inorg. Chem.* **1968**, *7*, 922-928.
(34) McArdell, C. S.; Stone, A. T.; Tian, J. *Environ. Sci. Technol.* **1998**, *32*, 2923-2930.
(35) Owens, G.; Ferguson, V. K.; Mclaughlin, M. J.; Singleton, I.; Reid, R. J.; Smith, F. A. *Environ. Sci. Technol.* **2000**, *34*, 885-891.
(36) Parkhurst, D. L.; Appelo, C. A. J. "User's Guide to PHREEQC (version 2) - A Computer Program for Speciation, Batch-Reaction, One-Dimensional Transport, and Inverse Chemical Calculations," U. S. Geological Survey, 1999.
(37) Brooks, S. C.; Carroll, S. L. *J. Contam. Hydrol.* **2002**, *58*, 191-207.
(38) Brooks, S. C.; Carroll, S. L. *Appl. Geochem.* **2003**, *18*, 423-433.
(39) Schweich, D.; Sardin, M. *J. Hydrol.* **1981**, *50*, 1-33.
(40) Jardine, P. M.; Parker, J. C.; Zelazny, L. W. *Soil Sci. Soc. Am. J.* **1985**, *49*, 867-873.
(41) Brooks, S. C.; Taylor, D. L.; Jardine, P. M. *Geochim. Cosmochim. Acta.* **1996**, *60*, 1899-1908.

Chapter 7

Analysis of Biodegradation Intermediates of Ethylenediaminetetraacetate and Nitrilotriacetate by High-Performance Liquid Chromatography

Zhiwen Yuan and Jeanne M. VanBriesen

Department of Civil and Environmental Engineering, Carnegie Mellon University, Pittsburgh, PA 15213

Anthropogenic chelating agents released to the environment can form degradation intermediates that are recalcitrant to further attenuation. Evaluation of the potential for environmental persistence requires new methods to detect these intermediates. Two HPLC methods were developed for measuring the biodegradation intermediates of EDTA and NTA. Ethylenediaminetriacetate (ED3A), N,N'-ethylenediaminediacetate (N, N'-EDDA) and 3-ketopiperazine-N,N'-diacetate (3KP) were measured concurrently by normal phase HPLC using a LiChrospher 100NH$_2$ column. Iminodiacetate (IDA) and glycine were measured concurrently by reversed phase HPLC after derivatization with 9-fluorenylmethyl chloroformate (FMOC-Cl). These two methods were simple, fast, accurate, specific, and applicable to the analysis of these intermediates in different matrices and their detection limits are all below 1μM using UV detector.

Introduction

EDTA and NTA are two important anthropogenic chelating agents able to form stable, water soluble complexes with metals. They are widely used in many industrial and consumer products and processes, such as detergents, photo processing, the paper and pulp industry, nuclear reactor decontamination, and many other industries. The estimated U.S. production of EDTA in 1992 was 7,241 tons (*1*). European usage of EDTA is much higher with an estimated 35, 000 tons in 1990 (*2*). European NTA consumption is about 17,350 tons in 1990 (*3*). The total worldwide use of aminopolycarboxylic acid (EDTA, DTPA and NTA) was 186,000 tons in 1994 (*4*). The usage of EDTA is expected to increase in the future, especially in the paper and pulp industry due to transition to chlorine free processes that require additional chelates.

Due to minimal biodegradation in natural and engineered systems and the low potential for photodegradation in many turbid waters, EDTA persists in surface waters at concentrations ranging from 1-100 μg/L (*2*). Aminocarboxylic chelating agents such as EDTA and NTA are generally regarded as nontoxic and used in food products and medicines. However, the effects of low persistent concentrations in the environment are unclear. Chelating agents may enhance eutrophication, remobilize heavy metals, and have long-term toxicological effects on microorganism (*5*).

Biodegradation of EDTA and NTA is a possible solution for their removal from the environment. Several species able to degrade NTA have been isolated. The well studied NTA degrader, *Chelatobacter heintzii* ATCC 29600 (*C. heintzii*), utilizes a monooxygenase pathway to catalyze NTA biodegradation (see Figure 1) (*5-9*). EDTA degrading organisms are not as thoroughly characterized. Nortemann and coworkers identified a mixed culture (BNC1/BNC2) and isolated a pure culture (BNC1, DSM 6780) that degraded low concentrations of EDTA (μM to mM) (*10-14*). Witschel et al (*15*) reported another pure strain DSM 9103 that is closely related to BNC1 and utilizes a similar degradation pathway to degrade EDTA. The proposed multi-step degradation pathway of EDTA by BNC1/BNC2 is shown in Figure 2 (*10-14, 16-18*).

Nortemann (*10*) reported that bacterium BNC1 initiates the breakdown of EDTA but nitrogen-containing intermediates were produced and persistent in the system. These intermediates may still have significant chelating ability. Intermediate persistence was also reported during NTA degradation by *C. heintzii (19, 20)*. If these intermediates persist, they sequester carbon and electrons, reducing cell yield, slowing the rate of removal of chelating agents from the system and continuing to influence the system. As shown in Figure 1, the anticipated intermediates of NTA biodegradation are iminodiacetic acid (IDA), glycine and glyoxylate. As shown in Figure 2, the anticipated intermediates of EDTA biodegradation may include ethylenediaminetriacetate (ED3A), N,N'-ethylenediaminediacetate (N,N'-EDDA), ethylenediaminemonoacetate (EDMA), ethylenediamine (ED) and glyoxylate. ED3A can spontaneously cyclize to form 3-ketopiperazine-N,N-diacetate (3KP), especially under acidic condition. This reaction is shown in Figure 3. 3KP is

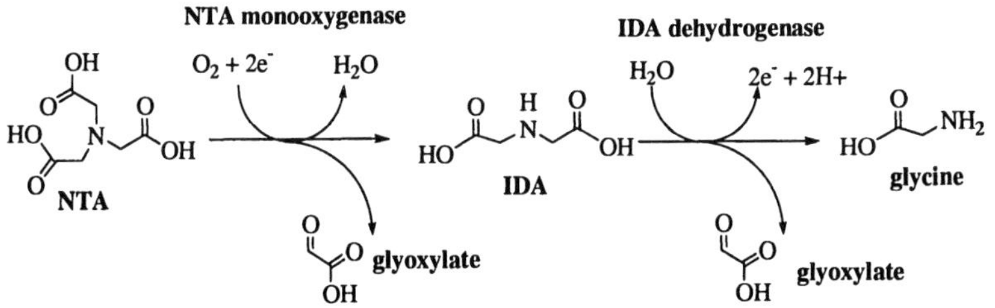

Figure 1. NTA degradation pathway by C. heintzii.

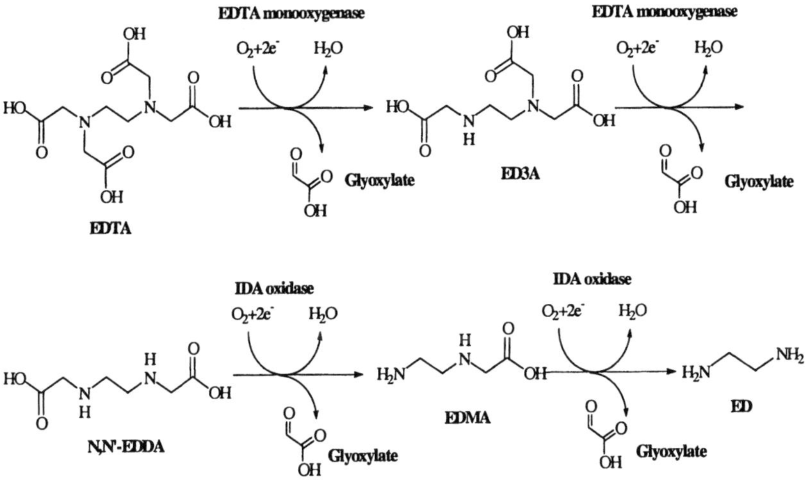

Figure 2. Proposed degradation pathway of EDTA by BNC1/BNC2 co-culture.

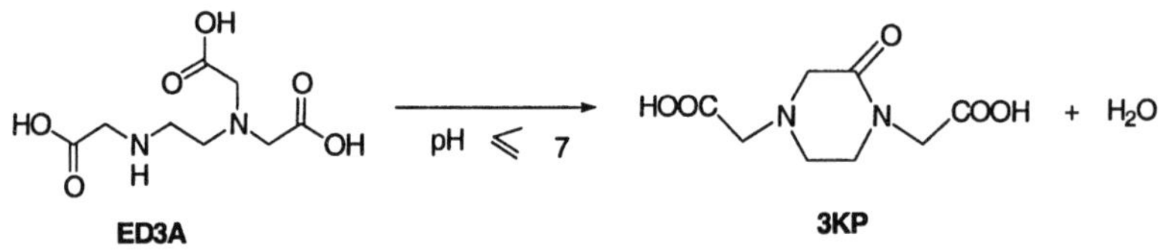

Figure 3. Spontaneous formation of 3KP from ED3A.

resistant to biodegradation and nothing is known about its toxic or ecotoxic behavior (*11, 21*).

Various analytical methods for measuring EDTA and NTA have been developed with different applications and detection limits (*2, 22-31*). However, few methods have been developed for measuring all the intermediates formed during degradation reactions. Glyoxylate has been previously measured by spectrophotometry (*32, 33*). Glycine was previously measured by High Performance Liquid Chromatography (HPLC) with fluorescence detector as an amino acid after derivatization with 9-fluorenylmethyl chloroformate (FMOC-Cl) (*34*). IDA was previously measured by HPLC with fluorescence detector as well (*19*). ED was previously measured by HPLC with fluorescence detector after derivatized with 5-dimethylamino-1-naphthalenesulfonyl chloride (*35*), though the derivatization process was time and labor intensive.

HPLC is well suited for separating EDTA, NTA and their degradation intermediates since they are non-volatile, highly soluble compounds. Kluner et al studied the biodegradation of EDTA and measured ED3A and 3KP concurrently by HPLC with UV detection. They measured N,N'–EDDA by another HPLC method (*14*).

The contribution of this work was to develop a method for ***concurrent*** measurement of EDTA biodegradation intermediates ED3A, N, N'-EDDA and 3KP and a method for ***concurrent*** measurement of NTA biodegradation intermediates IDA and glycine by HPLC. These two methods have been presented as an abstract previously (*36*). They were simple, fast, accurate, specific, and applicable to the analysis of these intermediates in different matrixes and their detection limits are all below 1 μM, which are low enough for most applications.

These two methods have been successfully applied to detect the possible persistent intermediates during NTA and EDTA biodegradation. Noticeable IDA accumulation and persistence during NTA degradation by *C. heintzii* and N,N'-EDDA accumulation and persistence during EDTA degradation by BNC1/BNC2 under high pH conditions have been observed and were monitored by these methods (Yuan and VanBriesen, unpublished results).

Materials and Methods

The HPLC was performed with a HP 1050 series quaternary pump and variable wavelength UV/Visible detector. A LiChrospher 100NH$_2$ column (250mm×4.0mm, 5μm particle size, Agilent Technologies, USA) was used for separating N,N'-EDDA, 3KP and ED3A. A Nucleosil 100-5 C18 Column (250mm×4.0mm, 5μm particle size, Agilent Technologies, USA) was used for separating IDA and glycine. HP Chemstation software (Revision A.07.01, Hewlett-Packard) was used for chromatogram integration and linear regression of the calibration table and curve. The concentration was calculated according to peak area.

EDTA mineral medium for BNC1/BNC2 contained (per liter): 0.9g $Na_2EDTA{\bullet}2H_2O$, 2.75g Na_2HPO_4, 1.0g KH_2PO_4, 1.0g $MgSO_4{\bullet}7H_2O$, 0.11g $CaCl_2$, 0.0005g $FeCl_3{\bullet}6H_2O$, 1mL trace element solution and 1mL vitamin solution. NTA mineral medium for *C. heintzii* contained (per liter): 1.5g $Na_3{\bullet}NTA{\bullet}H_2O$, 1.6g Na_2HPO_4, 0.4g KH_2PO_4, 0.2g $MgSO_4{\bullet}7H_2O$, 25mg $CaCl_2$, 2.5mg $FeCl_3{\bullet}6H_2O$, 1mL trace element solution, and 1mL vitamin solution.

$Na_3{\bullet}$NTA was purchased from Acros Organics, USA. Iminodiacetic acid was from Sigma Chemicals, USA. $Na_3{\bullet}$ED3A was a gift from Hampshire Chemicals, USA. 3KP was made from $Na_3{\bullet}$ED3A by adding HCl according to Figure 3. N,N'-EDDA was from Tokyo Kasei Kogyo, Japan. All other reagents were reagent grade or better and purchased from Fisher Scientific, USA.

After complexation with excess copper acetate, N,N'-EDDA, ED3A and 3KP were separated by LiChrospher 100NH$_2$ column, which has a layer of aminopropyl-silane chemically bonded to porous silica gel support and has special affinity for nitrogen containing compounds. The chromatography conditions were shown in the second column of Table I. All mobile phase solutions were filtered by 0.45μm membrane to remove particles and degas the solution.

IDA and glycine were measured as amino acids after derivatization with 9-fluorenylmethyl chloroformate (FMOC-Cl) under basic conditions. The reaction of FMOC-Cl with amino acid is shown in Figure 4. The derivative is highly fluorescent and stable and can be monitored by a fluorescence or UV detector. The derivatization procedures were as following: 3mL sample (or standard solution) and 1mL borate buffer (0.4mM boric acid, adjusted to pH 9.5 by 2.5M NaOH solution) were mixed in a 20mL reaction vial. Then 3mL 10mM FMOC-Cl reagent (dissolved in pure acetone) was added. After 1minute 10mL diethyl either was added into the mixture to extract excess reagent and acetone. The capped vials were shaken by hand for 1 minute and then kept still for five minutes to separate the ether and aqueous layers. Then the ether layer was removed by a Pasteur pipette and discarded. The extraction was repeated once and the aqueous samples were ready for injection. The chromatography conditions were shown in the third column of Table I.

Table I. Chromatography Conditions

Method	*N,N'-EDDA, ED3A and 3KP*	*IDA and glycine*
Temperature	Room temperature (about 20°C)	Room temperature (about 20°C)
UV detector wavelength	256nm	256nm
Mobile phase	0.7% glacial acetic acid in deionized H_2O, 1mM Cu $(CH_3COOH)_2 \cdot H_2O$, adjusted to pH 4.0 with NaOH solution.	Channel A: 0.3% glacial acetic acid in deionized H_2O, adjusted to pH 4.3 with NaOH solution. Channel B: 75% acetonitrile: 25% deionized water (v/v).
Separation mode	Isocratic elution	Gradient elution, 30%A: 70%B linearly changed to 70%A: 30%B from 0-6min, and then remained at 70%A: 30%B from 6-10min. The whole elution ended at 10min.
Flowrate	1.8mL/min	1.2mL/min
Running time	15min	10min
Post running time before next injection	0min	10min

FMOC-Cl

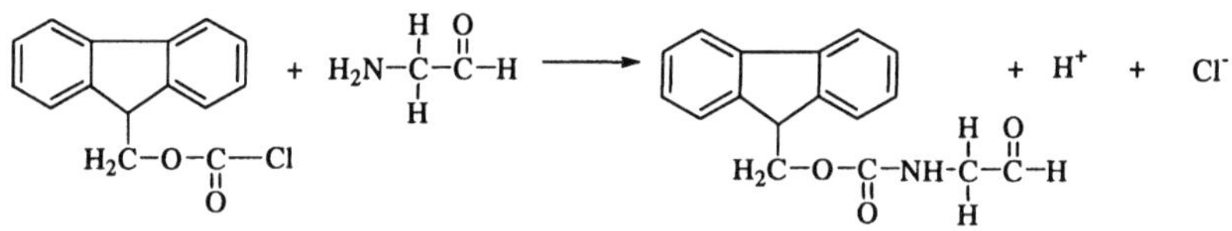

Figure 4. Derivatization of FMOC-Cl with amino acids.

Results and Discussion

The chromatogram of 30μM mixed standards of N,N'-EDDA, 3KP and ED3A spiked into EDTA mineral medium for bacteria BNC1/BNC2 is shown in Figure 5 (injection volume 100μL). The retention times for N,N'-EDDA, 3KP and ED3A are 2.9 min, 7.0 min and 11.2 min, respectively. This is the first report of concurrent measurement of these EDTA biodegradation intermediates. Further, this represents detection in a complex media.

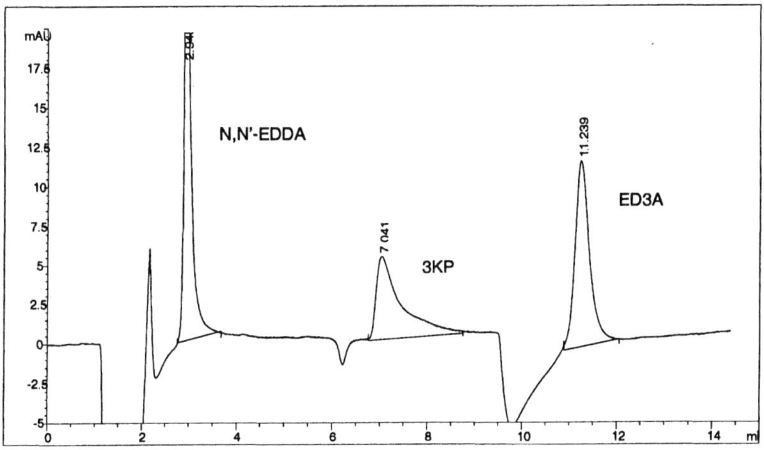

Figure 5. Chromatogram of 30μM N,N'-EDDA, ED3A and 3KP spiked into EDTA mineral medium for BNC1/BNC2 (injection volume 100μL).

The representative chromatogram of 100μM mixed standards of IDA and glycine spiked into NTA mineral medium for *C. heintzii* is shown in Figure 6 (injection volume 100μL). The retention times for IDA and glycine are 4.4min and 6.5min, respectively. This is the first report of concurrent measurement of these intermediates and the use of UV rather than fluorescence detection.

In summary, two HPLC methods were developed for measuring the degradation intermediates of EDTA and NTA, which were simple, fast, accurate, specific, and applicable to different matrices with detection limits below 1μM using UV detector. The retention times of all the degradation intermediates can be adjusted by varying the pH to avoid the possible interference caused by other compounds that might be present in environmental samples. The retention times of glycine and IDA specifically can also be adjusted by varying the gradient strength. The excess EDTA and NTA in the original media were not detectable by these two methods and they can be measured by previously described methods (*2, 22-31*). The ability to monitor for intermediates that form during biodegradation of chelates is expected to have significant utility in studying the fate of these compounds in the environment.

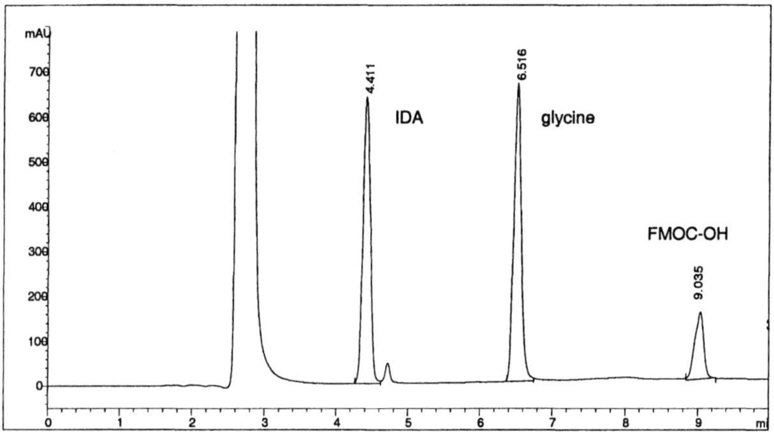

Figure 6. Chromatogram of 100μM glycine and IDA standards spiked into NTA mineral medium for C. heintzii *(injection volume 100μL).*

Acknowledgement

Support for this work was provided by the National Science Foundation (BES-0092463).

Reference

1. United States Trade Commission. *Synthetic Organic Chemicals - U.S. Production and Sales 1992.* USITC Publication 2720: Washington DC, 1994.
2. Frimmel, F. H. In *Detergents in the Environment;* Schwager, M. J., Ed.; Marcel Dekker: New York, 1997; pp 289-312.
3. Kiessling, D.; Kaluza, U. In *Detergents in the Environment;* Schwager, M. J., Ed.; Marcel Dekker: New York, 1997; pp 265-288.

4. Sisto, J. D.; Jacke, M.; Ishikawa, M. *Chemical Economics Handbook.* SRI Consulting: Menlo Park, CA, 1997.
5. Bucheli-Witschel, M.; Egli, T. *FEMS Microbiol. Rev.* **2001**, 25, 69-106.
6. Uetz, T.; Egli, T. *Biodegradation.* **1993**, 3, 423-434.
7. Uetz, T.; Schneider, R.; Snozzi, M.; Egli, T. *J. Bacteriol.* **1992**, 174, 1179-1188.
8. Firestone, M. K.; Tiedje, J. M. *Appl. Environ. Microbiol.* **1978**, 35, 955-961.
9. Firestone, M. K.; Tiedje, J. M. *Appl. Microbiol.* **1975**, 29, 758-764.
10. Nortemann, B. *Appl. Environ. Microbiol.* **1992**, 58, 671-676.
11. Nortemann, B. *Appl. Environ. Microbiol.* **1999**, 51, 751-759.
12. Henneken, L.; Nortemann, B.; Hempel, D. C. *J. Chem. Technol. Biotechnol.* **1998**, 73, 144-152.
13. Henneken, L.; Nortemann, B.; Hempel, D. C. *Appl. Microbiol. Biotechnol.* **1995**, 44, 190-197.
14. Kluner, T.; Hempel, D. C.; Nortemann, B. *Appl. Microbiol. Biotechnol.* **1998**, 49, 194-201.
15. Witschel, M.; Nagel, S.; Egli, T. *J. Bacteriol.* **1997**, 179, 6937-6943.
16. Payne, J. W.; Bolton, H. Jr.; Campbell, J. A.; Xun, L. **1998**, 180, 3823-3827.
17. Bohuslavek, J.; Payne, J. W.; Liu, Y.; Bolton, H. Jr.; Xun, L. *Appl. Environ. Microbiol.* **2001**, 67, 688-695.
18. Liu, Y.; Louie, T. M.; Payne, J.; Bohuslavek, J.; Bolton, H. Jr.; Xun, L. *Appl. Environ. Microbiol.* **2001**, 67, 696-701.
19. Bolton, H. Jr.; Girvin, D. C.; Plymale, A. E.; Harvey, S. D.; Workman, D. J. *Environ. Sci. Technol.* **1996**, 30, 931-938.
20. VanBriesen, J. M.; Rittmann, B. E. *Biotechnol. and Bioeng.* **2000**, 67, 35-52.
21. Ternes, T. A.; Stumpf, M.; Steinbrecher, T.; Brenner-Weiss, G.; Haberer, K. *Vom Wasser.* **1996**, 87, 275-290.
22. Lee, H.-B.; Peart, T. E.; Kaiser, K. L. E. *J. Chromatogr. A.* **1996**, 738, 91-99.
23. Bergers, P.; DeGroot, A. *Water Res.* **1994**, 28, 639-642.
24. Parkes, D. G.; Caruso, M. G.; Spradling, J. E. I. *Anal. Chem.* **1981**, 53, 2154-2156.
25. Schneider, R. P.; Zurcher, F.; Egli, T.; Hamer, G. *Anal. Biochem.* **1988**, 173, 278-284.
26. Geschke, R.; Zehringer, M. *Fresenius J. Anal. Chem.* **1997**, 357, 773-776.
27. Nowack, B.; Kari, F. G.; Hillger, S. U.; Sigg, L. *Anal. Chem.* **1996**, 68, 561-566.
28. Sillanpaa, M.; Sihvonen, M.-L. *Talanta.* **1997**, 44, 1487-1497.
29. Kloster, G. In *Detergents in the Environment;* Schwager, M. J., Ed.; Marcel Dekker: New York, 1996, pp65-121.

30. Nirel, P. M.; Pardo, P.-E.; Landry, J.-C.; Revaclier, R. *Wat. Res.* **1998**, 32, 3615-3620.
31. Owens, G.; Ferguson, V. K.; Mclaughlin, M. J.; Singleton, I.; Reid, R. J.; Smith, F. A. *Environ. Sci. Technol.* **2000**, 34, 885-891.
32. Kramer, D. N.; Klein, N.; Baselice, R. A. *Anal. Chem.* **1959**, 31, 250-252.
33. Trijbels, F.; Vogles, G. D. *Biochimica Et Biophysica Acta.* **1966**, 113, 292-301.
34. Einarsson, S.; Joseffsson, B.; Lagerkvist, S. *J. Chromatogr.* **1983**, 282, 609-618.
35. Chu, S.; Taliant, M. J.; Yang, R. S. H. *J. Chromatorgr.* **1983**, 280, 394-399.
36. Yuan, Z.; VanBriesen, J. M. In *American Chemical Society Annual Meeting Abstracts*; Boston, MA, 2002.

Biological Reactions

Chapter 8

Biodegradation of Chelating Agents: EDTA, DTPA, PDTA, NTA, and EDDS

Bernd Nörtemann

Institute of Biochemical Engineering, Technical University of Braunschweig, Gauss-Strasse, 17, D–38106 Braunschweig, Germany

The biodegradability of a chelating agent appears to depend on the nature of the organic ligand as well as on the metal ion complexed with it. The influence of the latter may be explained at least in part by the thermodynamic stability of the complex. Among the chelating agents of technical relevance, aminopolycarboxylates (APCs) are one of the most important groups. In general, APCs which form complexes with relatively low or moderately high stability constants (e.g., NTA, EDDS) are readily degradable, whereas those forming stronger complexes (e.g., EDTA, PDTA, DTPA) are relatively resistant to biodegradation. In all cases, however, the nature of the metal ion bound by the APC significantly determines the biodegradability of the ligand. This chapter gives a short overview on the biodegradability of some important APCs and their metal chelates, describes the biological degradation of EDTA, PDTA, and DTPA by specially enriched bacterial cultures, and introduces biological methods for the removal of these chelating agents from wastewaters.

Introduction and Background

Complexing agents from of the group of aminopolycarboxylates (APCs) are used in large quantities in a wide range of domestic products and industrial applications, e.g. for controlling the oxidation potential of metal ions, for adjusting metal ion concentrations, for masking free metal ions to avoid metal-catalyzed spoilage or the precipitation of sparingly soluble salts, and for cleansing processes where precipitates of salts in tubes, bottles and on membranes must be removed. For most technical purposes, those APCs are used as multidentate ligands ("ligare", Latin for "to bind") that contain two or more complexing groups forming bonds with the central metal ion. These chelating agents form stable and water soluble chelates ("chele"- Greek for "claw of crab") with most alkaline earth and heavy metal ions. APCs such as NTA (nitrilotriacetate), EDDS (ethylenediaminedisuccinate), EDTA (ethylene-diaminetetraacetate), DTPA (diethylenetriaminepentaacetate) and PDTA (1,3-propylenediaminetetraacetate) are the dominant products used for chelation worldwide, and are used almost exclusively for this purpose. The quantitatively most important areas of application are as cleansing compounds, in pulp and paper manufacture, photography, and agriculture.

In contrast to NTA and EDDS, most of the other technically important APCs - particularly EDTA, DTPA and PDTA (Figure 1) - are not eliminated by conventional biological and physico-chemical methods for the treatment of wastewater and the purification of drinking water. Of the measurable synthetic organic compounds, EDTA and DTPA have the highest concentration in many surface waters and drinking waters (*1-12*). As, in principle, drinking water should not contain xenobiotic compounds, the amount of recalcitrant APCs released into aquatic systems should be reduced as far as technically possible, e.g. by using substitutes that are readily biodegradable. However, EDTA and DTPA are the cheapest and most suitable complexing agents for many technical purposes, especially where strong chelating properties are required. As a rule, substitutes with chelating properties comparable to those of EDTA or DTPA are non-biodegradable as well, and could often cause more critical problems because of their (eco-) toxic properties. On the other hand, a high resistance of chelating agents to biodegradation is often desired or necessary for the stability of technical processes (e.g., photographic process baths) or products. In these cases, the elimination of the recalcitrant APC from the corresponding process water, or partial wastewater streams, by special treatment systems is a promising approach to reduce the amounts of APCs which are released into the aquatic environment.

Another reason to investigate the biodegradation of strong chelating agents is the fact that they can be used for the remediation of sites contaminated with heavy metals or radionuclides, and for the decontamination of nuclear power plant equipments (*13-15*). In addition, APCs in groundwater could enhance undesirably the migration of heavy metals or radionuclides from contaminated soils or disposal sites (*16, 17*). It is thus of interest to understand the biodegradation of APCs and their metal complexes, and to develop new approaches to stimulate biological degradation.

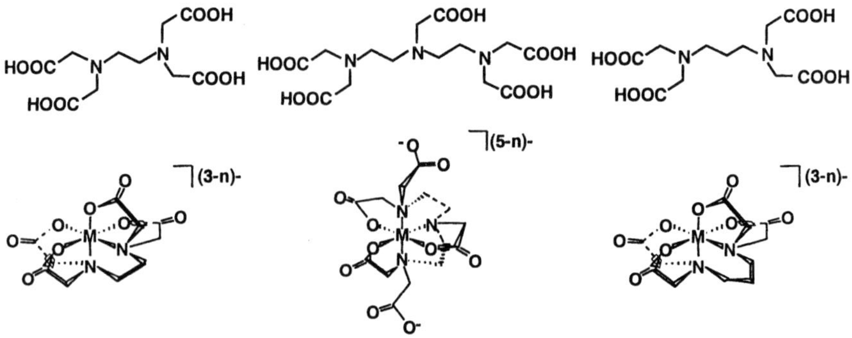

EDTA DTPA PDTA

Figure 1 Free acids and metal chelates of EDTA, DTPA, and PDTA (Dotted bonds to M are co-ordinate, otherwise dotted bonds are covalent).

Sykora et al. (*18*) have investigated the correlation between the structure or the functional groups of an APC and its biological degradability. They found, e.g., that the biodegradability of ethylene- and propylenediamine derivatives depends on the type and number of substituents; it decreases in the sequence of the substituents $-COCH_3$, $-CH_3$, $-C_2H_5$, $-CH_2CH_2OH$, $-CH_2COOH$, and with polysubstitution. Moreover, the biodegradability depends also on the kind and number of nitrogen atoms. Simple complexing agents with a single nitrogen atom in the molecule (e.g., NTA) are relatively readily subject to biodegradation. In contrast, the resistance to biodegradation increases with the number of tertiary amino groups. Interestingly, a lowering in the degree of substitution of the amino groups results in an increased susceptibility of the APC to biodegradation. This is valid, e.g., for the replacement of tertiary amino groups with secondary ones. These observations are in good agreement with most of the reports on relatively readily degradable APCs such as NTA or EDDS as well as with the investigations on the degradation of recalcitrant APCs (EDTA, PDTA, DTPA) by specially enriched and adapted bacterial cultures as described below in this chapter.

Microorganisms, Materials and Methods

Microorganisms.

The laboratory studies described in this chapter were carried out with an EDTA-degrading mixed culture and with two primary utilizers isolated from this community after different times of prolonged subcultivation with EDTA as sole source of carbon and energy. Strain BNC1 (DSM 6780) was originally found to be the only primary utilizer in the mixed culture (*19*). After prolonged subcultivation, however, another primary utilizer, strain ANP 11, was obtained from the community (*20-22*). In addition, some other EDTA-degrading organisms were obtained by enrichment from different habitats (20) that were continuously receiving EDTA containing wastewaters. Another EDTA-degrading axenic culture, strain DSM 9103, was described by Witschel et al. (*23-25*). All of these EDTA utilizing isolates are Alphaproteobacteria and belong to the same genus (and species) which is closely related to *Mesorhizobium* but will probably be described as a new genus (*26*). However, some biochemical differences destinguish strain ANP11 from the other isolates such as strain BNC1, and indicate that this organism represents a separate strain within the same common species of the EDTA-degrading isolates. This was also confirmed by the non-identical ranges of chelating agents that are degraded by whole cells or are converted by cell-free extracts of strain ANP11 and the other isolates.

Cultivation conditions.

Enrichment experiments, cultivation of bacteria, performance of turnover experiments, and the analysis of substrates and products were carried out as described elsewhere (*21, 22, 27-32*). The nutrient media usually contained 2.0 mmol/L $MgSO_4 \cdot 7\,H_2O$, 0.85 mmol/L $CaCl_2$, 0.02 mmol/L $FeCl_3 \cdot 6\,H_2O$, 1.0 mL/L trace element solution [described by Pfennig and Lippert (*33*)], 10 mL/L vitamin solution [described by Klüner (*34*)] and 25.0 mmol/L phosphate buffer (K_2HPO_4 and NaH_2PO_4, pH 7.75). The trickle-bed reactor described in this study consisted of a glass tube (height 40 cm, diameter 25 cm) filled with Raschig rings, a receiving flask and a peristaltic pump. 500 mL inoculum were added to 2L selective medium in the receiving flask, and pumped continuously from the base to the top of the reactor.

Readily Degradable Chelating Agents

NTA.

NTA possesses a tertiary amino group, but compared to other APCs of technical importance it is a simple complexing agent with a single nitrogen atom in the molecule. It generally binds bivalent metal ions in a ratio of 1:1, and the corresponding chelates have a relatively low or only moderately high stability constants. NTA is readily degraded by bacteria under both oxygenic and anoxic conditions, and many NTA-degrading bacteria could be obtained from a wide range of ecosystems including river waters, activated sludge, sediments and soils (*35-44*). The best characterized bacteria that degrade NTA under aerobic conditions are members of the genera *Chelatobacter* and *Chelatococcus*. Both of them belong to the Alphaproteobacteria, and are present in fairly high numbers in surface waters, soils and sewage treatment plants (*45-49*). Another organism, strain TE11, was found to be a member of the Gammaproteobacteria and degrades NTA under denitrifying conditions [44]. In particular, however *Chelatobacter heintzii* ATCC 29600 (*50*) has been subject to extensive physiological and biochemical investigations. This organism is a later subjective synonym of *Aminobacter aminovorans* (*51*) as described by Kämpfer et al. (*52*). The range of metal-NTA complexes that can be degraded by whole cells of *Chelatobacter heintzii* was investigated by Bolton et al. (*53*). The order of the rates of mineralization was: free NTA > FeNTA = CoNTA = ZnNTA > AlNTA > CuNTA > NiNTA. These differences could neither be explained by the different thermodynamic stability constants for the chelates, nor were they accounted for by the toxicity of the metal ions released from the complex.

The first catabolic step in aerobic NTA-degradation by *Chelatobacter heintzii* is catalyzed by nitrilotriacetate monooxygenase (NTA-MO) which hydroxylates NTA at an α-carbon atom and spontaneously forms iminodiacetate (IDA) and glyoxylate. The NTA-MO activity requires two component proteins, A and B. Component cA is a monooxygenase which uses $FMNH_2$ and O_2 to oxidize NTA; component cB is an NADH:FMN oxidoreductase that provides $FMNH_2$ for NTA oxidization by using NADH to reduce FMN to $FMNH_2$. In the following step, IDA is further converted to glycine and glyoxylate by iminodiacetate dehydrogenase [summary, see (25)].

The range and relative oxidation rates of NTA and its metal complexes by NTA-MO differed from that observed with whole cells of *Chelatobacter heintzii* (*54*). Free NTA and CaNTA were not oxidized by NTA-MO, while MgNTA, MnNTA, CoNTA, FeNTA, NiNTA, and ZnNTA were degraded, and had similar K_m-values. It was concluded that the metal associated with the NTA substrate is not transported into the cells and that biodegradation of various metal-NTA complexes is limited by the rate of transport into the cells.

Other readily degradable chelating agents.

In recent years, other newly synthesized or naturally occuring APCs such as L-alaninediacetate (L-ADA), iminodisuccinate (IDS), and EDDS (Figure 2) have been tested with regard to their technical applicability as chelating agents and to their biodegradability (*55-61*). L-ADA and IDS as well as serinediacetate (SDA), asparaginic acid diacetate (ASDA) and methylglycinediacetate (MGDA) were found to be readily degradable. Although they can be used for some technical purposes, the stability of their metal complexes is usually relatively low so that the field of possible applications is clearly restricted.

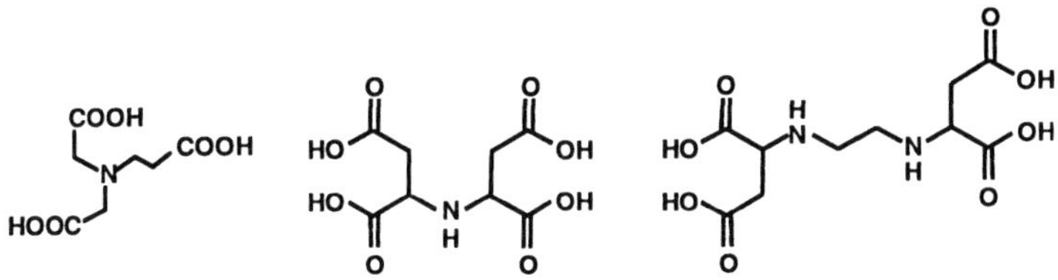

L-ADA **IDS** **EDDS**

Figure 2 Free acids of L-alaninediacetate (L-ADA), iminodisuccinate (IDS), and ethylenediaminedisuccinate (EDDS).

Ethylenediaminedisuccinate (EDDS) is a relatively efficient transition metal chelator and its [*S,S*]-stereoisomer is biologically produced as shown for *Amycolatopsis orientalis* (e.g., *62-64*). The EDDS structure contains two chiral carbon atoms, and has three stereoisomers ([*R,R*], [*R,S*]/[*S,R*], [*S,S*]). The biodegradation of these stereoisomers of EDDS was investigated extensively by various authors (*57-60, 62-64*). In these studies, the [*S,S*]-isomer was found to be subject to rapid and complete mineralization whereas the [R,R] remained undegraded in conventional test systems (e.g., OECD 301B), and the mixed stereoisomers were converted slowly and incompletely. As a consequence, the [*S,S*]-stereoisomer of EDDS has substituted traditional chelant agents in a number of consumer products. In addition, it can be applied efficiently for some technical purposes, e.g., for the extraction of heavy metals from soils using biodegradable chelating agents (*65, 66*). Tandy et al. (*66*) reported that for Cu at pH 7, the order of the binding efficiency for equimolar ratios of chelating agent to metal was EDDS > NTA> IDS > MGDA > EDTA and for Zn it was NTA > EDDS > EDTA >MGDA > IDS. Vandevivere et al. (*67*) fed activated sewage sludge with 1 mM/L pulses of [*S,S*]-EDDS and a number of its metal chelates as respective sole source of carbon and nitrogen. A ready biodegradability was observed for Ca-, Cr(III)-, Fe(III)-, Pb-, Al-, Cd-, Mg-, Na-, or ZnEDDS (the latter only after extensive lag phase). However, the Cu-, Ni-, Co-, and Hg-complexes remained essentially undegraded. Witschel and Egli (*68*) showed that the catabolism of EDDS by the EDTA-degrading strain DSM 9103 was initiated by a carbon-nitrogen lyase catalysing the non-hydrolytic cleavage of the C-N bond between the ethylenediamine part of the molecule and one of the succinyl residues without any cofactors being required. This reaction led to the formation of fumarate and N-(2aminoethyl) aspartate (AEA). The enzyme catalyzed the transformation of free [*S,S*]-EDDS and of [*S,S*]-EDDS-metal complexes with stability constants lower than 10^{10}, namely of Mg-, Ca-, Ba- and to a small extent also of MnEDDS. In contrast, Fe(III)-, Ni-, Cu-, Co-, and ZnEDDS were not transformed.

Whether or not EDDS could be used for the remediation of soil contaminated with heavy metals remains unknown. It is after all biodegradable. This biodegradability is a problem for many technical processes, so that the so called recalcitrant chelating agents will remain the agents of choice. Problems associated with their use can be overcome by the treatment of wastewater.

Recalcitrant Chelating Agents

EDTA.

Although the ferric chelate of EDTA can be oxidized to biodegradable metabolites by photodegradation (*69*), EDTA and DTPA have the highest concentration in many surface waters and drinking waters (*1-12*). This can be explained by the low potential for photodegradation in many turbid waters, and by the presence of other metal ions leading to metal-EDTA chelates that are resistant to photodegradation.

It was demonstrated that EDTA is efficiently degraded by cells of a specially enriched and adapted bacterial mixed culture in which the primary utilizer was strain BNC1 (*21, 70*), and by cells of the strain DSM 9103 described by Witschel et al. (*23-25*). During growth of an axenic culture of strain BNC1 with EDTA as sole C- and N-source, no organic metabolites were detectable by HPLC- or DOC-analyses. In the mixed culture, however, approximately 40 % of the colony forming units are represented by secondary organisms that can not metabolize EDTA. For prolonged cultivation with EDTA, cells of strain BNC1 require either the addition of vitamins (biotin, thiamine, and folic acid) or the presence of the secondary organisms. Therefore, a cross-feeding of carbon sources and vitamins between strain BNC1 and secondary organisms can be assumed (Figure 3).

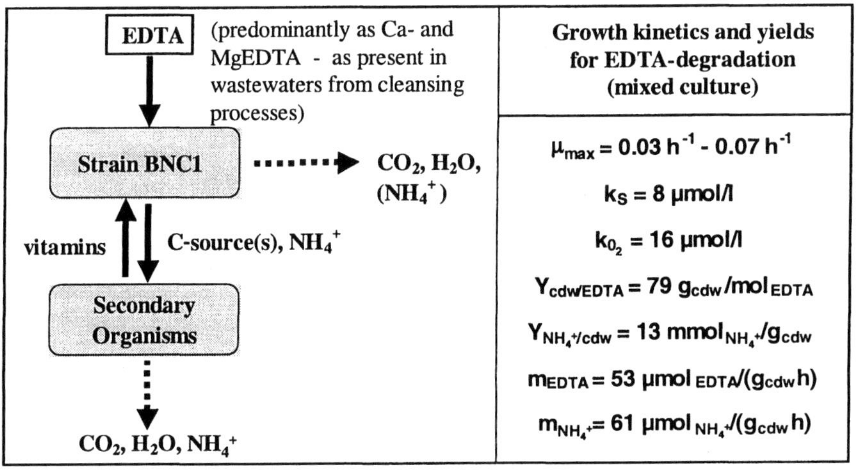

Figure 3 Degradation of EDTA by cells of the bacterial mixed culture: cross-feeding, kinetic data and yields (cdw, cell dry weight).

The first two catabolic steps in EDTA-degradation in both strain BNC1 (*34, 70-72*) and strain DSM 9103 (*24, 25*) are shown in Figure 4. The cleavage of the EDTA molecule to N,N'-ethylenediaminediacetate (N,N'-EDDA) via ethylenediaminetriacetate (ED3A) by a two-component monooxygenase occurs analogous to that in NTA catabolism. Component A is a monooxygenase which uses $FMNH_2$ and O_2 to oxidize EDTA and ED3A; and B is an NADH:FMN oxidoreductase that provides $FMNH_2$ by using NADH for FMN reduction.

The EDTA-degrading strain BNC1 also mineralizes NTA, and its EDTA monooxygenase oxidizes NTA to iminodiacetate (IDA). Recently, a soluble IDA oxidase was purified to apparent homogeneity from this strain by using a combination of eight purification steps (*73, 74*). The purified enzyme contained FAD as a prosthetic group and oxidized IDA to glycine and glyoxylate with the direct reduction of O_2 to H_2O_2. It also oxidized N,N'-EDDA to ethylenediamine and glyoxylate, probably via ethylenediaminemonoacetate (EDMA). Thus, IDA oxidase is obviously the second enzyme in both the NTA- and EDTA-degradation pathways in strain BNC1.

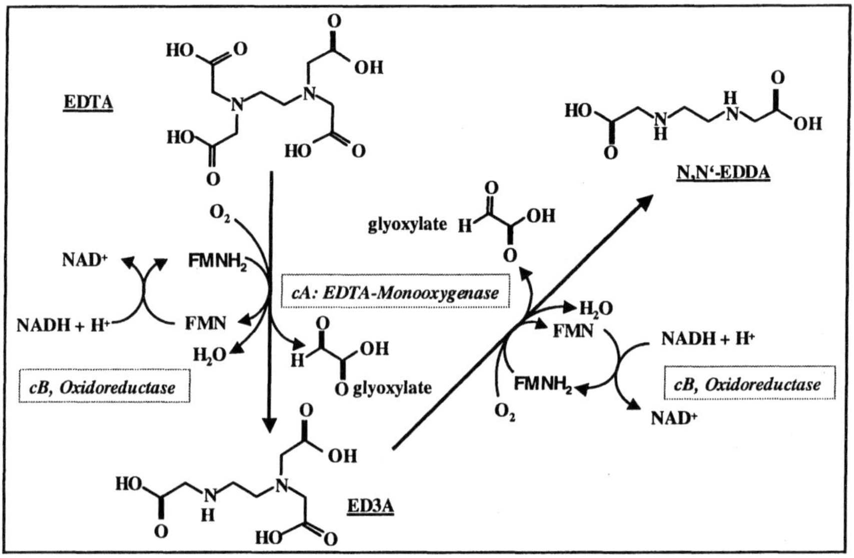

Figure 4 First catabolic steps in EDTA-degradation catalyzed by EDTA-monooxygenase (cA, Monooxygenase; cB, Oxidoreductase; ED3A, ethylenediaminetriactetate; N,N'-EDDA, N,N'-ethylenediaminediactate).

Influence of metal ions on the degradation of EDTA.

Metal speciation of an APC strongly influences both transport and subsequent metabolism. A clear dependency of the biodegradability of metal-EDTA chelates on the stability of the complex could only be demonstrated for whole cells, however, and not for the EDTA-monooxygenases. With the exception of ZnEDTA (conditional stability constant, $K = 10^{14.2}$), only metal-EDTA chelates with stability constants below 10^{12} were degraded by whole cells of strain BNC1. The maximum turnover rate for Ba- ($K = 10^{5.5}$), Mg- ($K = 10^{6.4}$), and CaEDTA ($K = 10^{8.4}$) was approximately 20 μmol · g^{-1} protein · min^{-1}. The EDTA chelate of Mn^{2+} ($K = 10^{11.7}$) was oxidized at a rate of approximately 15 μmol · g^{-1} protein · min^{-1}. Metal-EDTA chelates with stability constants above 10^{12} such as Fe(III)-, Co-, Cd-, Pb-, Ni-, and CuEDTA ($K = 10^{13.7}$, $10^{13.9}$, $10^{14.2}$, $10^{15.2}$, $10^{16.2}$, and $10^{16.2}$, respectively) were not metabolized. These strong chelates did not inhibit the oxidation of uncomplexed EDTA or degradable metal-EDTA complexes. In the presence of Fe-, Co-, Cd-, Pb-, Ni- or CuEDTA, a given surplus of uncomplexed EDTA was always consumed at the same rate as observed in the absence of the added complex. Comparable results were also obtained for the uptake of EDTA and its metal chelates by cells of strain DSM 9103 (*75-77*).

The carrier for the transport of EDTA into the cytoplasm of strain BNC1 is inducible, and appears to be a high affinity system because the Monod constant for EDTA-degradation by whole cells is only 8 µmol/L EDTA (see Figure 3) whereas under optimum conditions, the Michaelis constant of EDTA-oxidation by cell-free extracts is 120 µmol/L.

The MgEDTA chelate appears to be the best substrate for the EDTA-monooxygenase both of strain BNC1 and DSM 9103. In contrast to the results obtained with whole cells of these organisms, no dependency of the degradability of EDTA chelates on their stability constants was found with ultrafiltered cell-free extracts of strain BNC1 or with the purified EDTA-MO of strain DSM 9103. Surprisingly, CaEDTA ($K = 10^{8.4}$) is not a substrate for the first catabolic enzyme whereas CoEDTA ($K = 10^{13.9}$) is oxidized. Since CaEDTA but not CoEDTA can be degraded by whole cells of strain BNC1 and strain DSM 9103, it is probable that at least for some metal-EDTA chelates the uptake of the ligand into the cytoplasm is accompanied by a change in the metal speciation. This appears to be possible only for chelates with low stability constants and/or high dissociation rates.

Onsite treatment of industrial wastewaters containing EDTA.

Despite the low growth rate of the bacterial mixed culture with EDTA (μ_{max} = 0.03 to 0.07 h^{-1}), high degradation rates were achieved by immobilizing the cells on solid carriers in fluidized bed reactors (*78-80*). The low K_S-value of the mixed culture with only 8 µmol/L EDTA is advantageous for a wastewater treatment because it affords low residual EDTA-concentrations in the effluent. The practical applicability of a treatment system with immobilized cells of the mixed culture in an airlift-loop reactor was demonstrated for various wastewaters, e.g. from cleansing processes or the oxidative bleaching of pulp and paper (*27*). However, the release of metal ions as a result of the degradation of the organic ligand EDTA led to the formation of precipitates which could not be separated from the support material used in airlift-loop reactors. In most of the fixed-bed reactors tested, an uncontrollable growth of the bacteria on the support materials led to thick and partially inactive biofilms. These problems were avoided by use of a rotating disk reactor, or a trickle-bed reactor which was periodically fluidized to separate surplus biomass or precipitates (*28, 29, 32, 81, 82*).

A periodically fluidized trickle-bed reactor was used for the treatment of a photographic wastewater (from bleaching-fixing baths) containing EDTA in form of its Fe-chelate. In contrast to other metal-EDTA chelates relevant for industrial wastewaters (e.g., Ca-, Mg-, and MnEDTA), FeEDTA is a rather strong complex and cannot be biodegraded by the mixed culture, possibly because the complexation kinetics for the dissociation of the FeEDTA chelates are too slow and EDTA is sequestered in a form that is biologically unavailable (*83*). In a physico-chemical pretreatment stage, however, EDTA was readily rearranged from FeEDTA to CaEDTA which was degraded in the subsequent biological treatment stage (Figure 5). Consequently, EDTA was removed with a

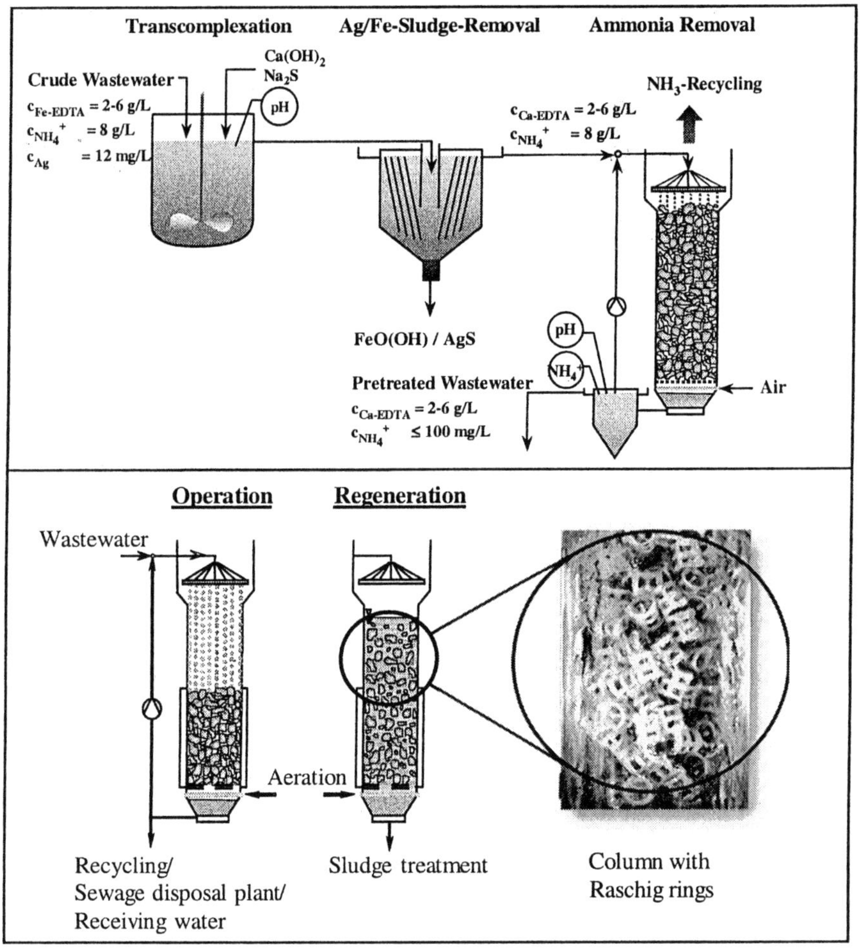

Figure 5 Two-stage process for the treatment of an EDTA-containing photographic wastewater: physico-chemical pretreatment (above); periodically fluidized trickle-bed reactor (below).

degradation degree of > 98 % from a photographic wastewater (c_{EDTA} = 2-6 g/L) in a two-stage treatment system (dilution rate, 0.125 h^{-1}) (*84-85*).

Because at present EDTA cannot be substituted in many photographic bleaching-fixing processes, this type of treatment of corresponding wastewaters is an appropriate approach to reduce the amounts of EDTA which are released into the aquatic environment. Moreover, comparable systems for the treatment of wastewaters containing other strong and recalcitrant chelating agents such as DTPA and PDTA might be established on the basis of the results obtained for the treatment of EDTA-containing wastewaters.

PDTA.

Presently, EDTA and PDTA are the main chelating agents in the bleaching baths of the photo finishing industry in Germany. Amounts of 153.5 and 28.1 tons respectively are used annually. Between 70 and 80 % of these amounts is released into the aquatic environment in process waters and washing waters. Although the complex stabilities of most metal-PDTA chelates are lower than the respective metal-EDTA chelates, the biodegradability of PDTA is as poor as that of EDTA. As a consequence, PDTA is found in concentrations up to about 10 µg/L in some rivers (*12*). The problem should be resolvable because the EDTA-degrading organisms described in this chapter, are also able to degrade PDTA as sole source of carbon, nitrogen, and energy.

In particular strain ANP11, which was isolated from the EDTA-degrading bacterial mixed community after prolonged subcultivation with EDTA (see Microorganisms, Material and Methods), exhibited relatively high activities for the degradation of PDTA. Batch experiments with strain ANP11 suggested that the substrate uptake system is the same for both EDTA and PDTA. In addition, it is likely that the EDTA-MO in strain ANP11 also catalyzes the first step in PDTA-degradation as reported for the EDTA-MO in strain DSM 9103 (*86*).

The chelating agents EDTA and PDTA in a molar ratio of 3 : 1 (1.2 mmol/L EDTA : 0.4 mmol/L PDTA, corresponding to the composition of the wastewater of one photo finisher), were used as mixed substrate in a batch growth experiment with cells of strain ANP11. A strict successive degradation of the complexing agents, without the occurrence of a diauxic phase, was observed (Figure 6). EDTA was the preferred substrate and the concentration of PDTA remained constant until EDTA (or its residual metal-species) was reduced to a concentration of approximately 0.1 mmol/L. This residual concentration could be attributed to the strong and non-biodegradable FeEDTA complex which was formed in the mineral salts medium. Other experiments in which EDTA and PDTA were supplied in equal stoichiometric amounts revealed similar results.

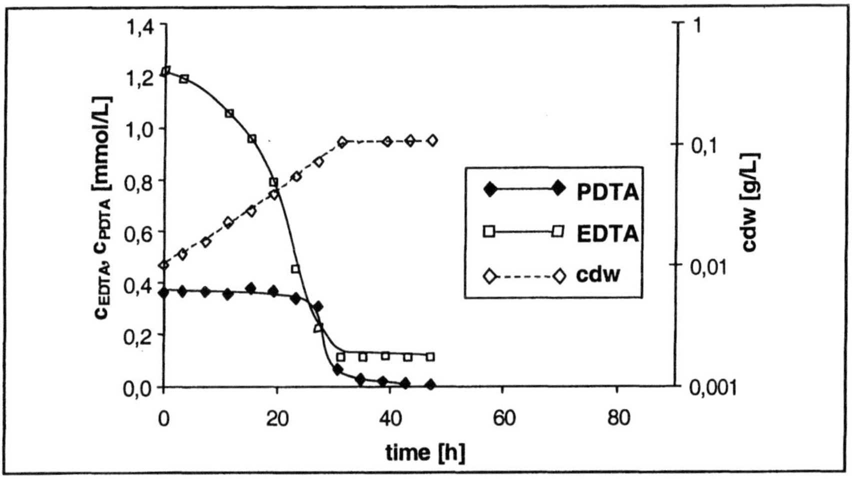

Figure 6 Growth of strain ANP11 with a substrate mixture of EDTA and PDTA (1.2 mmol/L EDTA; 0.4 mmol/L PDTA; pH 7.75; 36 °C; inoculum, cells of strain ANP11 grown on PDTA).

The EDTA-MO of strain DSM 9103 also catalyzes the first step in the degradation of PDTA (*86*). This has not been tested directly with the EDTA-MO from ANP11, but a similar situation seems probable. When resting cells of strain ANP11 (grown on PDTA) were incubated with uncomplexed EDTA or PDTA as sole substrate, the oxidation rates obtained were very similar (EDTA, 0.35 mmol EDTA/g_{CDW} h; PDTA, 0.33 mmol PDTA/g_{CDW} h). Similar results were obtained when EDTA and PDTA were used as their Mg- or Ca-chelate (22). The successive degradation of the chelating agents in the growth experiment (Figure 6) could thus be due to different substrate affinities of the uptake system or the EDTA-MO, but clearly additional experiments are required to confirm this.

The application of strain ANP11 in a mixed community for the removal of EDTA and PDTA in a biological wastewater treatment plant could probably be developed as has been for EDTA. In a trickle bed reactor degradation of the two complexing agents might occur at separate locations within the reactor. Where the complexing agents are used in different process baths, the reactor could be used successively with different loads of the corresponding wastewaters.

Photo finishing wastewaters contain EDTA and/or PDTA as their Fe(III)-chelates. As shown above, cells of strain BNC1 only degraded metal-EDTA complexes with relatively low stability constants and thus do not degrade Fe(III)EDTA. Strain ANP11 does not seem to degrade FeEDTA either (Figure 6). In order to investigate the influence of chemical speciation on the biodegradability of PDTA by cells of strain ANP11, resting cells of strain ANP11 (grown on PDTA) were incubated with 0.65 mmol/L of different metal-PDTA chelates, or with uncomplexed chelant as a control. FePDTA, CuPDTA, ZnPDTA, and MnPDTA were studied because of the range of their complex stabilities, and the maximum specific degradation rates are shown in Figure 7.

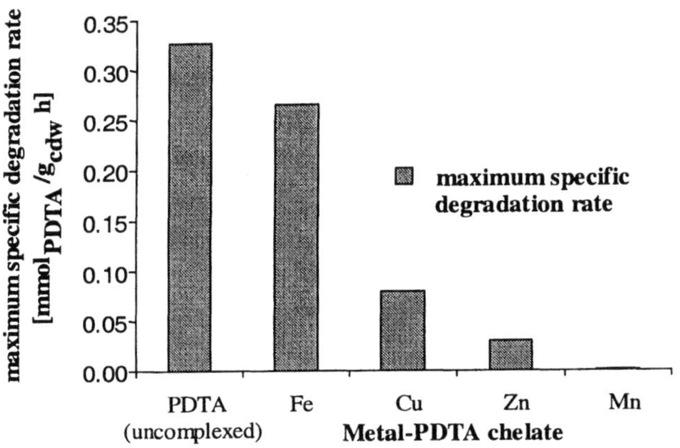

Figure 7 Maximum degradation rates of different metal PDTA-complexes by resting cells of strain ANP11 (0.65 mmol/L metal-PDTA; pH 7.75; 36 °C) (21).

Interestingly, the range of metal-PDTA chelates which are degradable by cells of strain ANP11 does not depend on their stability constants (Table 1). The relatively strong FePDTA chelate was oxidized almost as rapidly as free PDTA, and the less stable Cu and Zn complexes were degraded at a lower rate. The least strong complex, that of Mn, was not degraded at a significant rate.

More extensive studies on the influence of the aqueous speciation on the biodegradability of metal-EDTA and metal-PDTA chelates are under present investigation and will be described elsewhere.

DTPA.

Compared to EDTA and PDTA, DTPA forms even stronger chelates with most di- and trivalent metal ions. The main area of application of DTPA is in the pulp and paper industry where strong chelating agents are required to avoid metal-catalyzed decomposition of H_2O_2 which is used as a bleaching agent.

Table 1 Logarithmic thermodynamic stability constants of some metal chelates[a] of EDTA, PDTA, and DTPA

Metal ion	EDTA [b]	PDTA [b]	DTPA
Mg^{2+}	**8.8**	**6.0**	9.3
Ca^{2+}	**10.6**	**7.1**	10.8
Fe^{3+}	25.1	**21.4**	28.0
Zn^{2+}	**16.4**	**15.2**	18.4
Mn^{2+}	**13.8**	10.0	15.6
Cu^{2+}	18.8	**18.9** [c]	21.5

a) Data extracted from various publications (*87-90*).

b) Values in **bold-type**: chelates are degradable by cells of strain ANP11 (*21*).

c) In contrast to the results obtained with whole cells of strain ANP11 by Hille et al. (*21*), Baaß (*22*) did not observe CuPDTA-degradation by cell-free extracts of this organism.

To date no microorganisms have been described which are able to grow with DTPA as sole source of carbon and energy. However, DTPA can be unspecifically oxidized to ketopiperazinepolycarboxylates (KPPC), dead-end metabolites which can be detected in several surface waters and drinking waters (*8*). Extensive enrichment experiments carried out with various samples from natural habitats, or with sewage sludge from pulp and paper industry, did not result in the isolation of bacteria able to grow with DTPA as sole carbon source even though consumption of DTPA could be observed with inocula from different habitats. As shown in Figure 8, DTPA was metabolized by sewage sludge from a paper mill, but the DOC-concentration did not decrease in accordance with the consumption of DTPA.

As indicated by GC/MS-analysis, uncompletely substituted APCs (usAPCs) or their corresponding cyclization products, ketopiperazinepolycarboxylates (KPPCs) were accumulated as dead-end metabolites in the culture fluid. From this it can be concluded that only one acetyl group was cleaved from DTPA yielding usAPC(s) and glyoxylate as products (Figure 9). Although glyoxylate should be degradable as a source of carbon and energy, no significant bacterial growth was observed.

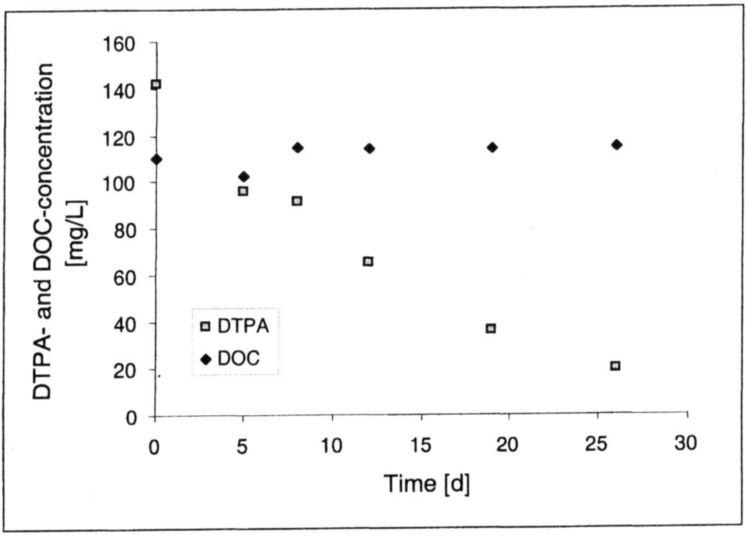

Figure 8 Removal of DTPA from by sewage sludge from a paper mill.

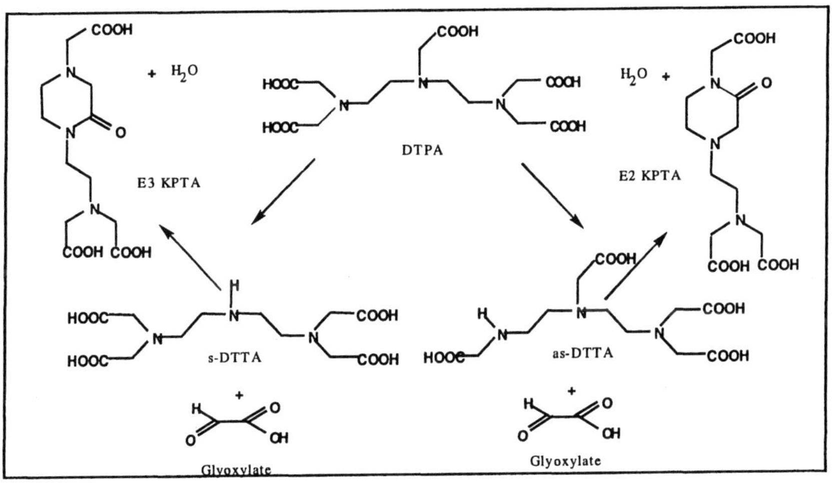

Figure 9 Postulated partial metabolism of DTPA in sewage sludge (E3 KPTA, ethyleneamine-3-ketopiperazinetriacetate; E2 KPTA; ethyleneamine-2-ketopiperazinetriacetate; s-DTTA, symmetric diethylenetriaminetetraacetate; as-DTTA, asymmetric diethylenetriaminetetraacetate).

EDTA- and PDTA-degrading bacteria such as strain ANP11 could not be adapted to grow with DTPA, and resting cells of strain ANP11 did not metabolize free DTPA or various metal-DTPA chelates tested. However, free DTPA as well as several metal-DTPA complexes were oxidized by cell-free extracts of this organism (relative rate: free DTPA > MgDTPA > CaDTPA > FeDTPA > ZnDTPA; no oxidation of Mn-, Cu-, and ZnDTPA). The range of metal-DTPA chelates which can be oxidized by cell-free extracts of strain ANP11 does not depend on their thermodynamic stability constants (see Table 1) because the rather strong FeDTPA was oxidized at a significant rate whereas MnDTPA with a relatively low complex stabilty was not degraded.

These results indicate that a complete biodegradation of DTPA could possibly be achieved either by the enrichment of DTPA-degrading bacteria, by the continued adaptation of EDTA-degrading organisms, or by the enrichment of bacteria which are capable to degrade other APCs or KPPCs, and which could be cultivated together with DTPA-converting organisms.

ACKNOWLEDGEMENTS

This study was supported by grants from the Deutsche Forschungsgemeinschaft (DFG, no. He 1515/5, No 226/2), the Bundesministerium für Bildung, Wissenschaft, Forschung und Technologie (BMBF, no. 0310458A), and the Bundesministerium für Wirtschaft via the Arbeitsgemeinschaft industrieller Forschungsvereinigungen „Otto von Guericke“ e.V. (AiF, no. 11031 N, 12060 N). I am indebted to Dietmar C. Hempel for his support and aid, and to Eric A. Barnsley for intensive and stimulating discussions.

References

1. Frimmel F.H.; Grenz R.; Kordick E.; Dietz F. *Vom Wasser* **1989**, *72*, 175-184.
2. Nusch, E.A.; Eschke, H.D.; Kornatzkis, K.H. *Korr. Abwasser* **1991**, *38*, 944-949.
3. Wanke, T.; Eberle; S.H. *Acta Hydrochim. Hydrobiol.* **1992**, *20*, 192–196.
4. Klopp, R.; Pätsch, B. *Wasser Boden* **1994**, *8*, 32-37.
5. Richardson, D.E.; Ash, G.H.; Harden, P.E. *J. Chromat. A* **1994**, *688*, 47-53.
6. Pietsch, J.; Schmidt, W.; Sacher, F.; Fichtner, S.; Brauch, H.-J. *Fresenius J. Anal. Chem.* **1995**, *353*, 75-82.
7. Stumpf, M.; Ternes; T.A.; Schuppert, B.; Haberer, K.; Hoffmann, P.; Ortner, H.M. *Vom Wasser* **1996**, *86*, 157-171.

8. Ternes, T.A.; Stumpf, M.; Steinbrecher, T.; Brenner-Weiß, G.; Haberer, K. *Vom Wasser* **1996**, *87*, 275-290.
9. Sillanpää, M. *Rev. Environ. Contam. Toxicol.* **1997**, *152*, 85-111.
10. Sillanpää, M.; Sihvonen, M.-L. *Talanta* **1997**, *44*, 1487-1497.
11. Sacher, F.; Lochow, E.; Brauch, H.-J. *Vom Wasser* **1998**, *90*, 31–41.
12. Knepper, T.P.; Weil, H. *Vom Wasser* **2001**, *97*, 193-232.
13. *Decontamination of Nuclear Reactors and Equipment*; Ayers J.A., Ed; The Ronald Press Company, New York, 1970.
14. Mcfadden, K.M. Organic components of nuclear wastes and their potential for altering radionuclide distribution when released to soil. PNL 2563. National Technical Information Service, Springfield, Va, 1980.
15. Macaski, L.E. *Crit. Rev. Biotechnol.* **1991**, *11*, 41-112.
16. Means, J.F., Crerar, D.A., Duguid, J.O. *Science* **1978**, *200*, 1477-1481.
17. Means, J.L.; Alexander, C.A. *Nucl. Chem. Waste Manage.* **1981**, *2*, 183-196.
18. Sykora, V.; Pitter, P.; Bittnerova, I.; Lederer, T. *Water Res.* **2001**, *35*, 2010-2016.
19. Nörtemann, B. *Appl. Environ. Microbiol.* **1992**, 58, 671-676.
20. Noll, A. Diploma thesis, Institute of Biochemical Engineering, Technical University of Braunschweig, Germany, 1996.
21. Hille, A.; Baaß, A.-C.; Goecke, Y.; Nörtemann, B.; Hempel, D.C. *Proc. 3rd Int. Conf. on Remediation of Chlorinated and Recalcitrant Compounds*, Monterey, USA, 2002.
22. Baaß, A.-C. In *ibvt-Schriftenreihe des Instituts für Bioverfahrenstechnik*; Hempel D.C., Ed., FIT, Paderborn, 2004; Vol. 18 (in press).
23. Witschel, M.; Weilenmann, H.-U.; Egli, T. Poster presented at the 54 Annual Meeting of the Swiss Society for Microbiology, Lugano, Switzerland, 1995.
24. Witschel, M.; Nagel, S.; Egli, T. *J. Bacteriol.* **1997**, *179*, 6937-6943.
25. Bucheli-Witschel, M.; Egli, T. *FEMS Microbiol. Rev.* **2001**, *25*, 69-106.
26. Nörtemann, B.; Cordes, C.; Lang, E.; Busse, H.-J. Poster no. PA011 presented at the Annual Meeting of the German Society of Microbiology VAAM, *Biospektrum* **2004**, *special edition*, 74.
27. Brüggenthies, A. In *ibvt-Schriftenreihe des Instituts für Bioverfahrens-technik*; Hempel D.C., Ed., FIT, Paderborn, 1996; Vol. 3.
28. Otto, P.; Nörtemann, B.; Hempel, D.C. *UTA- Umwelt Technologie Aktuell* **2000**, *1*, 52-53.
29. Otto, P.; Nörtemann, B.; Hempel, D.C. *Chem.-Ing.-Tech.* **2000**, *72*, 955-956.
30. Soßdorf, D.; Brenner-Weiß, G.; Kreckel, P.; Ternes, T.; Wilken, R.D. *Vom Wasser* **2000**, *94*, 121-134.

31. Baaß, A.-C.; Hille, A.; Göcke, Y.; Nörtemann, B.; Hempel, D.C. *Proc. 3rd Int. Conf. on Remediation of Chlorinated and Recalcitrant Compounds*, Monterey, USA, 2002.
32. Goecke, Y.; Baaß, A.-C.; Hille, A.; Otto, P.; Nörtemann, B.; Hempel, D.C. *Proc. 3rd Int. Conf. on Remediation of Chlorinated and Recalcitrant Compounds*, Monterey, USA, 2002.
33. Pfennig, N.; Lippert, K.D. *Arch. Microbiol.* **1966**, *55*,245-256.
34. Chemie und Biochemie des mikrobiellen EDTA-Abbaus; Klüner, T.; Cuvillier, Göttingen, 1996.
35. Focht, D.; Joseph, H. *Can. J. Microbiol.* **1971**, *17*, 1553-1556.
36. Cripps, R.E.; Noble, A.S. *Biochem. J.* **1973**, *136*, 1059-1068.
37. Enfors, S.; Molin, N. *Water Res.* **1973**, *7*, 881-888.
38. Liu, D.; Wong, P.; Dutka, B. *J. Water Pollut. Control Fed.* **1973**, *45*, 1728-1735.
39. Parks, S.; Stukus, P. *J. Sci. Lab. Denison Univ.* **1973**, *54*, 79-86.
40. Tiedje, J.M.; Mason, B.B.; Warren, C.B.; Malec, E.J. *Appl. Microbiol.* **1973**, *25*, 811-818.
41. Firestone, M.K.; Tiedje, J.M. *Appl. Environ. Microbiol.* **1978**, *35*, 955-961.
42. Kakii, K.; Yamaguchi, H.; Iguchi, Y.; Teshima, M.; Shirakashi, T.; Kuriyama, M. *J. Ferment. Technol.* **1986**, *64*, 103-108.
43. Egli, T.; Weilenmann, H.U.; El-Banna, T.; Auling, G. *Syst. Appl. Microbiol.* **1988**, *10*, 297-305.
44. Wanner, U.; Kemmler, J.; Weilenmann, H.-U.; Egli; T., El-Banna, T.; Auling, G. *Biodegradation* **1990**, *1*, 31-41.
45. Wilberg, E.; El-Banna, T.; Auling, G.; Egli, T. *Syst. Appl. Microbiol.* **1993**, *16*, 147-152.
46 Bally, M. Ph.D. Thesis No. 10821, Swiss Federal Institute of Technology, Zürich, Switzerland, 1994.
47. Bally, M.; Egli, T. *Appl. Environ. Microbiol.* **1996**, *62*, 133-140.
48 Xu, Y.; Mortimer, M.W.; Fisher, T.S.; Kahn, M.L.; Brockman, F.J.; Xun, L. *J Bacteriol.* **1997**, *179*, 1112-1116.
49. Bally, M.; Wilberg, E.; Kuhni, M.; Egli, T. *Microbiology* **1994**, *140*, 1927-1936.
50. Auling, G.; Busse, H.J.; Egli, T.; El-Banna, T.; Stackebrandt, E. *Appl. Microbiol.* **1993**, *16*, 104-112.
51. Urakami, T.; Araki, H.; Oyanagi, H.; Suzuki, K.I.; Komagata, K. *Int. J. Syst. Bacteriol*, **1992**, *42*, 84-92.
52. Kämpfer, P.; Neef, A.; Salkinoja-Salonen, M.S.; Busse H.-J. *Int. J. Syst. Evol. Microbiol.* **2002**; 52, 835-839.
53. Bolton, H.J.; Girvin, D.C.; Plymale, A.E.; Harvey, S.D.; Workman, D.J. *Environ. Sci. Technol.* **1996**, *30*, 931-938.

54. Xun, L.; Reeder, R.B.; Plymale, A.E.; Girvin, D.C.; Bolton, H. *Environ. Sci. Technol.* **1996**, *30*, 1753-1755.
55. Potthoff-Karl, B. *Seifen Öle Fette Wachse* **1994**, *120*, 104-109.
56. Grasshoff, A.; Potthoff-Karl, B. Tenside-Surfactants-Detergents **1996**, *33*, 278-288.
57. Schowanek, D.; Feijtel, T.C J.; Perkins, C.M.; Hartman, F.A.; Federle, T.W.; Larson, R. J. *Chemosphere* **1997**, *34*, 2375–2391.
58. Takahashi, R.; Fujimoto, N.; Suzuki, M.; Endo, T. *Biosci. Biotech. Biochem.* **1997**, *61*, 1957–1959.
59. Gangoda, C.K., PhD Thesis, University of Wales, Cardiff, UK, 1998.
60. Associated Octel: Octaquest E, an effective biodegradable chelating agent, Product literature, 2000.
61. Bayer: Iminodisuccinate, Iminodisuccinic acid sodium salt, Product literature (from www.bayer.com), 2000.
62. Cebulla, I. Ph.D. Thesis, Eberhard-Karls-Universität Tübingen, Tübingen, 1995.
63. Nishikiori, T.; Okuyama, A.; Naganawa, H.; Takita, T.; Hamada, M.; Takeuchi, T.; Aoyagi, T.; Umezawa, H.J. *Antibiot.* **1984**, *37*, 426-427.
64. Stegmann, E.; Pelzer, S.; Wilken, K.; Wohlleben, W.J. *Biotechnol.* **2001**, *92*, 195-204.
65. Kos, B.; Lestan, D. *Environ. Sci. Technol.* **2003**, *37*, 624-629.
66. Tandy, S.; Bossart, K.; Mueller, R.; Ritschel, J.; Hauser, L.; Schulin, R.; Nowack, B. *Environ. Sci. Technol.* **2004**; *38*, 937-44.
67. Vandevivere, P.C.; Saveyn, H.; Verstraete, W.; Feijtel, T.C.; Schowanek, D.R. *Environ. Sci. Technol.* **2001**, *35*, 1765-1770.
68. Witschel, M.; Egli, T. *Biodegradation* **1998**, *8*, 419-428.
69. Nowack, B.; Baumann, U. *Acta Hydrochim. Hydrobiol.* **1998**, *26*, 104-108.
70. Nörtemann, B. *Appl. Microbiol. Biotechnol.* **1999**, *51*,751-759.
71. Klüner, T.; Hempel, D.C.; Nörtemann, B. *Appl. Microbiol. Biotechnol.* **1998**, *49*, 194-201.
72. Payne, J.W.; Bolton, H.; Campbell, J.C.; Xun, L. *J. Bacteriol.* **1998**, *180*, 3823-3827.
73. Bohuslavek, J.; Payne, J.W.; Liu, Y.; Bolton, H.; Xun, L. *Appl. Environ. Microbiol.* **2001**, *67*, 688-695.
74. Liu, Y.; Louie, T.M.; Payne, J.; Bohuslavek, J.; Bolton, H.; Xun, L. *Appl. Environ. Microbiol.* **2001**, *67*, 696-701.
75. Witschel, M.; Egli, T. Zehnder, A.J.B.; Wehrli, E.; Spycher, M. *Microbiology* **1999**, *145*, 973-983.
76. Satroutdinov, A.D.; Dedyukhina, E.G.; Chistyakova, T.I.; Witschel, M.; Minkevich, I.G.; Eroshin, V.K.; Egli, T. *Environ. Sci. Technol.* **2000**, *34*, 1715-1720.
77. Egli, T. *J. Biosci. Bioeng.* **2001**, *92*, 89-97.

78. Henneken, L.; Nörtemann, B.; Hempel, D.C. *Appl. Microbiol. Biotechnol.* **1995**, *44*, 190-197.
79. Henneken, L.; Brüggenthies, A.; Nörtemann, B.; Hempel, D.C. *Chem.-Ing.-Tech.* **1996**, *68*, 310-314.
80. Hennecken, L.; Nörtemann, B.; Hempel, D.C. *J. Chem. Technol. Biotechnol.* **1998**, *73*, 144-152.
81. Otto, P.; Nörtemann, B.; Hempel, D.C. *Proc. 4th* GVC-VDI-Abwasser-Kongress, Bremen, 1999.
82. Otto, P.; Wäsche, S.; Horn, H.; Hempel, D.C. *Proc. IAWQ-Conference on Biofilm Systems*, New York, 1999.

Chapter 9

Microbial Degradation of EDTA: New EDTA-Degrading Bacterial strains

Aidar D. Satroutdinov, Tatiana I. Chistyakova, Emiliya G. Dedyukhina, and Igor G. Minkevich

Institute of Biochemistry and Physiology of Microorganisms, Russian Academy of Sciences, Pushchino, 142290, Russia

Novel EDTA-degrading bacterial strains (*Pseudomonas* sp. LPM-410 and EDTA-dependent strain LPM-4) were isolated. Resting cells of these new strains degraded EDTA and EDTA complexes with Mg, Ca, Mn, and Zn at higher rates (0.143–0.525 mmol EDTA/(g h)) than EDTA-degrading strain DSM 9103 (0.096–0.255 mmol EDTA/(g h)), but were unable to degrade EDTA complexes with Co, Cu, and Pb. By using continuous cell cultivation in pH-auxostat regime, kinetic and stoichiometric characteristics of growth and EDTA degradation by bacterial strains were studied. The data indicating partial utilization of energy released during oxidation of side groups of EDTA molecule were obtained.

The widely applied synthetic chelating agent, ethylenediaminetetraacetate (EDTA), is characterized by high persistence; no significant elimination of EDTA in conventionally operated actvated sludge plants has been revealed (*1*). The extensive environmental accumulation of EDTA observed in recent years (*2*) is highly undesirable, since it leads to remobilization of heavy and toxic metal ions from river sediments and contamination of drinking water. Recently, efficient EDTA elimination in an industrial wastewater treatment plant (*3*) and in semicontinuous activated sludge units (*4*) were achieved. However, only a few EDTA-degrading bacterial strains have been isolated as pure cultures: *Agrobacterium* sp. ATCC 55002 utilizing Fe(III)-EDTA (5), Gram-negative isolate BNC1 capable of degrading Mg–, Ca–, Mn–, and Zn–EDTA (*6, 7*), and a Gram-negative bacterium DSM 9103 (*8, 9*). Cell suspensions of DSM 9103 were capable of degrading EDTA and metal–EDTA complexes with low stability constants (Mg–, Ca–, and Mn–EDTA) to completion at constant rates, whereas more stable complexes (Zn–, Co–, Cu–, and Pb–EDTA) were degraded only partly and at lower rates (*10, 11*). Biochemistry of EDTA degradation by strain DSM 9103 was studied in detail (*9, 12, 13*); a two-enzyme system (monooxygenase) catalyzing EDTA oxidation was isolated from the cells and investigated (*12*). All the known EDTA-degrading strains grew well on complex media with organic nitrogen sources. No obligate EDTA-degraders has yet been known.

The isolation of new EDTA-degrading strains, characterization of their specificity towards different metal-EDTA complexes, and comparative studies on the efficiency of EDTA utilization by different strains is of much interest from the theoretical standpoint of understanding the microbial growth on such an unusual substrate as EDTA and from the practical viewpoint of developing approaches to remove EDTA from wastewater.

Materials and Methods

Isolation Procedure

EDTA-degrading bacterial strains were isolated from a municipal sewage sluge by enriching sewage sludge (10 g) in a 750-ml flask with 100 ml of the following EDTA-medium (g/l): EDTA, 1.0; $MgSO_4 \cdot 7H_2O$, 1.0; KH_2PO_4, 0.26; $CaCl_2 \cdot 2H_2O$, 0.4; $Na_2HPO_4 \cdot 12H_2O$, 0.83; yeast extract (Difco, United States), 0.1; trace elements (mg/l): $FeCl_2 \cdot 4H_2O$, 1.5; H_3BO_3, 0.06; $MnCl_2 \cdot 4H_2O$, 0.1; $CoCl_2 \cdot 6H_2O$, 0.12; $ZnCl_2$, 0.07; $NiCl_2 \cdot 6H_2O$, 0.025; $CuCl_2 \cdot 2H_2O$, 0.015; $Na_2MoO_4 \cdot 2H_2O$, 0.025; initial pH 7.0. The enrichment culture was cultivated in the EDTA-medium on a shaker (150—200 rpm) at 28°C. After two weeks, 5 ml of the enrichment culture was inoculated into a 750-ml flask with 100 ml of fresh EDTA-medium and cultivated on a shaker at 28°C for two weeks. After five such passages, 0.05 ml of the enrichment culture was spread onto a Petri dish with agar EDTA-medium and incubated at 28°C; individual colonies were isolated and tested for EDTA-degrading activity. The active isolates were maintained on slants with agar EDTA-medium.

Cell Cultivation and Degradation Assays

Batch cultivation of isolates was performed in 750-ml flasks with 100 ml of the above described EDTA-medium on a shaker (150–200 rpm) at 28°C. Cell growth was followed by measuring optical density of the cultures at 546 nm with a Specol 221 spectrophotometer (Carl Zeiss, Germany) after acidifying the cell suspension with 3% HNO_3 to pH 2.0 to dissolve precipitates which had been formed during bacterial growth. Biomass dry weight was estimated from optical density of cell suspensions by using calibration curves constructed for each isolate. To constract calibration curve, serial dilutions of acidified culture broth were used for measuring optical density at 546 nm and cell dry weight (CDW) as follows: aliquote (10 ml) of each dilution was centrifuged (8000 *g*, 3 min), washed with acidified distilled water (pH 2.0) and dried in weighed Eppendorf tubes at 80°C to constant weight.

Continuous cell cultivation in pH-auxostat regime was performed in an LKB fermentor (Sweden) with a working volume of 2.5 l equipped with a controller and computer. The medium input was controlled by a pH detector and performed by Gilson peristaltic pumps (France) through two channels: (1) the medium containing (mg l^{-1}), EDTA, 2000; $MgSO_4$ x $7H_2O$, 1000; $CaCl_2$ x $2H_2O$, 400; Na_2HPO_4 x 12 H_2O, 340; KH_2PO_4, 100; trace elements and vitamins (*14*) and (2) 0.2% (v/v) H_2SO_4;. pH value in a fermentor was maintained at 7.0 ± 0.1.

Aeration rate was 0.3 l (l min)$^{-1}$; agitation was 900 rpm. The pO_2 value was maintained at 80–90% (of air saturation) in the course of fermentation.

The temperature optimal for the growth of strains LPM-410 and LPM-4 (28 and 32°C, respectively) was maintained with an accuracy of ± 0.2°C.

EDTA concentration was analyzed by high-pressure liquid chromatogrphy (*15*) on a HPLC-chromatograph (Waters, Great Britain) equipped with a Nucleosil 100 C_6H_5 column (Machery und Nagel, Germany) at 285 nm.

Ammonium concentration in the medium was determined spectrophotometrically with Nessler reagent (*16*).

The ability of isolates to degrade EDTA and its complexes with Mg^{2+}, Ca^{2+}, Ba^{2+}, Mn^{2+}, Zn^{2+}, Cu^{2+}, Pb^{2+}, and Fe^{3+} was studied by incubating suspensions of exponential-phase cells with 1 mM of an appropriate EDTA species as described earlier (*10, 11*). The exponential-phase cells were harvested by centrifugation

(5500 *g*, 20°C, 30 min) and washed once with 2.0 mM of EDTA and 5.0 mM of $NaNO_3$ (pH 7.0). Assay mixture (100 ml) contained 10 mM HEPES buffer (pH 7.0), EDTA or metal–EDTA complex (1 mM), $NaNO_3$ (10 mM), and biomass (0.2–0.5 g dry cells/l). Initial pH of the assay mixture was 7.0. The mixture was incubated at 28°C on a rotary shaker (150—200 rpm) for 24 h. Aliquots of assay mixture (1 ml) were withdrawn every hour and after removing the cells by centrifugation (8000 *g*, 20°C, 3 min), the EDTA concentration was analyzed. The metal-EDTA complexes were prepared by mixing equimolar amounts of EDTA (as tetrasodium salt) and corresponding metal ion 30 min prior to usage for degradation assays. With the exception of $ZnSO_4$, all metal ions were used in the form of chloride salts.

Results and Discussion

Two EDTA-degrading bacterial strains designated as LPM-410 and LPM-4 were isolated from a municipal sewage sluge. Both isolates were unable to utilize the other widespread aminopolycarboxylic acid, nitrilotriacetate (NTA).

Characterization of Isolate LPM-410

Cells of isolate LPM-410 are single rods (0.6–0.8 by 1.0–3.0 μm), nonsporing, motile by polar or lateral flagella. Colonies grown on nutrient agar for 7 days are 2–4 mm in diameter, round, with even edge, smooth, transparent, pale yellow; on the Corynebacterium agar, colonies are golden-yellow. No diffusible pigments is produced. The organism is obligate aerobic chemoorganotroph with respiratory metabolism; growth factors were not required (*17*). Diagnostic tests that were carried out by routine techniques (*18*) revealed positive reactions of the strain for catalase and oxidase; oxidation of glucose; acid production from D-glucose, L-arabinose, D-xylose, and maltose; hydrolysis of gelatin, starch (weakly), and milk casein; reduction of nitrate to nitrite. The following reactions were negative: fermentation of glucose; arginine dihydrolase activity; levan production from sucrose; acid production from lactose, sucrose, trehalose, D-sorbitol, i-inositol, adonitol, mannitol, and glycerol. The temperature optimum for cell growth was 28–30°C; the cells did not grow in nutrient broth at 40°C. On the basis of these characteristics, strain LPM-410 was identified as *Pseudomonas* sp. (*19*).

Characterization of Isolate LPM-4

Cells of isolate LPM-4 grown in liquid EDTA-medium are rods (0.2–0.25 by 0.5–0.6 μm), single or in pairs, oxidase- and catalase-positive. According to 16S rDNA sequencing, this isolate belonged α-*Proteobacteria* and was closely related to the DNA sequences from uncultured bacteria extracted from soil samples of uranium wastes (accession numbers AJ536856, AJ296548, AJ582029, and AJ536869) (*20*).

Strain LPM-4 was unique in its nutrient requirements and grew only in the presence of EDTA; no cell growth was observed in nutrient broth, nutrient agar, or in media containing glucose, ethanol, or acetate as the sole carbon and energy sources and either inorganic (ammonium sulfate, potassium nitrate) or organic (urea, peptone, casein hydrolysate, amino peptide, yeast extract, ammonium acetate) nitrogen sources. Inorganic nitrogen compounds added to EDTA-medium showed no effect on the cell growth, whereas organic nitrogen sources, such as peptone, casein hydrolysate, amino peptide, or ammonium acetate, completely inhibited growth.

When glucose was added to EDTA-medium, glucose was utilized by strain LPM-4 after the EDTA was removed from the medium. The cells likely required EDTA as a specific nitrogen source.

Degradation of Metal–EDTA Complexes by New Isolates

Comparative data on the specific rates of degrading different EDTA compounds by the washed cell suspensions of two isolates and strain DSM 9103 (*10*) are given in Table I.

Uncomplexed EDTA and metal–EDTA chelates with low stability constants (log K below 16), such as Mg–, Ca–, Ba-, and Mn–EDTA, were degraded by the washed cell suspensions of both isolates at constant specific rates ranging from 0.310 to 0.525 mmol EDTA/(g cells h). Zn–EDTA with a higher stability constant (log K 18.3) was not completely degraded by resting cell suspensions of *Pseudomonas* sp. LPM-410 and LPM-4 after an incubation period of 24 h. Initially, the degradation rates of *Pseudomonas* sp. LPM-410 and LPM-4 were 0.195 and 0.143 mmol EDTA/(g cells h), respectively. After 8 h, degradation rates declined. No degradation of the stable chelates of Pb–, Co–, Cu–EDTA, and Fe(III)–EDTA was observed. By comparison, resting cells of strain DSM 9103 degraded EDTA and complexes Mg–, Ca–, and Mn–EDTA at lower specific rates ranging from 0.096 to 0.255 mmol EDTA/(g cells h), but were able to utilize a wider range of stable complexes, including Co–, Cu–, and Pb–EDTA (*10, 11*).

Table I. Specific Rates of EDTA Degradation (mmol EDTA/(g dry cells h)) by Different Bacterial Strains

EDTA complexes	*log* K_{MeEDTA}	*LPM-410*	*LPM-4*	*DSM 9103*[a]
EDTA		0.506 ± 0.034	0.453 ± 0.070	0.146 ± 0.008
Ba-EDTA	7.8	0.372 ± 0.076	0.423 ± 0.042	ND
Mg-EDTA	10.6	0.525 ± 0.026	0.501 ± 0.025	0.255 ± 0.022
Ca-EDTA	10.7	0.383 ± 0.069	0.364 ± 0.073	0.245 ± 0.006
Mn-EDTA	15.6	0.363 ± 0.070	0.310 ± 0.043	0.096 ± 0.002
Zn-EDTA	18.3	0.195 ± 0.030	0.143 ± 0.007	0.106 ± 0.030
Pb-EDTA	18.0	0	0	0.099 ± 0.034
Co-EDTA	18.1	0	0	0.048 ± 0.011
Cu-EDTA	20.5	0	0	0.021 ± 0.005
Fe(III)-EDTA	25.0	0	0	0

NOTE: [a] Data are from Reference 10; "ND" stands for "not determined"

To conclude, the isolated EDTA-degrading strains *Pseudomonas* sp. LPM-410 and LPM-4 exhibited higher degradation activities towards metal–EDTA complexes with comparably low stability constants than strain DSM 9103 but did not degrade the more stable metal–chelate complexes.

Application of Mass–Energy Balance Regularities for Description of Bacterial Growth on EDTA.

Quantities that characterize the mass–energy balance of cell metabolism are based on the generalized unit of reductivity, redoxon (RO) (*21, 22*), which is a generalization of the earlier used definition "available electron" (*23–26*). Redoxon is a discrete quantity estimated in equivalents. It is a new unit of amount of every compound reduced relative to zero level (carbon dioxide, water etc. (*21*)). Unlike mole, redoxon is the most close to biologically available energy content of organic compounds as measured by Gibbs standard free energy and enthalpy. The energetic level of RO is approximately the same for the majority of organic compounds as well as in dry cell biomass. It is shown using known physicochemical data for organic compounds as well as our and literature data for cell biomass (*21*). The mass–energy quantity estimating the efficiency of cell growth is the energetic yield of cell biomass from the substrate, $\eta_{x/s}$. By definition, it is the fraction of total number of substrate redoxons which is incorporated into biomass (*21, 23, 24, 27, 28*). This characteristic is independent of the energy content per mass unit of any organic substrate. Therefore, it allows a comparison of the real (energetic) growth efficiency of microorganisms on different substrates as well as comparative analysis of experimental values of growth efficiency with those predicted basing on bioenergetic properties of primary stages of substrate oxidation. The latter analysis was made for the strain DSM 9103 growing on EDTA as a source of carbon, nitrogen and energy (see the next section).

The mass-energy (based on redoxons) and usual (based on mass) characteristics are proportional to each other as follows: $\eta_{x/s} = (\sigma_B\gamma_B/\sigma_S\gamma_S)Y_{X/S}$, where σ is a mass fraction of carbon in an organic substance, γ is a degree of reductivity of carbon (eq. RO per C-mol) (*22–24, 28, 29*); subscripts S and B denote substrate and biomass, respectively. The reductance degree of biomass with elemental composition of $CH_pO_nN_q$ is calculated as follows (*24*): $\gamma_B = 4 + p - 2n - 3q$. For all non-oleaginous microorganisms for which the biomass elementary composition was measured, $\sigma_B\gamma_B \approx 2$ (*23, 30*). We assumed this value for DSM 9103. The elementary composition of EDTA is $C_{10}H_{16}O_8N_2$, from which the RO-number, N_{RO} and other mass-energy quantities of EDTA molecule are as follows: $N_{RO} = 4\times10 + 1\times16 - 2\times8 - 3\times2 = 34$; $\sigma_S = 0.411$; $\gamma_S = 34/10 = 3.4$; $\sigma_S\gamma_S = 1.4$. Therefore, energetic yield of EDTA-grown cells may be calculated as follows: $\eta_{x/s} = (2/1.4)Y_{X/S}$, or $\eta_{x/s} = 1.43Y_{X/S}$. Correspondingly, the value of $\sigma_S\gamma_S$ for glucose is equal to 1.6 and the $\eta_{x/s}$ value is calculated as 1.25 $Y_{X/S}$.

Comparative Studies on the Efficiency of EDTA Utilization by Different Strains

By using continuous cell cultivation, the kinetic and stoichiometric characteristics of growth and EDTA degradation by strains LPM-410 and LPM-4 were studied. Strains *Pseudomonas* sp. LPM-410 and LPM-4 were cultivated in EDTA-medium in pH-auxostat regime, which allows cell growth at the maximal specific growth rate (μ_{max}) (*31*). Under optimal conditions for the growth of *Pseudomonas* sp. LPM-410 (28—30°C) the value of μ_{max} was 0.087 ± 0.002 h^{-1}. The optimum temperature for the growth of strain LPM-4 was higher (32—34°C) with a μ_{max} of 0.095 ± 0.001 h^{-1}. The maximal specific growth rates of new isolates were considerably higher than the μ_{max} obtained for batch- and chemostat-grown strain DSM 9103 (0.050 and 0.053 h^{-1}, respectively) (*9, 32*). The maximal attained values of mass cell yield ($Y_{X/S}$) for strains *Pseudomonas* sp. LPM-410 and LPM-4 were 0.267 ± 0.008 and 0.219 ± 0.001, respectively, and the maximal attained energetic yield from EDTA were 0.382 ± 0.011 and 0.313 ± 0.002, respectively (Table II). Chemostat-grown strain DSM 9103 exhibited higher values of maximal mass and energetic cell yields (0.313 and 0.448, respectively).

Table II. Growth Characteristics of EDTA-Degrading Strains

Parameters	*LPM-410*	*LPM-4*	*DSM 9103*[a]
μ_{max}, h^{-1}	0.087 ± 0.002	0.095 ± 0.001	0.053
$Y_{x/s}$	0.267 ± 0.008	0.219 ± 0.001	0.313
$\eta_{x/s}$	0.382 ± 0.011	0.313 ± 0.002	0.448

NOTE: [a]Data are from Reference 32.

The maximal energetic growth efficiency ($\eta_{x/s}$) is assumed to be obtained during growth of microorganisms on glucose (*27, 33*). For a variety of microbes, mass growth yield on glucose is equal to $Y_{X/S} \approx 0.5$, which corresponds to $\eta_{x/s} \approx$ 0.62–0.63 (*23, 33*). This is the largest reliable value of cell energetic yield among those obtained for any microorganisms. Therefore, $\eta_{x/s} \approx 0.6$ is assumed to be a standard for comparison of the obtained $\eta_{x/s}$ with its possible maximum. The energetic yield may be reduced from the standard value of 0.6 due to high energy expenditures for cell maintenance or the involvement of oxygenase or oxidase reactions in substrate metabolism. These reactions are entirely or partly uncoupled with transformation of substrate energy into form suitable for further metabolic utilization (transmembrane electrochemical potential, high-energy bonds) (*27, 33, 34*). For example, yeasts do not utilize energy during oxidation of methanol to formaldehyde. At the same time, this oxidation in bacteria is coupled with ATP formation, although it is less efficient compared with the main respiratory chain.

Metabolism of EDTA by strain DSM 9103 utilized a monooxygenase (*12*). This enzyme requires one O_2 molecule per reaction, which eliminates four RO of substrate molecule. If all four side chains of EDTA are oxidized, 16 of 34 redoxons are transferred to oxygen via the oxygenase path. Under the assumption that energy from these 16 RO dissipates completely, the energetic cell yield from EDTA $\eta_{x/s}$ should be no more than 0.6 × (34–16)/34 = 0.32. The obtained $\eta_{x/s}$ value for strains *Pseudomonas* sp. LPM-410 and DSM 9103 are higher (0.382 and 0.448, respectively). These values are close to a middle of 0.6 (a maximum) and 0.32 (see above). From this it follows that oxidation of the side chains of EDTA is partly coupled with the utilization of energy of electrons transferred to oxygen due to the oxygenase reactions. Evidently, this process, like the main respiratory chain, includes proton transfer across the cytoplasm membrane against a H^+ gradient, but the number of coupling sites is lower than that for the main chain. Probably, the paths of electron transport connected with oxygenase operation coincide in part with the main respiratory chain.

Efficiency of EDTA and Glucose Utilization by Isolate LPM-4

As mentioned above, the EDTA-dependent strain LPM-4 was capable of glucose utilization during growth in EDTA-medium supplemented with glucose.

The time courses of batch growth of strain LPM-4 in EDTA-medium (control) and in the same medium supplemented with glucose are given in Figure 1. In both media, EDTA was consumed in 3 days. Glucose utilization started after EDTA exhaustion. In both variants, ammonium accumulation in the culture broth occurred during EDTA utilization and reached the maximum (33–45 mg/l) by the 3rd day of cultivation. No appreciable uptake of ammonium ions was observed during glucose utilization. Therefore, cell growth on glucose appeared to occur at the expense of intracellular nitrogen supply or nitrogen-containing metabolites other than ammonium.

The biomass (0.287 g/l) in EDTA-medium was formed from 1.28 g/l EDTA, i.e., mass cell yield ($Y_{x/s}$) was 0.22. In EDTA-medium supplemented with glucose, biomass produced from 1.14 g/l EDTA should be equal to 0.256 g/l and, therefore, the amount of biomass produced from glucose (1.03 g/l) should be equal to 0.519 – 0.256 = 0.263 g/l. Thus, mass cell yield from glucose was 0.26 that was considerably lower than the values of Y_s known in literature for different glucose-grown microorganisms (0.43–0.51) (*35*).

The mass cell yield from glucose in our experiment may be underestimated since glucose uptake occurred under strong nitrogen limitation of cell growth. Since energy capacities of EDTA and glucose are different, it was reasonable to compare the values of energetic cell yield ($\eta_{x/s}$) from these substrates. The calculated values of the energetic cell yields from EDTA and glucose in the case of strain LPM-4 were similar (0.32). However, it is necessary to take into account that, in our experiment, the obtained value of the energetic cell yield (similar to mass cell yield) from glucose may be underestimated because of nitrogen shortage.

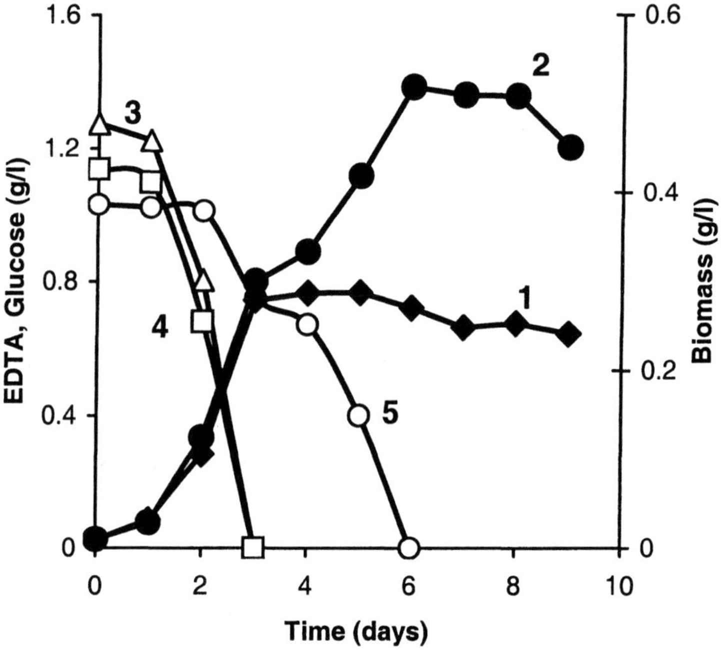

Figure 1. Time courses of the growth of isolate LPM-4 in (1) EDTA-containing medium (control) and (2) EDTA-containing medium supplemented with glucose (experiment). Residual EDTA concentration in (3) control and (4) experiment; residual glucose concentration (5).

To conclude, the new EDTA-degrading strains represent a practical interest as means for the removal of EDTA from polluted environment. The intrinsic feature of isolate LPM-4 is its EDTA-dependence. The elucidation of metabolic peculiarities responsible for a specific EDTA requirement for cell growth invites further investigations.

References

1. Bucheli-Witschel, M.; Egli, T. *FEMS Microbiol. Rev.* **2001,** *25,* 69–106.
2. Sillanpäa, M. *Rev. Environ. Contamin. Toxicol.* **1997,** *152,* 85–111.
3. Kaluza, U.; Klingelhöfer, P.; Taeger, K. *Water Res.* **1998,** *32,* 2843–2845.
4. van Ginkel, C.G.; Vandenbroucke, K.L.; Stroo, C.A. *Bioresour. Technol.* **1997,** *59,* 151–155.

5. Lauff, J.J.; Steele, D.B.; Coogan, L.A.; Breitfeller, J.M. *Appl. Environ. Microbiol.* **1990,** *56,* 3346–3353.
6. Nörtemann, B. *Appl. Environ. Microbiol.* **1992,** *58,* 671–676.
7. Klüner, T.; Hempel, D.C.; Nörtemann, B. *Appl. Microbiol. Biotechnol.* **1998,** *49,* 194–201.
8. Witschel, M.; Weilenmann, H.-U.; Egli, T. 54th Annual Meeting of the Swiss Society of Microbiology **1995.**
9. Witschel, M. Ph.D. thesis, Swiss Federal Institute of Technology, Zurich, 1999.
10. Satroutdinov, A.D.; Dedyukhina, E.G.; Chistyakova, T.I.; Witschel, M.; Minkevich, I.G.; Eroshin, V.K.; Egli, T. *Environ. Sci. Technol.* **2000,** *34,* 1715–1720.
11. Satroutdinov, A.D.; Dedyukhina, E.G.; Chistyakova, T.I.; Minkevich, I.G.; Eroshin, V.K.; Egli, T. *Mikrobiologiya* (Moscow) **2003,** *72,* 14–18.
12. Witschel, M.; Nagel, S.; Egli, T. *J. Bacteriol.* **1997,** *179,* 6937–6943.
13. Witschel, M.; Egli, T.; Zehnder, A.J.B.; Wehrli, E.; Spycher, M. *Microbiology* **1999,** *154,* 973–983.
14. Egli, T.; Weilenmann, H-U.; El-Banna, T.; Auling, G. *Sys. Appl. Microbiol.* 1988, *10,* 297–305.
15. Kluener, T.; Hempel, D.C.; Nörtemann, B. *Appl. Microbial. Biotechnol.* **1998,** *49,* 194–201.
16. *Manual of Methods for General Bacteriology;* Gerhardt, P.; Murray, R. G. E.; Costilow, R. N.; Nester, E. W.; Wood, W. A.; Krieg, N. R., Eds.; American Society for Microbiology: Washington, DC, 1981; p 355.
17. Chistyakova, T.I.; Belikova, V.L.; Satroutdinov, A.D.; Dedyukhina, E.G.; Eroshin, V.K. *World J. Microbiol. Biotechnol.* **2003,** *19,* 977–980.
18. Smibert, R. M.; Krieg, N. R. In *Methods for General and Molecular Bacteriology;* Gerhardt, P.; Murray, R.G.E.; Wood, W.A.; Krieg, N.R., Eds.; American Chemical Society: Washington, DC, 1994; pp 607–654.
19. *Bergey's Mannual of Determinative Bacteriology;* 9th ed.; Holt, J.; Krieg, N.; Seath, P.; Staley, J.; Williams, S., Eds.; Williams & Wilkins: Baltimore, 1994.
20. Pearson, W.R.; Lipman, D.J. *Proc. Natl. Acad. Sci.* **1998,** *85,* 2444–2448.
21. Minkevich, I.G. *J. Theor. Biol.* **1982,** *95,* 569–590.
22. Minkevich, I.G. *Biotechnol. Bioeng.* **1983,** *25,* 1267–1293.
23. Minkevich, I.G.; Eroshin, V.K. *Fol. Microbiol.* **1973,** *18,* 376–385.
24. Erickson, L.E.; Minkevich, I.G.; Eroshin, V.K. *Biotechnol. Bioeng.* **2000,** *67,* 748–773.
25. Erickson, L.E.; Minkevich, I.G.; Eroshin, V.K. *Biotechnol. Bioeng.* **1979,** *21,* 575–591.
26. Mayberry, W.R.; Prochazka, G.J.; Payne, W.J. *Appl. Microbiol.* **1967,** *15,* 1332–1338.

27. Minkevich, I.G.; Eroshin, V.K. *Stud. Biophys.* **1975,** *49,* 43–52.
28. Minkevich, I.G. *Appl. Biochem. Microbiol.* **1996,** *32,* 91–99.
29. Minkevich, I.G.; Eroshin, V.K. *Uspekhi Sovremennoi Biologii (Adv. Mod. Biol.) (Moscow)* **1976,** *82,* 103–116.
30. Minkevich, I.G.; Eroshin, V.K.; Alekseeva, T.A.; Tereshchenko, A.P. *Mikrobiologicheskaya Promyshlennost (Microbiol. Ind.) (Moscow)* **1977,** *2,* 1–4.
31. Mutafov, S.B.; Minkevich, I.G.; Eroshin, V.K. *pH-Auksostat: Teoriya i Praktika (pH-Auxostat: Theory and Practice);* Scientific Centre of Biological Researches, USSR Academy of Sciences: Pushchino (Russia), 1985.
32. Minkevich, I.G.; Satroutdinov, A.D.; Dedyukhina, E.G.; Chistyakova, T.I.; Eroshin, V.K. *Proc. Biochem.* **2003,** *38,* 1559–1564.
33. Minkevich, I.G. *Biotechnol. Bioeng.* **1985,** *27,* 792–799.
34. Antony, C. *The Biochemistry of Methylotrophs;* Academic Press: New York, 1982.
35. Pirt, S.J. *Principles of Microbe and Cell Growth;* Blackwell: Oxford, London, Edinburgh, Melbourne, 1975, pp 81–101.

Chapter 10

Biodegradation of L-Glutamatediacetate by Mixed Cultures and an Isolate

Cornelis G. van Ginkel, Roy Geerts, and Phuong D. Nguyen

Akzo Nobel Chemicals Research, Velperweg 76, 6824 BM Arnhem, The Netherlands (kees.vanginkel@akzonobel-chemicals.com)

The aerobic biodegradation of both enantiomers of glutamate-*N,N*-diacetate (GLDA) was studied according to official OECD test guidelines. L-GLDA was found to be readily biodegradable. Degradation of D-GLDA was demonstrated in an inherent biodegradability test. Analysis of L-GLDA as well as monitoring the change of dissolved organic carbon demonstrate that L-GLDA degrades extensively following a short acclimatization period in activated sludge treatment plants. A bacterium strain BG-1 was isolated from activated sludge on the basis of its capacity to use L-GLDA as sole nitrogen, carbon and energy source. The isolated strain was identified as a *Rhizobium radiobacter*. Strain BG-1 also utilized glyoxylate, L-glutamate, and NTA. D-GLDA, *S,S*-EDDS, and EDTA did not support growth of the strain. Significant oxygen uptake rates by L-GLDA-grown washed cell suspensions were observed with glyoxylate, L-glutamate, and oxoglutarate. Iminodiacetate did not stimulate oxygen consumption by L-GLDA-grown cells. This substrate utilization pattern suggests that L-GLDA is degraded by the successive removal of the two carboxymethyl groups, resulting in the formation of L-glutamate. The results of this study strongly indicate that L-GLDA readily undergoes complete degradation by microbial cultures.

Introduction

Anthropogenic chelating agents are of great importance for different purposes. They have a wide variety of uses, from household cleaners, pulp and paper bleaching, water treatment, industrial and institutional cleaning to photography. L-Glutamate-*N,N*-diacetate (L-GLDA), a new chelating agent, is composed of two carboxymethyl groups linked to the nitrogen atom of L-glutamate. L-GLDA is primarily used in detergents. After use this chelating agent will end up in wastewater. Therefore, it is important to understand whether or not L-GLDA is biodegradable in the environment and technosphere.

When the issue of biodegradation became a matter of public interest, tests to assess the biodegradability were developed. These biodegradability tests were internationally harmonized by the Organization of Economic Cooperation and Development (OECD). Not only the biodegradability of the parent compound is of importance but also the possible formation of recalcitrant metabolites. Biodegradation without the formation of recalcitrant metabolites can be demonstrated by determining the biodegradation pathway. The use of pure cultures of microorganisms is a valuable tool in elucidating the biodegradation pathway of chelating agents. Many bacteria that degrade chelating agents, i.e. nitrilotriacetate (NTA), ethylenediaminetetraacetate (EDTA), and *S,S*-ethylenediaminedisuccinate (*S,S*-EDDS), have been isolated (*1, 2*). Two metabolic pathways have been described by which microorganisms degrade aminocarboxylates. Evidence on microbial degradation of *S,S*-EDDS demonstrates that microorganisms catalyze the cleavage of fumarate from ethylenediamine (*3, 4, 5*). Biodegradation of EDTA and NTA is initiated by a cleavage of the C-N bond, releasing glyoxylate as product (*1, 2*).

We report here the biodegradation of L-GLDA and D-GLDA in wastewater treatment plants, using OECD biodegradability tests. In addition, L-GLDA degradation by an isolate is described to provide evidence of complete mineralization.

Materials and methods

Chemicals: L-Glutamate-*N,N*-diacetate (L-GLDA) (Dissolvine®) and D-GLDA were obtained from Akzo Nobel Chemicals, BU Functional Chemicals, Amersfoort, the Netherlands. All other chemicals used were purchased. The metal-chelate solutions were prepared by adding equimolar amounts of the respective metals as salts to L-GLDA or NTA solutions with a concentration of 2 g/L. Subsequently, the metal-chelate solutions were diluted to a concentration of 1 g/L of L-GLDA.

Inocula and bacteria: The activated sludge for the biodegradation experiments was from the aeration tank of the municipal wastewater treatment plant Nieuwgraaf, Duiven, the Netherlands. The plant consists of mechanical and biological stages for the treatment of mainly domestic wastewater. *Rhizobium radiobacter* DSMZ 30147, *Aminobacter aminovorans* DSMZ 6449, and *Chelatococcus asaccharovorans* DSMZ 6461 were purchased from DSMZ, Braunschweig, Germany.

Biodegradability tests: The capability of microorganisms to degrade L-GLDA and D-GLDA was determined, using OECD biodegradability tests. The closed-bottle test was performed according to OECD Test Guideline 301 D with some minor modifications (*6, 7*). Activated sludge from a municipal treatment plant and an SCAS unit used as inocula were diluted to a concentration of 2 mg/L dry weight in the bottles.

The SCAS tests were performed in accordance with the OECD Test Guideline 302 A (*8*). The daily fill and draw cycle involved the addition of domestic wastewater spiked with 110 mg/L of GLDA. The concentration of the activated sludge was initially 2 g/L dry weight. The removal of GLDA in the SCAS units was followed by nonpurgeable organic carbon (NPOC) and GLDA analyses.

A standardized method was used for evaluating the behavior of L-GLDA in continuous activated sludge (CAS) systems (*9*). The reactor consisted of an aeration vessel capable of holding 0.35-Liter from which the liquor was passed continuously to a settler of 0.15-Liter capacity. Aeration was achieved through a capillary on the bottom of the aeration vessel at a rate of approximately 10 L/hour. Domestic wastewater spiked with 50 mg/L of L-GLDA was supplied with a pump. An SRT of 10 days and an HRT of 10 hours were maintained in the CAS reactor. The liquor passed through the aeration vessel and settler was analyzed for NPOC, and L-GLDA. Removal of L-GLDA was assessed at 10 and 20°C.

Isolation and characterization: The bacterium used in this study was isolated by selective enrichment on L-GLDA as the sole source of nitrogen, carbon, and energy. Activated sludge from a plant treating primarily domestic wastewater was used as inoculum. After growth was obtained, dilutions of the cell suspensions were streaked on agar plates containing L-GLDA in a nitrogen-free mineral salts medium (*10*). Colonies began to appear on these plates upon incubation at 30°C. One colony was picked and streaked to purity on L-GLDA agar plates. Analyses to identify the isolate were carried out by DSMZ (Braunschweig, Germany), using fatty acid methyl ester analysis and 16S rDNA sequencing.

Growth, and washed cell suspensions: The isolate, *Rhizobium radiobacter* DSMZ 30147, *Aminobacter aminovorans* DSMZ 6449, and *Chelatococcus asaccharovorans* DSMZ 6461, were grown in 5-L Erelnmeyer flasks with

500 mL of nitrogen-free mineral salts medium with 1 g/L L-GLDA, or NTA. The isolate was also grown on various other organic compounds (1 g/L) in a mineral salts medium containing ammonium as nitrogen source (10). The cultures were shaken at 200 rpm in an orbital incubator at 30°C. Growth on L-GLDA was followed by removing samples at various time intervals and analyzed for L-GLDA and ammonium.

Cell suspensions of the batch cultures were harvested by centrifugation at 10,000 x g for 5 min at 4°C, washed three times with 100 mM phosphate buffer, pH 7.0 or with 50 mM HEPES buffer, pH 7.0. The washed cell suspensions were stored at 4°C.

Respiration experiments: Oxygen uptake was measured with a Biological Oxygen Monitor (Yellow Springs Instruments, Yellow Springs, Ohio), which consisted of an electrode and a water-jacketed vessel (5 mL). Washed cell suspensions were incubated in the vessel at 30°C for at least five minutes to allow determination of the endogenous respiration rate. Subsequently, 0.1 mL of a substrate solution (1 g/L) was injected, and the increase in the respiration rate was determined. Influence of the counter ions on the oxidation of L-GLDA and NTA was assessed, using a HEPES buffer (*11*).

Analysis: L-GLDA was determined by high performance liquid chromatography (HPLC). The HPLC system consisted of a high precision pump model 300 (Separations, H.I. Ambacht, the Netherlands) a Spark autosampler model basic marathon (Separations, H.I. Ambacht, the Netherlands), an Ionpac AS7 column with an AG7 guard-column (Dionex, Bavel, the Netherlands), and a UV/VIS detector model ABI 759A (Separations, H.I. Ambacht, the Netherlands). The concentration of L-GLDA was measured at a wavelength of 330 nm. The mobile phase was 50 mM nitrate in demineralized water containing 50 mM sodium acetate, pH = 2.75 ± 0.2. This solution was filtered over a 0.45 μm cellulose nitrate filter prior to use. The flow rate was 0.4 mL/min. Samples were prepared by mixing 4 mL of the sample with 1 mL of an iron nitrate solution (12.4 mM $Fe(NO_3)_3$ solution containing 151.2 mM nitric acid). Samples were shaken and allowed to stand for 15 minutes. Samples (50 μl) were injected after particles were removed by filtration through a 0.45-μm filter.

The accumulation of ammonium in the medium during the transformation of L-GLDA was determined colorimetrically at 690 nm by the formation of indophenol blue with hypochlorite and salicylate in the presence of sodium nitroferricyanide as catalyst (*12*). Samples were passed through a membrane filter (0.45-μm) prior to analysis.

Dissolved organic carbon was determined using a TOC analyzer (Shimadzu Corporation, Kyoto, Japan). Samples were passed through a membrane filter

(0.45 μm pore diameter) prior to analysis. Samples were acidified prior to injection in the TOC apparatus.

Results and discussion

Assessment of the biodegradability: The most important aspect with regard to the environmental fate of GLDA is its biodegradability. GLDA contains one asymmetric carbon atom. Although only L-GLDA is marketed, the biodegradability of both enantiomers, i.e., L-GLDA and D-GLDA, was studied because of known enantio-selective degradation of other chelates (*13, 14*). The closed-bottle test (OECD 301 D) has been established for the investigation of the ready biodegradability by mixed microbial cultures. For degradation to occur in the closed-bottle test, microorganisms must be present that are capable of utilizing the compound as an energy and carbon source. In the closed-bottle test, growth of microorganisms on L-GLDA is accompanied by oxygen consumption, exhibiting lag, exponential, and stationary phases (Figure 1).

A high number of L-GLDA-degrading microorganisms present in the activated sludge probably account for a lag period of approximately 10 days. L-GLDA exhibited a high degradation percentage exceeding the pass level within 22 days (Figure 1).

L-GLDA does therefore meet the OECD test criteria for ready biodegradability. D-GLDA did not degrade when incubated with activated sludge in the closed-bottle test. Failure to achieve biodegradation in the closed-bottle test may either be due to recalcitrance of the test substance or to the severe conditions imposed during the ready biodegradability testing. Inherent biodegradability tests, therefore, are to satisfy conditions required to enable biodegradation, thus permitting the assumption that any failure to observe biodegradation is due to recalcitrance. The SCAS test (OECD 302 A) was employed to demonstrate inherent biodegradability of D-GLDA. In the SCAS test, D-GLDA was degraded after an incubation period of approximately 8 weeks (Figure 2). For comparison an SCAS test was also carried out with L-GLDA. L-GLDA and D-GLDA were shown to be almost completely removed within 3 and 10 weeks, respectively (Figure 2). After acclimatization in the SCAS units more than 95% of both L- and D-GLDA-carbon was removed from the influent. High removal of the organic carbon demonstrates that no recalcitrant water-soluble intermediates are formed during the degradation of both L-GLDA and D-GLDA.

Inoculation of the closed-bottle test with acclimated sludge, which was derived from the SCAS unit fed with D-GLDA, resulted in 80% biodegradation of D-GLDA within 4 weeks. Acclimatization of sludge to D-GLDA also affected the biodegradation of L-GLDA. No lag phase was detected and biodegradation of L-GLDA up to 80% took place within 2 weeks (data not shown).

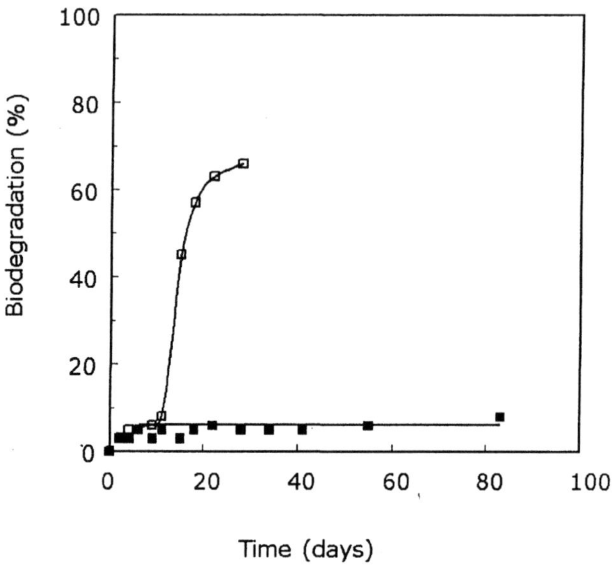

Figure 1. Biodegradation of L-GLDA (□) and D-GLDA (■) in Closed Bottle tests inoculated with activated sludge

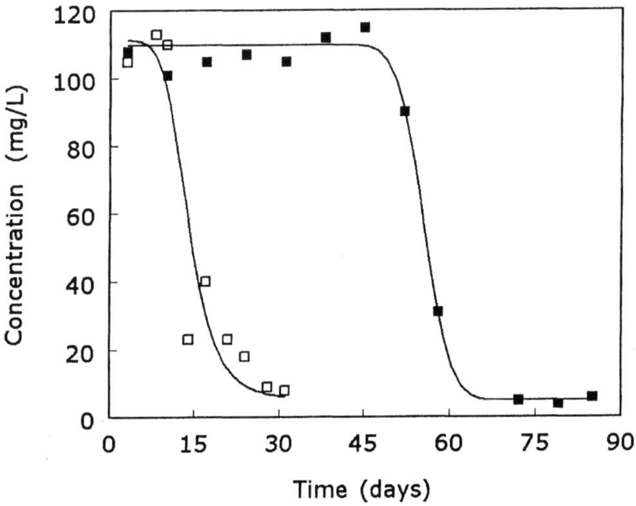

Figure 2. Biodegradation of L-GLDA (□) and D-GLDA (■) in SCAS tests

The findings demonstrate that the biodegradability of L- and D-GLDA differs. In the literature many examples of the preferred usage of one of the enantiomers are found. Examples of enantio-selective biodegradation of chelates are EDDS and iminodisuccinate (IDS). The three enantiomers of IDS are biodegradable; *R,R*-IDS being the least biodegradable enantiomer (*14*). One enantiomer of EDDS, i.e., *R,R*-EDDS, is not biodegradable, whereas *S,S*-EDDS is readily biodegradable (*13*).

Finally, a test simulating conventional activated sludge treatment was performed. In continuously-fed activated sludge (CAS) test, activated sludge was fed domestic wastewater spiked with L-GLDA. Biodegradation was followed by specific analysis of L-GLDA and by monitoring the change of dissolved organic carbon present in the effluent. At a temperature of 20°C, during an initial 10-day period little or no degradation of L-GLDA was observed due to the need for acclimatization of microorganisms. Lowering the temperature in the CAS test from 20° to 10°C, increased the lag period and the time required to obtain almost complete removal significantly (Figure 3).

The temperature did not influence the extent of biodegradation (Figure 3). Additional analysis of the dissolved organic carbon demonstrated that L-GLDA was not converted into a recalcitrant organic substance. Consequently, L-GLDA will be removed almost completely under conditions prevailing in conventional activated sludge plants.

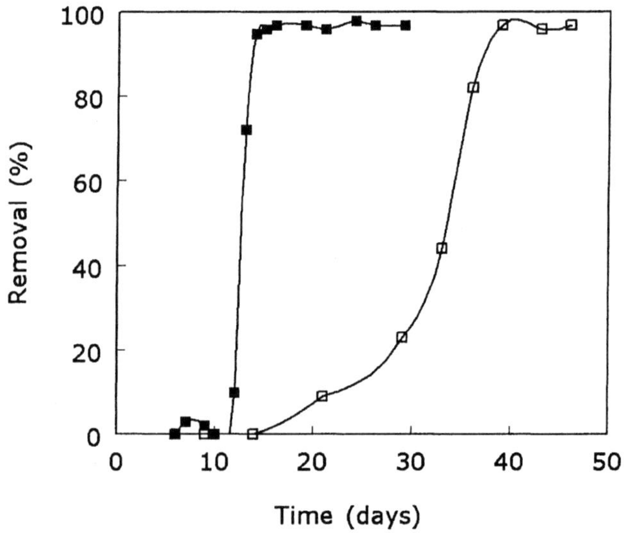

Figure 3. Removal of L-GLDA in continuously-fed activated sludge units operated at an HRT and SRT of 10 hours and 20 days, respectively. The tests were carried out according to OECD guideline 303 A at incubation temperatures of 10℃ (□)) and 20℃ (■)

Pure culture studies: Of general importance is the determination of the complete (ultimate) biodegradability of a substance as already indicated by the OECD test results. To prove complete degradation a bacterium designated as strain BG-1 was isolated from activated sludge. The isolation of strain BG-1 capable of degrading L-GLDA from activated sludge without a history of L-GLDA discharge suggests that L-GLDA degrading microorganisms are widespread in the environment. Strain BG-1 is an aerobic, Gram-negative, rod-shaped bacterium. The isolated strain was identified as a *Rhizobium radiobacter*. The type strain was unable to utilize L-GLDA. L-GLDA served as sole nitrogen, carbon and energy source for *Rhizobium radiobacter* BG-1. L-GLDA biodegradation was observed by virtue of its total disappearance from the growth medium as detected by HPLC. The consumption of L-GLDA by *Rhizobium radiobacter* strain BG-1 led to the release of ammonium. At an initial substrate concentration of 4.0 mM the molar ratio of L-GLDA degraded to ammonia produced was found to be 1:0.5. The observed maximum doubling time of *Rhizobium radiobacter* growing on L-GLDA was 8 h. *Rhizobium radiobacter* strain BG-1 was also able to grow on the aminocarboxylate NTA. *Aminobacter aminovorans* DSMZ 6449 and *Chelatococcus asaccharovorans* DSMZ 6461 (*15*) capable of utilizing NTA as sole carbon and energy source were also capable of growing on L-GLDA. Other aminocarboxylates did not support growth of *Rhizobium radiobacter* strain BG-1. These included D-GLDA, EDTA, *S,S*-EDDS and diethylenetriaminepentaacetate.

Oxidation of L-GLDA complexes was studied with washed cell suspensions of L-GLDA-grown *Rhizobium radiobacter* BG-1. Only limited oxidation of L-GLDA complexed with Ni Co, Cu, Fe, and Zn with washed cell suspensions of L-GLDA-grown cells was detected. Washed cell suspensions of L-GLDA grown *Rhizobium radiobacter* BG-1 were capable of oxidizing L-GLDA in the form of Ca, Mg, and Mn complexes as well as uncomplexed L-GLDA. Mg-NTA, Ca-NTA and Mn-NTA were oxidized by washed cell suspensions of NTA-grown *Rhizobium radiobacter* BG-1 at high rates. The presence of these counter ions increased the oxygen uptake rate. Metal-NTA complexes with Ni, Co, Cu, Fe and Zn were not or slowly oxidized (Table I). Firestone and Tiedje (*16*) found comparable results with a bacterium capable of

Table I. Increase of oxygen uptake rates in percentages by washed cells suspensions grown on L-GLDA, and NTA at 30°C in the presence of the respective growth substrate and various counter ions (1:1 molar ratio). Rates of oxygen uptake are expressed as percentages increase compared to the endogenous respiration

	Substrate	
Counter ion	*L-GLDA*	*NTA*
none	570	220
Ca^{2+}	660	860
Mg^{2+}	520	530
Mn^{2+}	460	660
Zn^{2+}	0	90
Fe^{3+}	50	50
Cu^{2+}	20	0
Co^{2+}	30	20
Ni^{2+}	20	0

degrading NTA. For example, they found no oxygen uptake with Cu-NTA and with Ni-NTA and an intermediate rate with Zn-NTA. High rates were found for complexes with Ca and Mn. VanBriessen *et al* (17) found the concentration of Ca-NTA was rate limiting for biodegradation of NTA by *C. heintzii* and removal of other species was dependent on the concentration of Ca-NTA in equilibrium with other metal-chelate forms.

Rhizobium radiobacter BG-1 was grown on a number of different substrates as sole source of carbon, after which respiration rates of washed cell suspensions were determined with a variety of substrates. Cells grown on L-GLDA respired D-GLDA and NTA. Washed cell suspensions of NTA-grown *Rhizobium radiobacter* strain BG-1 were able to oxidize L-GLDA but not D-GLDA. Further substrates oxidized by strain BG-1 grown on NTA were intermediates of NTA catabolism, i.e. iminodiacetate (IDA), glycine, and glyoxylate. Washed cell suspensions of L-GLDA-grown cells oxidized L-GLDA, glutamate, oxoglutarate, and glyoxylate. Trans-ketoglutaconate and IDA did not increase the rate of oxygen uptake by the washed cells above the endogenous rate. Glycine enhanced the respiration of L-GLDA-grown cells slightly (Table II). This substrate utilization pattern indicates strongly that *Rhizobium radiobacter* BG-1 has evolved enzymes that convert L-GLDA into glyoxylate and L-glutamate. By analogy with NTA metabolism, a pathway for L-GLDA degradation involving the successive removal of the carboxymethyl groups is proposed (Figure 4). Acetate-grown cells were not able to oxidize NTA and L-GLDA at high rates demonstrating that the enzymes catalyzing the degradation of the aminocarboxylates are not constitutively expressed.

Table II. Increase of oxygen uptake rates of washed cells suspensions grown on L-GLDA, NTA, and acetate at 30°C. Rates of oxygen uptake are expressed as percentages increase compared to the endogenous respiration

	Growth substrate		
Substrate	*L-GLDA*	*NTA*	*Acetate*
L-GLDA	540	70	40
D-GLDA	100	0	10
L-Glutamate	420	800	600
D-Glutamate	0	0	0
trans Ketoglutaconate	0	0	0
IDA	0	1500	0
Glyoxylate	30	20	30
Oxoglutarate	80	120	120
Glycine	50	80	160
Acetate	80	150	470
NTA	50	1400	0

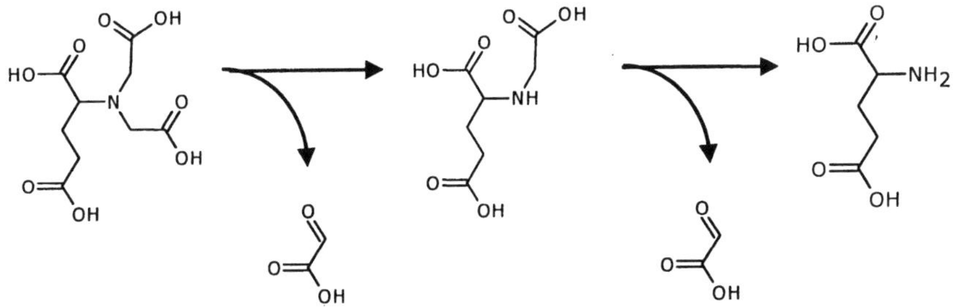

Figure 4. Proposed biodegradation route of L-GLDA

In conclusion, biodegradability test results and the isolation of strain BG-1 from activated sludge without a history of L-GLDA discharge demonstrate that L-GLDA degrading microorganisms are widespread in the environment and biological treatment systems. These microorganisms degrade L-GLDA completely as strongly indicated by the *Rhizobium radiobacter* strain BG-1.

References

1. Nortemann, B. *Appl. Microbiol. Biotechnol.* **1999,** *51*, 751-759.
2. Bucheli-Witschel, M.; Egli, T. *FEMS Microbiol. Rev.* **2001**, *25*, 69-106.
3. Takahashi, R.; Fujimoto, N.; Suzuki, M.; Endo, T. *Biosci. Biotechnol. Biochem.* **1997**, *61*, 1957-1959.
4. Takahashi, R.; Yamayoshi, K.N.; Fujimoto, N.; Suzuki, M. *Biosci. Biotechnol. Biochem.* **1999**, *63*, 1269-1273.
5. Bucheli-Witschel, M.; Egli, T. Biodegradation. **1998**, *8*, 419-428.
6. OECD. Guidelines for testing of chemicals; Degradation and accumulation, No. 301: Ready biodegradability, Paris, France. 1992.
7. Ginkel, C.G. van; Stroo, C.A. *Ecotoxicol. Environ. Saf.*, **1992**, *24*, 319-327.
8. OECD Guidelines for testing chemicals; Degradation and accumulation No 302 A, Inherent biodegradability; modified SCAS test, Paris, France. 1981.
9. OECD Guidelines for testing chemicals; Simulation test - Aerobic sewage treatment. No 303 A, Paris, France. 1981.
10. Kroon, A.G.M.; Ginkel, C.G. van. *Environ. Microbiol.* **2001**, *3*, 131-136.
11. Kluner, T.; Hempel, D.C.; Nortemann, B. *Appl. Microbiol. Biotechnol.* **1998,** *49*, 194-201.

12. Verdouw, H.; Echteld van, C.J.A; Dekkers E.M.J. *Water Res.* **1978**, *12*, 399-402.
13. Schowanek, D.; Feijtel, T.C.J.; Perkins, C.M.; Harman, F.A.; Ferderle, T.W.; Larson, R.J. Chemosphere. **1997**, *34*, 2375-2391.
14. Reinecke, F.; Groth, T.; Heise, K.P.; Joentgen, W.; Muller, N.; Steinbuchel, A. *FEMS Microbiol. Lett.* **2000,** *18*, 41-46.
15. Auling, G.; Busse, H.J.; Egli, T.; El-Banna, T.; Stackebrandt, E. *System Appl. Microbiol.* **1993**, *16*, 104-116.
16. Firestone, M.K.; Tiedje, J.M. *Appl Microbiol.* **1975**, *29*, 758-764.
17. vanBriessen, J.M.; Rittmann, B.E.; Girvin, D.; Bolton, H. *Environ. Sci. Technol.* **2000,** 3346-3353.

Chapter 11

Full-Scale Biological Treatment of Industrial Effluents Containing EDTA

Cornelis G. van Ginkel and Roy Geerts

Akzo Nobel Chemicals Research, Velperweg 76, 6824 BM Arnhem, The Netherlands (kees.vanginkel@akzonobel-chemicals.com)

In Europe, the ability of microorganisms to biodegrade EDTA aerobically under slightly alkaline conditions in existing full-scale activated sludge plants has been applied. The use of alkaline conditions in activated sludge plants is spurred by increasingly stringent environmental regulations. This process has been implemented in plants treating wastewater from pulp and paper mills at negligible costs. High removal of EDTA has also been observed in full-scale activated sludge plants treating wastewater from the dairy and beer industry. The results of monitoring studies of these full-scale activated sludge plants demonstrate removal efficiencies of more than 80%. The effectiveness of the activated sludge treatment is not only dependent on the pH level but also on the metal profile of the wastewater and the sludge retention time.

Introduction

Ethylenediaminetetraacetic acid (EDTA) has numerous applications based on its ability to control the action of different metal ions. The pulp and paper industry and industrial and institutional cleaning are very important application areas. EDTA is used in the pulp and paper industry to stabilize the action of hydrogen peroxide on pulp by complexing with metals that catalyze the decomposition of peroxide. In industrial and institutional cleaners, EDTA is used to prevent precipitation of calcium and magnesium. As EDTA is water-soluble and not volatile, it is released mainly with wastewater effluents. In Europe, treatment of wastewaters containing EDTA is increasingly required because environmental regulations are becoming more stringent.

Biological treatment of many industrial wastewaters has traditionally been based on methods successfully employed for processing domestic sewage. Aerobic processes, such as activated sludge systems, are the principle way in which naturally occurring microorganisms convert organic material into environmentally benign substances. EDTA has been reported to pass through conventionally operated biological treatment plants without notable degradation (*1, 2, 3*). However, the activated sludge process can be used to biodegrade EDTA under alkaline conditions (*4, 5*). Removal of EDTA in activated sludge systems also depends on sludge retention time (SRT) to ensure that EDTA-utilizing microorganisms do not wash out (*4*). Finally, it has been shown that the counter ions determine the treatability of EDTA (*6, 7, 8, 9*).

Biological treatment of EDTA-containing wastewater in activated sludge plants operated under alkaline conditions is now being implemented. Monitoring studies conducted in full-scale activated sludge plants are reported to disclose the potential of biological EDTA removal in existing treatment plants. A simple laboratory set-up is presented to assess the possibilities of EDTA removal in existing plants.

Materials and methods

Chemicals: EDTA (Dissolvine®) was obtained from Akzo Nobel Chemicals, BU Functional Chemicals, Amersfoort, the Netherlands. All other chemicals used were purchased.

Continuously-fed activated sludge test: Laboratory-scale continuously-fed activated sludge (CAS) reactors permit evaluation of the treatability of EDTA-containing effluents (10). The CAS reactor consisted of an aeration vessel with a capacity of 0.35-L, from which the liquor was passed continuously to a settler of 0.10-L capacity. The wastewater flow through the system was maintained by a peristaltic pump with a throughput of 0.35 L/day, to give a hydraulic retention time of 1 day. Treated effluent from the reactor was collected in a vessel. Aeration was achieved by means of a capillary fitted on the bottom of the aeration section, delivering air at a rate of approximately 10 L/hour. Aeration was provided to operate the air-lift and to maintain a dissolved oxygen concentration in excess of 2 mg/L. Daily 17.5 ml of sludge was removed from the aeration tank to maintain a sludge retention time of 20 days. The pH in the units was monitored and maintained at 8.0 by automatic addition of 1 M NaOH. At the start, the unit was filled with activated sludge at a concentration of 3 g/L of suspended solids. Treated effluent was collected in a vessel over 24-hour periods to serve in monitoring process efficiency by measurement of the EDTA content.

Monitoring studies: Five activated sludge plants treating wastewater from the dairy and beer industry, including publicly owned (*I, II, III*) and privately owned plants (*IV, V*) were evaluated. The influent samples were taken from the discharge of the primary clarifier (*IV, V*) or the influent of the aeration tank (*I, II, III*) and the effluent samples from the discharge of the secondary clarifier. Plant IV and V received a mix of domestic wastewater and industrial wastewater from a dairy plant. Plant IV experienced two seasonal changes characterized by a cold winter (5 °C) and a summer (20 °C). Two activated sludge plants receiving wastewater of pulp and paper mills were monitored. The influent samples were taken from the discharge of the industrial plants or the primary clarifier of the activated sludge system. The effluent samples were taken from the discharge of the secondary clarifier.

Analysis: EDTA was determined by high performance liquid chromatography (HPLC). The HPLC system consisted of a high precision pump model 300 (Separations, H.I. Ambacht, the Netherlands) a Spark autosampler model basic marathon (Separations, H.I. Ambacht, the Netherlands), an Ionpac AS7 column with an AG7 guard-column (Dionex, Bavel, the Netherlands) and an UV/VIS detector model ABI 759A (Separations, H.I. Ambacht, the Netherlands). The concentration of EDTA was measured at a wavelength of 330 nm. The mobile phase was 50 mM nitrate in demineralized water containing 50 mM sodium acetate, pH = 2.75 ± 0.2. This solution was filtered over a 0.45 μm cellulose nitrate filter prior to use. The flow rate was 0.5 ml/min. Samples were prepared by mixing 4 ml of the sample with 1 ml of an iron nitrate solution (12.4 mM $Fe(NO_3)_3$ solution containing 151.2 mM nitric acid). Samples were shaken and allowed to stand for 15 minutes. Samples (50 μl) were injected after particles were removed by filtration through a 0.45-μm filter.

Suspended solids (SS) concentrations in the CAS units were measured by filtering samples through a preweighed 12-μm pore size (cellulose nitrate) membrane filter (Schleicher and Schüll, Darmstadt, Germany), drying the filter at 105°C overnight, and measuring the weight increase.

Results and discussion

Dairy and beer industry: Monitoring studies of full-scale studies should provide the necessary real-world confirmation of the laboratory results by giving special attention to the SRT and the pH. Five activated sludge plants treating only or predominantly wastewater from the dairy and beer industries were evaluated (Table I).

Table I. Overview of monitoring results from five full-scale activated sludge plants treating wastewater from the dairy and beer industry

Plant	*Wastewater*	*pH*	*SRT (days)*	*Removal (%)*
I	dairy	7.5-8.1	~20	~90
II	beer	7.3-7.7	~23	~50
III	dairy	7.8-8.4	~9	~30
IV (5°C)	dairy and domestic	7.5-7.8	~40	~35
IV (20°C)	dairy and domestic	7.8-8.0	~40	~95
V	dairy and domestic	6.9-7.1	~20	0
VI	dairy	8.7-8.9	20	95[a]

[a] *measured in a laboratory-scale activated sludge unit*

Excellent performance was obtained from plant I, with effluent EDTA concentrations in the 2 to 8 mg/L range. The average EDTA removal was ~90% and the minimum and maximum removals were 78% and 95%, respectively. The pH in the activated sludge treatment system was slightly alkaline due to the use of NaOH during Cleaning In Place (CIP). The SRT in plant I was approximately 20 days. This SRT enables EDTA-degrading micro-organisms to maintain themselves in the activated sludge system (*4*). Plant I also removed over 95% of the COD. Another plant (*II*) removed over 90% of the COD and ~50% of the EDTA. The EDTA effluent concentrations ranged from 7 to 12 mg/L. The microbial degradation of EDTA present in the wastewater from this beer producing plant was therefore only partially successful. The pH values measured ranged from 7.3 to 7.7, which is not optimal for biodegradation of EDTA (*4*). The activated sludge of plant III was capable of removing ~30% of the EDTA present in the wastewater resulting in effluent concentrations varying from 14 to 21 mg/L. This partial removal can probably be attributed to an SRT of approximately 9 days in the activated sludge system. The pH in this activated sludge plant was within the optimal range. During the monitoring study, Plant IV experienced two seasonal changes characterized by a very cold winter (temperature of 5°C in the aeration tank) and a summer (temperature of 20°C in the aeration tank). The SRT of this plant was approximately 40 days and the pH measured ranged from 7.5 to 8.0. In the summer, the removal of EDTA was ~95%. Under these conditions, effluent EDTA concentrations of the treatment plant were only 0.2 to 1.6 mg/L. However, in the winter, the extent of elimination of EDTA was only 35%. At 5°C, an SRT of approximately 40 days probably still results in a wash-out of the competent EDTA-utilizing micro-organisms (Table I). A comparable well-known phenomenon is the low conversion of ammonium into nitrate by activated sludge at low temperatures due to wash-out of nitrifying bacteria (*11*).

Wastewaters from the dairy, soft drink, and beer industry are usually alkaline. This is also true for the wastewater from plant V. In contrast to the other plants monitored, wastewater of plant V was neutralized before being discharged into a publicly-owned activated sludge plant. Under the neutral conditions no EDTA removal was detected in this activated sludge system (Table I). The treatability of the wastewater from plant V was studied in a laboratory-scale CAS reactor operated at an SRT of 20 days and a hydraulic retention time of 1 day. An alkaline pH of 8.7 to 8.9 was obtained in the CAS unit fed with only wastewater from dairy plant V without any additional measures. The EDTA concentration of the wastewater flowing into the unit was 30 mg/L. Results from the CAS test demonstrated that the activated sludge acclimatized to EDTA within three weeks. After this period, removal percentages in excess of 95 were found, which are attributed to biodegradation (Figure 1). The steady-state effluent EDTA concentrations from the CAS unit were between 0.5 and 1.5 mg/L. The CAS reactor operated at pH of 8.7 to 8.9 also effectively reduced the COD and BOD_7 of the wastewater.

Paper and pulp industry: Many pulp and paper mills have installed conventional biological treatment systems, such as activated sludge plants to treat their effluents. The pH in these activated sludge plants is often slightly acidic. Under these conditions reduction of EDTA present in pulp and paper mill wastewater was not observed in activated sludge plants (*12*). In contrast, the results obtained in laboratory-scale experiments demonstrate that up to 80% of EDTA can be degraded if the pH level is raised to 8-9 (*5*). Because full-scale applications of activated sludge plants degrading EDTA are still relatively rare, there is a lack of information regarding the potential for enhanced removal under modified conditions. Two full-scale activated sludge plants were monitored to elucidate the factors that affect the performance.

Mill A produces newsprint using primarily thermomechanical pulp. The wastewater of this mill is treated in two activated sludge units in series. The first activated sludge unit primarily treats wastewater from the pulp mill. The second activated sludge unit not only receives the effluent from the first bioreactor but also effluent from the paper mill. The first unit has a high sludge load of 0.5 g COD g^{-1} SS day^{-1}, and the second unit has a sludge load of only 0.13 g COD g^{-1} SS day^{-1}. These sludge loads result in SRTs of 6 days (first) and >20 days (second). Part of the sludge of the second activated sludge plant is introduced into the first activated sludge plant. This system achieves a COD removal of approximately 80%. BOD_7 analyses show over 98% removal. The SRT of this

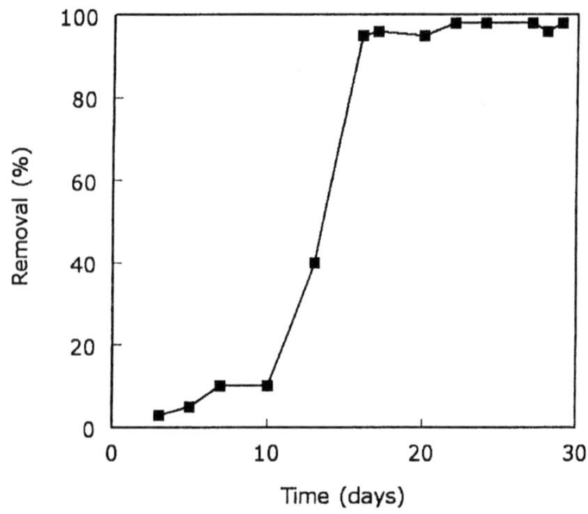

Figure 1. Removal of EDTA from wastewater derived from a dairy plant (V) in a CAS unit maintained at a pH of 8.7 to 8.9

activated sludge system could bring about EDTA removal. Under neutral conditions EDTA present in the effluent from the pulp and paper mill at concentrations ranging from 45 to 60 mg/L was not removed. However, at a pH of only 7.5 to 8.0 EDTA concentrations decreased to 2 mg/L at negligible costs. This represents >80% EDTA removal. The removal percentage can not be calculated with only the influent and effluent concentrations given because effluent from the paper mill is fed to the second reactor. The high EDTA removal percentages achieved at pH 7.5 to 8.0 have thus far only been obtained with wastewater from a dairy plant (*4*).

Mill B producing TCF (totally chlorine free) bleached softwood paper, treats its wastewater in a reconstructed lagoon. The construction of a settling tank transformed the original biological treatment system into an activated sludge system. Since the reconstruction of the biological treatment plant was completed, discharges of COD and BOD_7 have fallen. The COD removal of the activated sludge system averaged 77%. BOD_7 removed amounted to 98 to 99%. The SRT of the plant is 30 to 50 days. The high SRT of the system allows EDTA removal (*4*). To obtain EDTA removal Mill B realizes alkaline conditions by reducing the concentration of sulfite in the influent of the activated sludge plant. At first a pH of approximately 8 was achieved using this approach. Under these conditions an average EDTA reduction of 70% was achieved. A further increase in EDTA removal was achieved by maintaining a pH of 8.5, resulting in an average EDTA effluent concentration of 4.5 mg/L corresponding to ~85% removal of EDTA (Figure 2). Although the response of EDTA removal to the pH differs from one wastewater to another, an increase in the pH up to 8.5 usually leads to higher removal percentages (*5*). Mill B demonstrates that removal can be easily obtained at negligible costs.

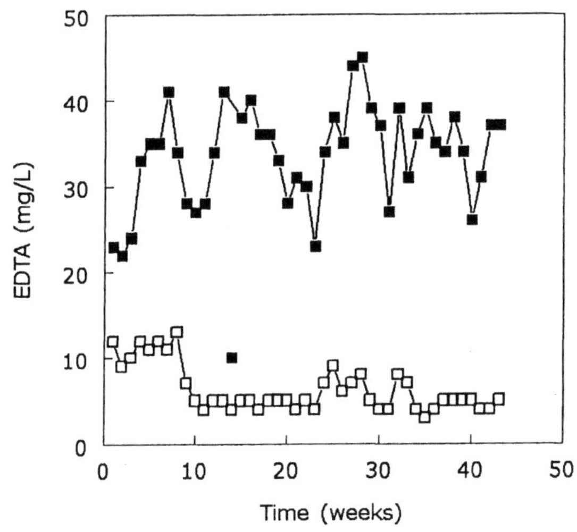

Figure 2. Changes in EDTA concentration of influent (■) and effluent (□) samples analyzed during full-scale activated sludge treatment of mill B effluent. The pH was maintained at 8 from week 0 to week 8. From week 8 to week 43 the pH was to 8.5

EDTA removal in activated sludge plants treating pulp and paper wastewater (approximately 80%) is lower than removal obtained with wastewater from dairy plants (over 90%). The different performance with respect to EDTA removal may be caused by different ratios of Fe and EDTA in the effluents since Fe-EDTA complexes have been shown to be recalcitrant (*6, 7, 8, 9*) and to re-speciate slowly to the biodegradable Ca-EDTA form (13). To demonstrate that extensive removal of EDTA can also be obtained with pulp and paper mill effluent, effluent from a bleaching plant (Q-filtrate) was investigated. This effluent contained 95 mg/L of EDTA and 0.4 mg/L of Fe giving an EDTA and Fe molar ratio of 50 : 1. The removal of EDTA present in the effluent of the bleaching plant was assessed in a CAS unit. The CAS unit was operated at an SRT and HRT of 20 days and 1 day, respectively. Sludge in the CAS unit maintained at pH 8.0 was capable of removing >95% of the EDTA. The residual EDTA concentrations were therefore less than 5 mg/L of which, maximal 2 mg/L was complexed with Fe. This result demonstrates that high EDTA removal can be achieved by treating flows from bleaching plants separately.

Conclusions

A few observations can be made from the monitoring studies. Slightly alkaline conditions are crucial to obtaining EDTA removal in activated sludge plants. The use of alkaline conditions proved to be a cost-effective and dependable technique for treatment of EDTA containing industrial effluents. Alkaline conditions exhibited no adverse effects on COD and BOD_7 removal during treatment. Available data suggest that the SRT should be more than approximately 20 days. Higher SRTs are a prerequisite when temperatures in the aeration tanks are lower than 15°C. EDTA removal may be determined by the ratio of EDTA and iron present in the wastewaters derived from the pulp and paper industry. Treatment options to reduce EDTA concentrations to acceptable levels in existing plants can be easily assessed in laboratory-scale CAS tests.

References

1. Gardiner, J. *Water Res.* **1976**, 10, 507-514.
2. Kari, F.G.; Giger, W. *Water Res.* **1996**, 30, 122-134.
3. Alder, A.C.; Siegrist, H.; Fent, K.; Egli, T.; Molnar, E.; Poiger, T.; Schaffner, C.; Giger, W. *Chimia* **1997**, 51, 922-928.
4. Ginkel, C.G. van; VandenBroucke, K.L.; Stroo, C.A. *Bioresource Technol.* **1997**, 59, 151-155.
5. Ginkel, C.G. van; Virtapohja, J.; Steyaert, J.A.G.; Alen, R. *Tappi J.* **1999**, 82, 138-142.
6. Henneken, L.; Nortemann, B.; Hempel, D.C. *Appl. Microbiol. Biotechnol.* **1995**, 44, 190-197.
7. Henneken, L.; Kluner, T.; Nortemann, B.; Hempel, D.C. *J. Chem. Technol. Biotechnol.* **1998**, 73, 144-152.
8. Kluner, T.; Hempel, D.C.; Nortemann, B. *Appl. Microbiol. Biotechnol.* **1998**, 49, 194-201.

9. Willett, A.I.; Rittmann, B.E. *Biodegradation* **2003**, 14, 105-121.
10. OECD Guidelines for testing of chemicals. Simulation Test - Aerobic Sewage Treatment. Guideline 303 A. Paris, France, 1981.
11. Birch, R.R. *J. Chem. Tech. Biotechnol.* **1991**, 50, 411-422.
12. Saunamäki, R. *Tappi J.* **1995**, 78, 185-192.
13. Xue, H.; Sigg, L.; Kari, F.G. *Environ. Sci. Technol.* **1995,** 29, 59-68.

Chapter 12

Effects of Chelating Agents on Trace Metal Speciation and Bioavailability

Kathrin Wenger, Susan Tandy, and Bernd Nowack

Institute of Terrestrial Ecology, Swiss Federal Institute of Technology (ETH), Grabenstrasse 3, CH–8952 Schlieren, Switzerland

The effect of chelating agents on metal bioavailability is discussed in terms of interactions between metals, chelating agents, microorganisms, aquatic organisms and plants. The statement "chelating agents are not bioavailable" was found not to hold true for all organisms. Microorganisms that are able to degrade chelating agents have specific, species-dependent transporters for their uptake. Some chelates are also as toxic as the free metal ion to microorganisms (e.g. HgEDTA). Exceptions to the free metal ion activity model (FIAM) are also observed for some higher aquatic organisms. Results from hydroponics experiments also clearly showed that metal uptake into plant shoots did not follow the FIAM model in the presence of chelants. Several possible explanations for the observed departures from the FIAM are discussed: the presence of kinetically distinct transport systems for solutes in higher plants including high-affinity ion specific transport systems and indiscriminate apoplastic transport of solutes via breakages in the endodermis, the possible existence of non-specific uptake sites, as well as the ability of plants to control and regulate ion uptake via shoot-root communication and modifications in the root plasma membranes.

Introduction

The majority of studies examining trace metal bioavailability have employed models such as the free ion activity model (FIAM) *(1, 2)* in an attempt to quantitatively relate chemical speciation to biological effects. FIAM is a conceptual model that describes how the effect of metals can be explained on the basis of metal speciation and metal interactions with the organism. The following key assumptions underpin the FIAM model *(2)*: i) the cell membrane is the primary site for metal interactions with living organisms; ii) the interaction of a metal with the cell surface can be described as a surface complexation reaction; iii) rapid equilibrium is established between metal species in the solution and those at the cell surface, i.e. steady-state equilibrium is attained if the internalization of the metal, the transport across the biological membrane, is the rate-limiting step; iv) biological response, whether expressed as uptake, nutrition or toxicity is directly proportional to the concentration of the metal cell surface complexes or to the metal internalization fluxes; v) the concentration of free surface sites on the cell membrane remains virtually constant, i.e. the cellular ligands or carrier ligands of the cell membrane remain under saturated, and the concentration of the metal surface complex is directly proportional to the concentration of free metal in solution; vi) internalization of metals, if it occurs, is first order or pseudo first order; vii) the membrane surface is chemically homogeneous (i.e. binding sites are independent of one another); viii) no significant modification of the plasma membrane is assumed to occur (e.g. no degradation or synthesis of carrier ligands), the metal does not induce any changes in the nature of the plasma membrane.

According to this view, the role of ligands present in soil or nutrient solution is limited to participating in complexation reactions with trace metals, resulting in a decreased free metal ion activity and hence in a decreased biological response *(2)*. This implies that chelating agents have no direct physiological effect and that it does not matter whether the metal is complexed by either weak or strong organic ligands. Data that do not fit this postulate are typically regarded as 'apparent exceptions' to FIAM *(2)*. Exceptions to the FIAM model documented so far include the transport of lipophilic complexes *(3)*, the transportation of low molecular metabolizable complexes *(4, 5)*, or the formation of ternary surface complexes *(6)*. Other exceptions to the FIAM have been reported for higher plants, e.g. for metal complexes with chloride *(7)*, and for some low-molecular weight organic acids *(8)*. In experiments with the same calculated free metal activity, higher plant metal concentrations were observed from chelant buffered solution which had the higher total metal concentration indicating that the free metal ion activity alone might not predict the phytoavailability of metals *(9, 10)*.

In this paper we discuss the effect of chelating agents on metal bioavailability, considering the existing literature about interactions between metals, chelating agents, microorganisms, aquatic organisms and higher plants. Our objective is to critically reexamine the validity of the FIAM model with special focus on chelating agents and their effect on metal uptake by higher plants. Chelating agents have been widely used in many nutrient solutions to buffer the free metal ion concentration. The underlying assumption of these studies is that the metal-chelating agent complex itself is not available. But, if this general assumption is not true, it would have considerable implications for the present concept of bioavailability and risk assessment.

Interactions between Trace Metals, Chelating Agents and Microorganisms

Organisms require on the one hand trace amounts of several metals for normal growth and metabolism while on the other hand metals may also be toxic to them. Chelating agents may therefore act in two ways: the chelating agent can detoxify the metal by binding it in an unavailable form or it can increase the bioavailability when the complexes are biologically active. Trace metals such as copper and zinc are important micronutrients for algae but are toxic at higher concentrations. Experimental evidence indicates that growth and toxicity of these metals are a function of the free metal ion concentration.

The methodology using chelation to buffer the free metal ion concentration at very low levels was introduced by Hutner et al. in 1950 *(11)*. The use of speciation calculations allows setting the free metal ion concentration at varying pH and total metal and ligand concentration, taking into account all side reactions. These buffers have permitted the demonstration that it is the free metal ion activity that controls biological activity of aquatic microorganisms *(12-14)*. The underlying assumption of these studies is that the metal-chelating agent complex itself is not available. Some experimental evidence verifies that EDTA does not cross the cellular membrane of phytoplankton *(15)*. The metal-EDTA complexes of Cu, Zn, Mn and Fe(III) are all unavailable and uptake or toxicity is only dependent on the free metal ion concentration *(12, 16)*. The first investigations were carried out with marine algae but the same relationship was also found for freshwater species *(17)*.

Addition of uncomplexed chelating agents to a water sample will therefore reduce the growth of algae if the free metal ion concentration is reduced below the threshold needed for a sufficient supply of metals to the algae. On the other hand, if the chelating agent reduces the free metal ion concentration into a range that is more favorable by alleviating a toxic effect, then an increase in growth can be observed. In addition, the chelating agent may also solubilize metals from particulate fractions in the water and increase the dissolved fraction that is more easily available compared to the solid fraction *(18)*. Biostimulation has been observed for both EDTA and NTA *(19)*, e.g. in effluents from wastewater treatment plants *(20)*. It has been hypothesized that the main stimulating influence of EDTA addition maybe to increase the amount of dissolved iron *(15)*. Iron is often the limiting element in water despite the presence of particulate iron. EDTA or other chelating agents form strong complexes with Fe(III) that are soluble at the pH range of natural waters. Also in soil and groundwater the availability of Fe(III) for iron-reducing microorganisms can be greatly enhanced using chelating agents *(21, 22)*.

However, there is a special group of microorganisms to which chelating agents are fully bioavailable. These are organisms that are able to take up chelating agents and mineralize them (*23, 24*). These organisms have energy-dependent carriers for the uptake of the chelating agents that are specific to a certain chelating agent (*25*). The EDTA-carrier of bacterial strain DSM 9103 for example is specific to EDTA and to a lesser extent to DTPA (*25*). Other similar compounds such EDDS or NTA are not taken up by this strain. The uptake of the metal complexes is species dependent. CaNTA for example is the NTA-complex that is taken up by *Chelatobacter heintzii* (*26*). The bacterial strain DSM 9103 is able to take up free EDTA and metal-EDTA with low log K value (*25*) whereas the Zn, Ni, Cu and Fe(III) complexes are not taken up.

Interactions between Trace Metals, Chelating Agents and Higher Aquatic Organisms

In general it appears that also for higher organisms only the free metal ion can cross the biological membrane and that chelating agents reduce the toxicity and the tissue concentration of metals *(2)*. It has been demonstrated in many studies that differences in metal availability or toxicity for fish or other aquatic organisms in the presence of chelating agents are well explained on the basis of free metal ion activities. This has been observed e.g. for fish *(27-31)*, *Daphnia (32)*, tadpoles *(33)*, and crustaceans *(13, 34, 35)*.

However, some studies indicate that metal-chelating agent complexes are bioavailable under certain conditions and that an increase in the uptake of metals can occur: i) Complexes of divalent metals such as Cd, Zn or Cu are to some extent bioavailable to some organisms. Uptake greater than predicted based on the free metal ion concentration has been observed for CdEDTA with the common mussel *(36)*, for SrEDTA and SrNTA with the common carp *(37)* and for CuEDTA with fish *(38)*. These studies indicate that a direct uptake of metal-chelating agent complexes may take place. Other explanations are that on the biologically active surface (e.g. the fish gill) a dissociation of the complexes takes place, in some cases supported by mucus secretion *(38)*. ii) Fe(III)NTA was found to have the same toxicity as uncomplexed Fe(III) for *Daphnia (39)*. Fe(III)NTA is neutral and may therefore cross the cell membrane like a lipophilic compound. iii) Mercury complexes are also toxic to organisms. DTPA for example did not change the toxicity of Hg to *Daphnia (40)* and EDTA did not influence Hg-toxicity to yeast *(41)*. Higher toxicity of the Hg-complex than of inorganic Hg was found for EDTA with *Daphnia (40)* and for some fish species with NTA *(42)*. Whether this is due to the very strong binding of Hg to sulfur containing proteins that are able to out compete EDTA for Hg is not clear.

Interactions between Trace Metals, Chelating Agents and Higher Plants

Natural chelating agents

In order to maintain the concentration of essential metals within physiological limits and to minimize the detrimental effects of non-essential metals, plants, like all other organisms, have evolved a complex network of homeostatic mechanisms that serve to control the uptake, accumulation, trafficking and detoxification of metals. The reactivity and limited solubility of most metal ions means they require constant chelation once they are taken up into the cell. Metal ions are bound by chelators and chaperons. Chelators contribute to metal detoxification by buffering cytosolic metal concentrations and chaperons specifically deliver metal ions to organelles and metal-requiring proteins. Organic and amino acids present in the xylem sap in different concentrations, could serve as ligands for the cations transported in the transpiration stream *(43)*. Mugineic acid (MA), a plant phytosiderophore (PS), and nicotianamine (NA) are two structurally similar molecules synthesized by plants. Although both compounds have roles in the acquisition and transport of Fe, their species distribution and physiological functions are distinct. MA are only made by graminaceous monocotyledonous plants and are secreted from the roots of Fe deficient grasses to mobilize Fe from insoluble sources *(44)*, and the Fe(III)-PS complex is probably the form in which Fe is taken up by the roots of grasses *(45, 46)*. In contrast, NA is made by all plants and is present in various plant organs *(47, 48)* but is not secreted. It is thought to have a role in the internal transport of Fe and other metals *(47, 49, 50)*. Another group of plant synthesized compounds are the Phytochelatines (PCs), which are structurally related to glutathione *(51, 52)*. PC synthesis can be induced by a range of metal ions in both intact plants and plant cell cultures *(52)* and seems to have a role in heavy metal detoxification within the plant. It has been shown for tobacco plants that PC-Cd complexes are sequestered to the vacuole *(53)*.

Considerable debate exists concerning whether some plants are able to take up undissociated metal-siderophore complexes. It has been shown that equimolar ratios of ^{59}Fe to [^{14}C]phytosiderophores were taken up by maize plants indicating that the Fe(III)-phytosiderophore complexes entered the root undissociated *(46)*. Other studies *(45, 54)* also support the uptake of the entire complex.

Plants of either Strategy I or II have developed mechanisms for utilizing chelated Fe *(45, 55, 56)*, but they differ widely in their abilities to obtain Fe from either microbially produced siderophores *(55, 57)* or from plant produced siderophores *(45, 56, 58)*.

Because phytosiderophores form chelates not only with Fe(III) but also with zinc, copper and manganese, they therefore also mobilize other micronutrients from soils and make them more bioavailable for plants *(59-61)*. Phytosiderophores have even been proposed to facilitate the influx of all micronutrients *(62)*. Copper phytosiderophores were found to be taken up preferentially by Fe-deficient barley as opposed to Cu as CuEDTA *(63)*. Uptake studies with labeled metal-phytosiderophores indicated that barely is able to uptake phytosiderophore-complexed zinc and copper *(64, 65)* and that maize roots not only absorb Fe(III)-phytosiderophores, but also the Zn(II)-phytosiderophore *(66)*. Because uptake rates by iron-deficient barley roots from Fe(III)-phytosiderophores was found to be much higher than corresponding complexes with Cu(II), Zn(II), Co(II) and Co(III), two different transporter proteins were proposed for the metal-phytosiderophore complexes with only the Fe(III)-phytosiderophore to be specifically transported *(65)*.

New insights in plant mineral homeostasis were gained using an approach based on molecular genetics. The gene *Yellow stripe1* (YS1) was predicted to encode a Fe(III)-phytosiderophore transporter of maize plants based on the phenotype of *ys1* mutants lacking the ability to retrieve iron from Fe(III)-phytosiderophores. Transposon-tagging of the *Yellow Stripe1* (YS1) gene in maize allowed the identification of a membrane protein (ZmYS1) that mediates Fe(III)-phytosiderophore uptake *(67)*. YS1 was found to be a proton-coupled broad-range metal-phytosiderophore transporter *(68)* that additionally transports Fe(II)-nicotianamine *(68, 69)* and Ni-nicotianamine through membranes *(68)*. Yellow Stripe-Like (YSL) proteins were also found in the dicotyledonous plant *Arabidopsis*. Because phytosiderophores are neither made nor used by non-grass species like *Arabidopsis*, the role of the YSL proteins in Arabidopsis was suggested to be in the transport of metals complexed by the phytosiderophore-related compound nicotianamine *(67)*. Indeed, transport of Fe(II)- and Cu(II)-nicotianamine through membranes was shown for the *Arabidopsis* YSL2 *(70)*. As a possible physiological reason for the presence of strategy II transporter proteins in strategy I plants, a complimentary action to the predominant iron transporters or a role in heavy metal transport and detoxification was suggested *(68)*.

Anthropogenic chelating agents and plants

Plant uptake

Similar to microorganisms or aquatic animals, plant uptake of metals also shows a marked dependence on the chemical speciation of the metal in solution. It is noted that plant response also generally correlates best with the activity of the free, uncomplexed metal ion in solution *(71)*. Chelating agents are used in nutrient solution to buffer the free metal ion concentration similarly to algal culture media *(72)*. Chelator-buffered nutrient solutions are now used routinely for studying plant-micronutrient interactions (e.g. Fe, Cu, Zn, or Mn) at realistically low micronutrient concentrations, mimicking the situation occurring in soils *(73)*. In such a system, micronutrient deficiencies could be induced in a predictable manner. Concentrations that are normally used range from 25 to 100 μM. Uptake of the chelated metals is normally considered to play no role *(74)*. These systems are therefore believed to conform to the FIAM model.

However, there are numerous observations that chelating agents are taken up by plants. The research into chelating agents and plants started in 1951 when Jacobson found that Fe(III)EDTA is a good source of available iron to plants in nutrient solution *(75)*. Just a few years later it was shown by radiotracer studies that FeEDTA actually enters the plant, at least partially, as an intact complex *(76, 77)*. Not only Fe(III)EDTA but also the complexes with DTPA *(78)*, HEDTA *(78)*, CDTA *(79)* and EDDHA *(80)* are effective in enhancing iron uptake. However, it was only in 1998 that a direct analytical determination of an intact metal-EDTA complex inside the plant proved that the metal and the chelating agent form a complex *(81)*. The EDTA complexes of Mn, Zn, Al, Cd and Cu have been detected inside the plant using chromatography/ mass-spectrometry *(82, 83)*. The presence of PbEDTA in leaves was also confirmed with EXAFS measurements *(84)*.

Because in most of the studies that detected chelate uptake high concentrations of chelating agents were used (e.g. several hundred micromolar), it was speculated that the uptake of the complexes is due to physiological damage of the cell walls *(85)*. This damage and therefore the inadvertent uptake of chelants would not occur at the low concentrations used in conventional nutrient solutions or especially in the environment. However, apoplastic uptake of solution through the endodermis at the site of secondary root formation is a normal process in plants *(86)*. Uptake of chelants may therefore occur also at low concentration but is not detected due to analytical methods that are not sensitive enough.

Chelant-enhanced Phytoextraction

Most of the initial studies dealing with chelating agents and plants were done with the background of plant nutrition and therefore focused to a large part on the treatment of the deficiency of the essential nutrient metals Fe, Mn, Cu and Zn. The first published study on PbEDTA uptake was using this complex for a plant physiological study and has been used to follow water uptake and the water pathways in plants *(87)*. EDTA or DTPA-chelated radionuclides e.g. fission products such ^{91}Y, ^{144}Ce, and ^{210}Po *(88, 89)* and transuranics such as ^{241}Am *(90)* and ^{239}Pu *(91)* were also found to be taken up by plants.

First results with the uptake of common heavy metals from soils were obtained with Pb *(92, 93)*, Hg *(93)* and Cr *(89)* chelated to DTPA or EDTA. Increased uptake of the heavy metal was observed in most cases. Research on the influence of chelating agents on metal uptake by plants increased dramatically after the initial reports by Jorgensen *(94)* and Huang and Cunningham *(95)* that addition of EDTA and HEDTA to soils increased Pb concentration in crop plants to such an extent that they might be used for clean-up of Pb-contaminated soils. The following investigations showed that addition of chelating agents increases Pb accumulation in shoots of various plants by factors as high as 265 *(95-98)*. Increased uptake was not only observed in nutrient solution and in pot experiments but also in the field *(99, 100)*.

Nutrient solution experiments

In this section we will discuss experimental data from 4 hydroponics experiments conducted in the absence and presence of chelating agents. The experimental details can be seen below in Table 1. In all experiments the plants were grown for 3 weeks prior to the experimental time in nutrient solution. All the experiments were carried out in 0.1 strength Hoagland solution with the FeEDTA replaced by $FeSO_4$. At the start of the experimental time the nutrient solutions were modified as seen in Table 1. The time for all experiments was 6 days. All experiments were carried out in controlled environment with a 16 hour day of between 21 and 25 °C and an 8 hour night of between 15 and 16°C.

Table 1. Hydroponics experiments, experimental set up.

Exp.	Chelant	Plant	Metals	Extra Factors	Ref.
1	NTA (500 μM)	Sunflower	Cu (126 μM), Zn (61, 122, 245 μM)	+ P	unpub.
2	EDDS (500 μM)	Sunflower	Cu (126μ M), Zn (122 μM), Pb (126 μM)	– P	unpub.
3	NTA (500 μM)	Tobacco	Cu (126 μM)	+ P	*(101)*
4	NTA (500 μM)	Tobacco	Cu, Zn, Pb (100 μM)	± P	*(102)*

All plants were exposed to trace metals prior to the experimental time as they were grown in nutrient solution containing trace metals. Except for Figure 3, the total metal content of the roots and shoots was taken and the values for the control plants subtracted from this to compensate for trace metal exposure prior to the experiment and to obtain a true value of metals taken up during the experimental time. The total metal concentrations were different in different experiments. Free metal concentrations of the experimental solutions were calculated using solution metal concentrations and the speciation program CHEMEQL. The results were plotted to investigate if these results conformed to the FIAM model.

Figure 1 and 2 show free metals (Cu, Zn and Pb) in the hydroponics solutions from the experiments in Table 1 and how they relate to the shoot and root metal uptake. The results are depicted as the range of free metals (x axis) or shoot and root metal (y axis) over all the experiments. Treatments with chelating agents and in the absence of them are plotted separately as are results for the 3 different metals. It can be seen from Figure 1 that uptake of Cu, Zn and Pb into shoots did not follow the FIAM model, where uptake is correlated to free ion concentration. Metal uptake in the absence of chelating agents varied much more than in the presence of them, but is generally lower and not higher as expected by the FIAM.

Root uptake seems to follow in general the FIAM model and is related to free metal concentration (Figure 2). The presence of chelating agents caused a decrease in the free metal ion concentration which in turn reduced root uptake of Cu, Zn, and Pb to a large extent.

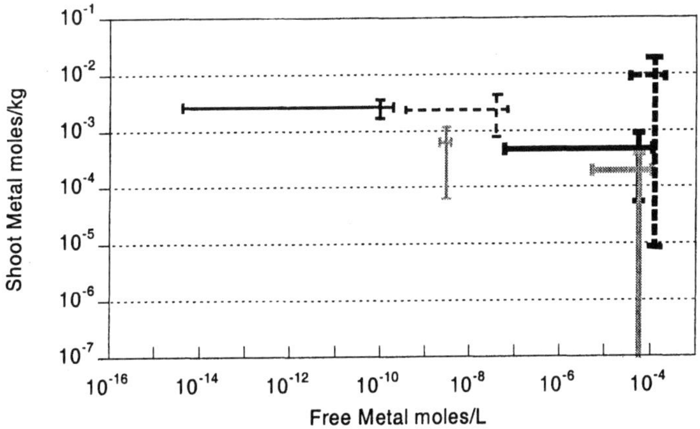

Figure 1. Shoot metal uptake from hydroponics experiments as a function of free metal concentration for all experiments shown in Table 1. Cu unchelated, black thick line; Cu chelated, black thin line; Zn unchelated, dashed black thick line; Zn chelated, dashed black thin line; Pb unchelated, grey thick line, Pb chelated, grey thin line. The lines show the range of observed concentrations.

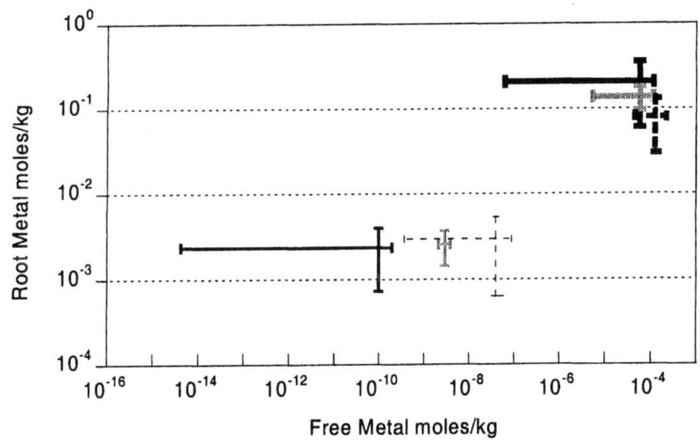

Figure 2. Root metal uptake from hydroponics experiments as a function of free metal concentration for all experiments shown in Table 1. Cu unchelated, black thick line; Cu chelated, black thin line; Zn unchelated, dashed black thick line; Zn chelated, dashed black thin line; Pb unchelated, grey thick line, Pb chelated, grey thin line. The lines show the range of observed concentrations.

The conformation of the root uptake of Cu, Zn and Pb to the FIAM model i.e. a relationship between root uptake and free metal concentration, can be explained by the nature of root metal uptake. In general most of the metals found in the roots are adsorbed to cation exchange sites on the external surfaces of the root and in the cell wall within the root. Negatively charged metal-chelant complexes will not be adsorbed and uptake is therefore related to the free ion concentration in solution around the roots.

Figure 3 a and b show the effect of competing Cu and metal chelation on Zn uptake and translocation into sunflower shoots. The highest amount of Zn was translocated in the unchelated Zn treatment without Cu competition. Although free Zn concentrations were similar in the treatments with and without competing Cu, and root Zn uptake was about the same as in the treatments without Cu (Figure 3b), only a small part of the free Zn was translocated into the shoots in the presence of Cu (Figure 3a). It seems that the plant reacted to high Cu concentrations by effectively blocking the highly efficient zinc transport system, while increasing Zn concentrations alone did not cause this response. A small amount of Zn was transported to the shoots despite the Cu induced blockage of the highly efficient Zn translocation system.

Although, much less Zn was taken up into the roots in presence of NTA (Fig. 3b), Zn was nevertheless translocated into the shoots in the same range as in the presence of Cu with and without NTA (Figure 3a). It may be that a certain amount of Zn must be bound to specific sites on the cell surfaces in the apoplast of roots in order to activate the high efficiency Zn specific transport system, and while Cu might outcompete Zn for these specific binding sites, NTA might prevented the binding of Zn to these specific sites due to the formation of negatively charged Zn-NTA complexes. Two different pathways for the uptake of Zn by maize roots have been proposed: a preferential uptake of the free Zn^{2+} and an additional uptake in form of Zn(II)-phytosiderophore complexes *(66)*.

According to Grusak et al. *(103)*, the control of certain processes, e.g. acquisition, distribution, utilization and storage of micronutrient ions may be coordinated at the whole-plant level via shoot-root communication. This coordination is important, because an individual organism has to ensure adequate nutrition of those minerals needed for essential functions, but it also must prevent excess accumulation of those micronutrient metals that can have deleterious effects on cellular processes such as high Cu concentrations.

In no case was a significant growth reduction or any other visible sign of metal or NTA toxicity found in the presence of NTA in these experiments. Indeed, NTA mitigated the toxicity of the accumulated metals on root and shoot growth (Figure 4). The chelated Zn treatments with and without Cu competition did not significantly differ in total dissolved Zn, free Zn or root / shoot dry weight. Therefore, only the data of the chelated Zn treatment with Cu competition is shown.

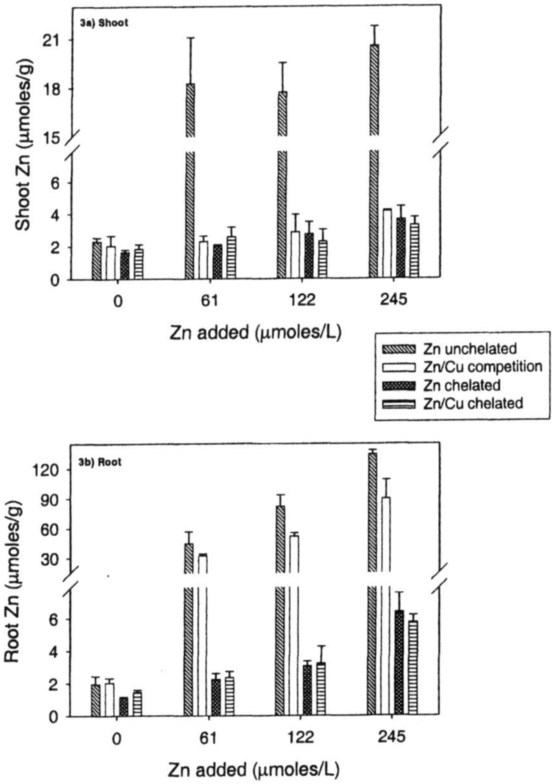

Figure 3. Zn uptake by sunflower shoots (3a) and roots (3b) under hydroponics conditions in presence or absence of NTA and competing Cu for various Zn treatments. Zn unchelated, Zn in competition with Cu (addition of 126 μM Cu), chelated Zn (addition of 500 μM NTA), Zn and Cu chelated (addition of 126 μM Cu and 500 μM NTA.

The lack of cohesion to the FIAM model for uptake of Cu, Zn and Pb into plant shoots supports the hypothesis that chelated metals can be taken up by plants and translocated to the above ground biomass. This is further backed up by the measurement of EDDS in sunflower shoots and xylem sap in experiment 2 (Figure 5). Calculations using EDDS concentrations measured in the xylem sap and loss of nutrient solution during the experiment (transpirational flow) also give values in the same range to that measured in the leaves giving rise to the theory that uptake could be through an apoplastic (passive) pathway.

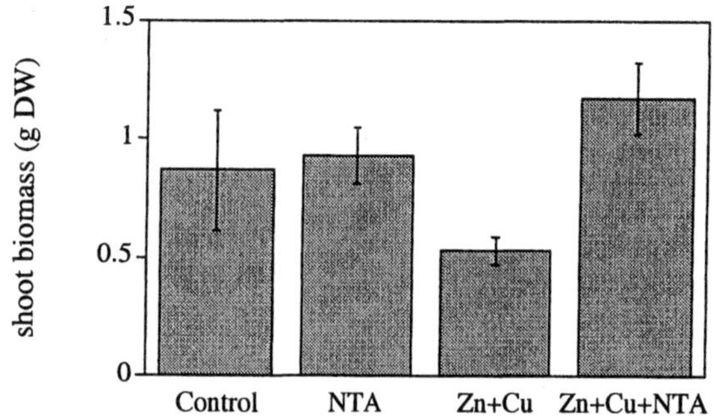

Figure 4: Influence of Zn+Cu and NTA on shoot biomass. 245 μM Zn+Cu, 500 μM NTA.

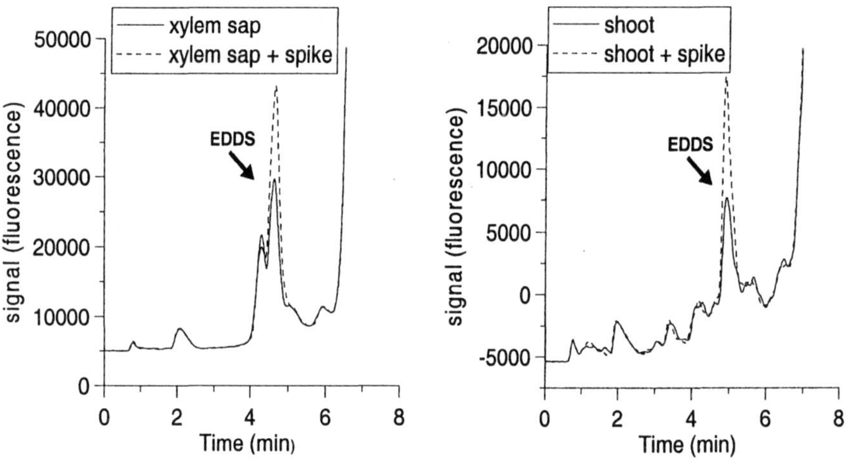

Figure 5. HPLC chromatograms showing EDDS in shoot extract (right) and xylem sap (left) after derivatization.

Uptake of chelating agents

There are two parallel transport pathways of ions (solutes) and water across the root cortex towards the stele: a pathway of passive transport through the apoplasm (cell walls and intercellular spaces), and a pathway of active transport from cell to cell in the symplasm. In general, the apoplasm of the root cortex is directly accessible to solutes from the external solution, although it does not provide a free space for movement of charged solutes. Within the cell wall, carboxylic groups and other moieties act as cation exchangers. Cations are preferentially bound to these sites, whereas anions are repelled. Cation binding in the apoplasm can significantly contribute to the total cation accumulation of roots *(104)*. The high metal concentrations found in the roots in the treatments without ligand may have resulted primarily from binding in the apoplasm of the root cortex, rather than from 'true' uptake into the root cells. The reduction of metal concentrations in the roots by addition of ligand may then simply be explained by the prevention of metal sorption to these cation exchange sites in the apoplast due to formation of negatively charged metal complexes. Indeed, the retention of unchelated Pb and unchelated Cu in cell walls of roots, particularly around intercellular spaces was observed in an ultra-structural study using transmission electron microscopy for Pb *(105)* and for Cu by means of EDX microanalysis (energy dispersive x-ray microanalysis) *(106)*. The main barrier against passive movement into the stele via the apoplastic pathway is the endodermis, the innermost layer of cells of the cortex. In the radial and transverse walls of the endodermis, hydrophobic incrustations (suberin), i.e. the Casparian band, constitute an effective barrier against passive solute transfer into the stele. The endodermis is not a perfect barrier for apoplastic transport, however. In addition to passage cells, there appear to be sites distributed along the root axis where this barrier is 'leaky'. At the root apex for instance the Casparian strip is not yet fully developed and, thus, allows apoplastic transport to reach the stele *(107)*. Crowdy and Tanton *(87)* showed that the uptake of Pb-EDTA was restricted to a region between 3 and 140 mm behind the root tip where the suberization of the cell walls had not yet occurred. This pathway would allow metal-chelates to by-pass the impermeable Casparian strip through the apoplast. To some degree, higher plants take up nutrients indiscriminately and presumably with little selectivity as to free or chelated metals. This occurs through breaks in the endodermis where primordial lateral roots penetrate. There, soil or nutrient solution can spill into the xylem and phloem vascular bundle without the usual barrier at the Casparian strip *(86, 108)*. In limited nutrient supply, where competition with chelators would inhibit the normal loading of the plants own carriers or would reduce Fe(III) to Fe(II) reduction, the uptake of micronutrients via breaks in the endodermis might be more important in the overall uptake of micronutrients. Breaks in the endodermis promoting indiscriminate uptake have also been found for widely different plants such as maize (*Zea mays*) *(109)*, broad bean (*Vicia faba*) *(86)*, and Norway spruce (*Picea abies*) *(108)*.

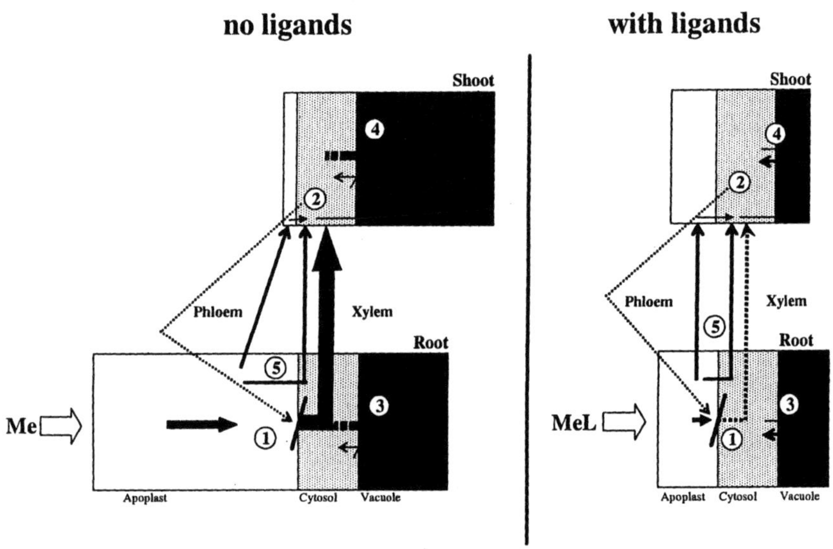

Figure 6. Conceptual model for micronutrient uptake and distribution within plants: without and with addition of ligands. (1) specific acquisition/uptake site for free metal ions; (2) communication of shoot micronutrient status via signal molecules; (3) storage / waste disposal of abundant metal in root vacuoles; (4) storage / waste disposal of metals in shoot vacuoles; (5) proposed additional pathway via apoplast. The size of the boxes indicates the size of the pools, the thickness of the arrows the size of the metal flux.

Aminopolycarboxylic acids (APCAs) of biological origin, e.g. rhizobactin, rhizoferrin, staphyloferrin (bacteria origin), nicotianamine, mugineic acid, avenic acid (plant origin), are mainly involved in metal acquisition by organisms. The identification of the Yellow stripe (YS) and the Yellow stripe like (YSL) transporter protein family in cell membranes opens now the field for speculation about metal-ligand transport through membranes. It has been shown that these transporters were not metal specific and do in fact transport other metals than iron, even so they preferentially transport iron under iron deficient conditions *(68-70)*. Instead, it has been proposed that this transporter family is ligand specific *(65, 66)*, and indeed, Fe(II)- and Fe(III)-EDTA were only transported to a negligible amount by these transporters *(68)*.

Based on our experiments and on extensive literature studies we suggest the following conceptual model for the effect of ligands on metal uptake by plants (Figure 6). For micronutrients required by the plant, an active high-affinity transport system for free metal ions (1) can be activated and regulated by the plant (2). Besides this effective and specific uptake and transport system, there might also be an additional pathway for micronutrients and eventually also for other solutes. This additional pathway might be fully apoplastic via breakages in the endodermis allowing a broad spectrum of solutes to cross the membrane at these sites (5).

General discussion

Uptake of metal chelates as it is proposed in this paper, is difficult to detect, because compared to the total concentration in the nutrient solution, the uptake via an apoplastic pathway or the uptake as a consequence of co-transport phenomena of the chelate-metal anion via non-specific uptake sites would have little or no effect on the concentration of the metal in solution. In fact, uptake via a different system might also be overlooked, because a high-affinity system super-imposes on the less effective low-affinity system. The existence of two kinetically distinct uptake components, a high-affinity system exhibiting Michaelis-Menten kinetics and a linear component has been described for iron *(46)*. Saturation kinetics in ion absorption of roots, tissues or single cells generally indicates the involvement of an active carrier- or channel-mediated transport system *(110)*. Biphasic kinetics are also known from uptake of other mineral nutrients. In nitrate uptake for example, the linear component was attributed to the activity of a constitutive low-affinity transport system that occurs independently of a high-affinity transport system *(111)*. Plants, which cannot change their location, might strongly depend on different transport systems to maintain the concentration of essential metals within physiological limits.

Several studies cited in this paper concluded that chelates were not bioavailable, because an amelioration of metal toxicity due to the addition of a chelating agent was observed. But, the fact that most chelating agents indeed mitigate metal toxicity does not exclude the uptake and sequestration of the intact metal-complex into a storage or waste compartment within the plant (e.g. vacuoles). Chelation and sequestration of metals in the vacuoles are natural plant defense mechanisms. Measuring only the expression of a toxicity response by a single organism without considering food chain pathways might therefore not be enough to determine the bioavailability of a metal. Chelants probably mitigate the toxicity due to the fact that the metal complex is not the form that is "recognized" by the plants own transporters (chaperons) and is therefore, as long as the metal-complex remains intact, not bioavailable for plant organelles or its metal-requiring proteins.

The exceptions to the FIAM reported in this paper, however, shall not infuse pessimism that accurate speciation of trace metals in soil solution might not be a useful tool to improve our ability to predict phytoavailability, toxicity and food-chain transfer and a better assessment of risks associated with metals in soils. Accurate speciation will always be necessary for our understanding of the chemical mechanisms in uptake and transport of metals and chelates. Our results suggest that some assumptions of the FIAM have to be reconsidered and adapted to plant physiology and the molecular mechanism of plant metal tolerance and homeostasis, e.g. the membrane surface is most probably not chemically homogeneous as it is assumed by the FIAM, and opposite to the underpinning assumption of the FIAM, it seems very likely that plants are able to significantly modify their plasma membrane, and even more importantly, our observations in conjunction with those of other authors e.g. *(8, 9, 68, 83, 84)* suggest that intact chelates can be taken up by plants.

References

(1) Morel, F. M. M. *Principles of Aquatic Chemistry*; Wiley-Interscience: New York, 1983.

(2) Campbell, P. G. C. In *Metal speciation and bioavailability in aquatic systems*; Turner, D. R., Ed.; John Wiley and Sons, New York, 1995; pp 45-102.

(3) Phinney, J. T.; Bruland, K. W. *Environ. Sci. Technol.* **1994**, *28*, 1781-1790.

(4) Errécalde, O.; Seidl, M.; Campbell, P. G. C. *Water Res.* **1998**, *32*, 419-429.

(5) Errécalde, O.; Campbell, P. G. C. *Journal of Phycology* **2000**, *36*, 473-483.

(6) Wilkinson, K. J.; Bertsch, P. M.; Jagoe, C. H.; Campbell, P. G. C. *Environ. Sci. Technol* **1993**, *27*, 1132-1138.

(7) Smolders, E.; McLaughlin, M. J. *Plant Soil* **1996**, *179*, 57-64.

(8) Parker, D. R.; Pedler, J. F.; Ahnstrom, Z. A. S.; Resketo, M. *Environ. Toxicol. Chem.* **2001**, *20*, 899-906.

(9) Mclaughlin, M. J.; Smolders, E.; Merckx, R.; Maes, A. In *Plant Nutrition for Sustainable Food Production and Environment*; Ando, T., al., e., Eds.; Kluwer: Dordrecht, The Netherlands, 1997; pp 113-118.

(10) Bell, P. F.; Chaney, R. L.; Angle, J. S. *Plant Soil* **1991**, *130*, 51-62.

(11) Hutner, S. H.; Provasoli, L.; Schatz, A.; Haskins, C. P. *Proc. Am. Phil. Soc.* **1950**, *94*, 152-170.

(12) Manahan, S. E.; Smith, M. J. *Environ. Sci. Technol* **1973**, *7*, 829-833.

(13) Sunda, W. G.; Engel, D. W.; Thuotte, R. M. *Environ. Sci. Technol* **1978**, *12*, 409-413.

(14) Sunda, W. G.; Guillard, R. R. L. *J. Mar. Res.* **1976**, *34*, 511-529.

(15) Jackson, G. A.; Morgan, J. J. *Limnol. Oceanogr.* **1978**, *23*, 268-282.

(16) Brand, L. E.; Sunda, W. G.; Guillard, R. R. L. *Limnol. Oceanogr.* **1983**, *28*, 1182-1198.

(17) Knauer, K.; Behra, R.; Sigg, L. *Environ. Toxicol. Chem.* **1997**, *16*, 220-229.

(18) Johnston, R. *J. Mar. Biol. Ass. U.K.* **1962**, *43*, 427-456.

(19) Yentsch, C. M.; Yentsch, C. S.; Owen, C.; Salvaggio, M. *Environ. Lett.* **1974**, *6*, 231-238.

(20) Eklund, B.; Bruno, E.; Lithner, G.; Borg, H. *Environ. Toxicol. Chem.* **2002**, *21*, 1040-1051.

(21) Lovley, D. R.; Woodward, J. C.; Chapelle, F. H. *Nature* **1994**, *370*, 128-131.

(22) Lovley, D. R.; Woodward, J. C. *Chem. Geol.* **1996**, *132*, 19-24.

(23) Egli, T. *J. Biosci. Bioeng.* **2001**, *92*, 89-97.

(24) Bucheli-Witschel, M.; Egli, T. *FEMS Microbiol. Rev.* **2001**, *25*, 69-106.

(25) Witschel, M.; Egli, T.; Zehnder, A. J. B.; Wehrli, E.; Spycher, M. *Microbiol.* **1999**, *145*, 973-983.

(26) Vanbriesen, J. M.; Rittmann, B. E.; Xun, L.; Girvin, D. C.; Bolton, H. *Environ. Sci. Technol.* **2000**, *34*, 3346-3353.

(27) Pärt, P.; Wikmark, G. *Aquat. Toxicol.* **1984**, *5*, 277-289.

(28) Muramoto, S. *J. Environ. Sci. Health A* **1982**, *17*, 313-319.

(29) Shaw, T. L.; Brown, V. M. *Water Res.* **1974**, *8*, 377-382.

(30) Khangarot, B. S. *Current Science* **1981**, *50*, 246-248.

(31) van Ginneken, L.; Chowdhury, M. J.; Blust, R. *Environ. Toxicol. Chem.* **1999**, *18*, 2295-2304.

(32) Tevlin, M. P. *Water Res.* **1978**, *12*, 1027-1034.

(33) Khangarot, B. S.; Sehgal, A.; Bhasin, M. K. *Acta Hydrochim. Hydrobiol.* **1985**, *13*, 121-125.

(34) Sunda, W. G.; Tester, P. A.; Huntsman, S. A. *Mar. Biol.* **1987**, *94*, 203-210.

(35) Daly, H. R.; Campbell, I. C.; Hart, B. T. *Environ. Toxicol. Chem.* **1990**, *9*, 997-1006.

(36) George, S. G.; Coombs, T. L. *Mar. Biol.* **1977**, *39*, 261-268.

(37) Chowdhury, M. J.; Blust, R. *Aquat. Toxicol.* **2002**, *58*, 215-227.

(38) Tao, S.; Long, A.; Pan, B.; Xu, F.; Dawson, R. *Ecotoxicol. Environ. Safety* **2002**, *53*, 317-322.

(39) Biesinger, K. E.; Andrew, R. W.; Arthur, J. W. *J. Fish. Board Can.* **1974**, *31*, 486-490.

(40) Sorvari, J.; Sillanpää, M. *Chemosphere* **1996**, *33*, 1119-1127.

(41) Kungolos, A.; Aoyama, I.; Muramoto, S. *Ecotoxicol. Environ. Safety* **1999**, *43*, 149-155.

(42) Eisler, R.; Gardner, G. R.; Hennekey, R. J.; LaRoche, G.; Walsh, D. F.; Yevich, P. P. *Water Res.* **1972**, *6*, 1009-1027.

(43) Senden, M. H. M. N.; Wolterbeek, H. T. *Acta Botanica Neerlandica* **1990**, *39*, 297-303.

(44) Sugiura, Y.; Nomoto, K. *Struct Bonding* **1984**, *58*, 107-135.

(45) Römheld, V.; Marschner, H. *Plant Physiol.* **1986**, *80*, 175-180.

(46) von Wirén, N.; Marschner, H.; Römheld, V. *Physiol. Plant* **1995**, *93*, 611-616.

(47) Stephan, U. W.; Schmidke, I.; Pich, A. *Plant Soil* **1994**, *165*, 181-188.

(48) Walter, A.; Pich, A.; Scholz, G.; Marschner, H.; Römheld, V. *J. Plant. Nutr.* **1995**, *18*, 1577-1593.

(49) Stephan, U. W.; Schmidke, I.; Stephan, V. W.; Scholz, G. *Biometals* **1996**, *9*, 84-90.

(50) Pich, A.; Scholz, I. *J. Exp. Bot.* **1996**, *47*, 41-47.

(51) Cobbett, C. S. *Plant Physiol.* **2000**, *123*, 825-833.

(52) Rauser, W. E. *Plant Physiol.* **1995**, *109*, 1141-1149.

(53) Voegeli-Lange, R.; Wagner, G. J. *Plant Physiol.* **1990**, *92*, 1086-1093.

(54) Mori, S.; Nishizawa, N.; Hayashi, H.; Chino, M.; Yoshimura, E.; Ishihara, J. In *Iron Nutrition and Interactions in Plants*; Chen, Y., Hadar, Y., Eds.; Kluwer Academic: Dordrecht, The Netherlands, 1991; Vol. 130, pp 175-188.

(55) Cline, G. R.; Reid, C. P. P.; Powell, P. E.; Szansizlo, P. J. *Plant Physiol.* **1984**, *76*, 36-40.

(56) Takagi, S.; Nomoto, K.; Takemoto, T. *J. Plant. Nutr.* **1984**, *7*, 469-477.

(57) Miller, G. W.; Pushnik, J. C.; Brown, J. C.; Emery, T. E.; Jolley, V. D.; Warnik, K. Y. *J. Plant. Nutr.* **1985**, *8*, 249-264.

(58) Jolley, V. D.; Brown, J. C. *J. Plant. Nutr.* **1991**, *14*, 45-58.

(59) Treeby, M.; Marschner, H.; Römheld, V. *Plant Soil* **1989**, *114*, 217-226.

(60) Awad, F.; Römheld, V. *Journal of Plant Nutrition* **2000**, *23*, 1847-1855.

(61) Shenker, M.; Fan, T. W.-M.; Crowley, D. E. *J Environ Qual* **2001**, *30*, 2091-2098.

(62) Crowley, D. E.; Reid, C. P. P.; Szaniszlo, P. J. In *Iron Transport in Microbes, Plants and Animals*; Winkelmann, G., Van der Helm, D., Neilands, J. B., Eds.; VCH Publishing: New York, 1987; pp 375-386.

(63) Marschner, H.; Treeby, M.; Römheld, V. *Z.Pflanzenernaehr.Bodenk.* **1989**, *152*, 197-204.

(64) Zhang, F.-S.; Römheld, V.; Marschner, H. *Soil Sci. Plant Nutr.* **1991**, *37*, 671-678.

(65) Ma, J. F.; Kusano, G.; Kimura, S.; Nomoto, K. *Phytochemistry* **1993**, *34*, 599-603.

(66) von Wirén, N.; Marschner, H.; Römheld, V. *Plant Physiol.* **1996**, *111*, 1119-1125.

(67) Curie, C.; Panaviene, Z.; Loulergue, C.; Dellaporta, S. L.; Briat, J.-F.; Walker, E. L. *Nature* **2001**, *409*, 346-349.

(68) Schaaf, G.; Ludewig, U.; Erenoglu, B. E.; Mori, S.; Kitahara, T.; von Wirén, N. *The Journal of Biological Chemistry* **2004**, *279*, 9091-9096.

(69) Roberts, L. A.; Pierson, A. J.; Panaviene, Z.; Walker, E. L. *Plant Physiol.* **2004**, *135*, 112-120.

(70) DiDonato, R. J.; Roberts, L. A.; Sanderson, T.; Bosler Eisley, R.; Walker, E. L. *The Plant Journal* **2004**, *39*, 403-414.

(71) Parker, D. R.; Pedler, J. F. *Plant and Soil* **1997**, *196*, 223-228.

(72) Chaney, R. L.; Bell, P. F.; Coulombe, B. A. *HortScience* **1989**, *24*, 565-572.

(73) Parker, D. R.; Chaney, R. L.; Norvell, W. A. In *Chemical equilibrium and reaction models*; Goldberg, S., Ed.; SSSA, 1995; Vol. 42, pp 163-200.

(74) Halvorson, A. D.; Lindsay, W. L. *Soil Sci. Soc. Am. J.* **1977**, *41*, 531-534.

(75) Jacobson, L. *Plant Physiol.* **1951**, *26*, 411-413.

(76) Wallace, A.; North, C. P.; Mueller, R. T.; Shannon, L. M.; Hemaidan, N. *Proc. Am. Soc. Hort. Sci.* **1955**, *65*, 9-16.

(77) Stewart, I. *Ann. Rev. Plant Physiol.* **1963**, *14*, 295-310.

(78) Wallace, A.; North, C. P.; Mueller, R. T.; Hemaidan, N. *Proc. Am. Soc. Hort. Sci.* **1953**, *62*, 116-118.

(79) Wallace, A.; Mueller, R. T.; Lunt, O. R.; Ashcroft, R. T.; Shannon, L. M. *Soil Sci.* **1955**, *80*, 101-108.

(80) Tiffin, L. O.; Brown, J. C.; Krauss, R. W. *Plant Physiol.* **1959**, *35*, 362-367.

(81) Vassil, A. D.; Kapulnik, Y.; Raskin, I.; Salt, D. E. *Plant Physiol.* **1998**, *117*, 447-453.

(82) Collins, R. N.; Onisko, B. C.; Mclaughlin, M. J.; Merrington, G. *Environ. Sci. Technol.* **2001**, *35*, 2589-2593.

(83) Collins, R. N.; Merrington, G.; Mclaughlin, M. J.; Knudsen, C. *Environ. Toxicol. Chem.* **2002**, *21*, 1940-1945.

(84) Sarret, G.; Vangronsveld, J.; Manceau, A.; Musso, M.; D'Haen, J.; Menthonnex, J. J.; Hazemann, J. L. *Environ. Sci. Technol.* **2001**, *35*, 2854-2859.

(85) Rengel, Z. *Plant and Soil* **1999**, *215*, 193-202.

(86) Peterson, C. A.; Emanuel, M. E.; Humphreys, G. B. *Can. J. Bot.* **1981**, *59*, 618-625.

(87) Crowdy, S. H.; Tanton, T. W. *J. Exp. Bot.* **1970**, *21*, 102-111.

(88) Essington, E.; Nishita, H.; Wallace, A. *Soil Sci.* **1963**, *95*, 331-337.

(89) Athalye, V. V.; Ramachandran, V.; D'Souza, T. J. *Environ. Pollut.* **1995**, *89*, 47-53.

(90) Wallace, A. *Health Phys.* **1972**, *22*, 559-562.

(91) Ballou, J. E.; Price, K. R.; Gies, R. A.; Doctor, P. G. *Health Phys.* **1978**, *34*, 445-450.

(92) Patel, P. M.; Wallace, A.; Romney, E. M. *Comm. Soil Sci. Plant Anal.* **1977**, *8*, 733-740.

(93) Hale, V. Q.; Wallace, A. *Soil Sci.* **1970**, *109*, 262-263.

(94) Jorgensen, S. E. *Ecol. Eng.* **1993**, *2*, 89-100.

(95) Huang, J. W.; Cunningham, S. D. *New Phytol.* **1996**, *134*, 75-84.

(96) Blaylock, M. J.; Salt, D. E.; Dushenkov, S.; Zakharova, O.; Gussman, C.; Kapulnik, Y.; Ensley, B. D.; Raskin, I. *Environ. Sci. Technol.* **1997**, *31*, 860-865.

(97) Huang, J. W.; Chen, J.; Berti, W. R.; Cunningham, S. D. *Environ. Sci. Technol.* **1997**, *31*, 800-805.

(98) Epstein, A. L.; Gussman, C. D.; Blaylock, M. J.; Yermiyahu, U.; Huang, J. W.; Kapulnik, Y.; Orser, C. S. *Plant Soil* **1999**, *208*, 87-94.

(99) Liphadzi, M. S.; Kirkham, M. B.; Mankin, K. R.; Paulsen, G. M. *Plant Soil* **2003**, *257*, 171-182.

(100) Kayser, A.; Wenger, K.; Keller, A.; Attinger, W.; Felix, H. R.; Gupta, S. K.; Schulin, R. *Environ. Sci. Technol.* **2000**, *34*, 1778-1783.

(101) Wenger, K.; Gupta, S. K.; Furrer, G.; Schulin, R. *J. Environ. Qual.* **2003**, *32*, 1669-1676.

(102) Wenger, K.; Gupta, S. K.; Schulin, R. In *Bioavailability of Soil Pollutants*; Naidu, R., Ed., 2004; Vol. I.

(103) Grusak, M. A.; Pearson, J. N.; Marentes, E. *Field Crops Research* **1999**, *60*, 41-56.

(104) Marschner, H. *Mineral nutrition of higher plants*; Academic Press: London, 1995.

(105) Jarvis, M. D.; Leung, D. W. M. *Plant Sci.* **2001**, *161*, 433-441.

(106) Jung, C.; Maeder, V.; Funk, F.; Frey, B.; Sticher, H.; Frossard, E. *Plant Soil* **2003**, *252*, 301-312.

(107) Huang, C. X.; Van Steveninck, R. F. M. *J. Plant Physiol.* **1989**, *135*, 554-558.

(108) Häussling, M.; Jorns, C. A.; Lehmbecker, G.; Hecht-Buchholz, C.; Marschner, H. *J. Plant Physiol.* **1988**, *133*, 486-491.

(109) Rasmussen, H. P. *Planta* **1968**, *81*, 28-37.

(110) Stein, W. D. *Channels, carriers and pumps: An introduction to membrane transport*; Academic Press: San Diego, 1990.

(111) Glass, A. D. M.; Shaff, J. E.; Kochian, L. V. *Plant Physiol.* **1992**, *99*, 456-463.

Transport and Fate

Chapter 13

Distribution and Fate of Chelating Agents in the Environment

Mika E. T. Sillanpää

Department of Environmental Sciences, University of Kuopio, P.O. Box 181, 50101 Mikkeli, Finland

This chapter discusses the behaviour of the chelating agents EDTA and DTPA in the receiving aquatic environment. EDTA and DTPA are used in high amounts in different industrial and household applications and this has raised concern about their ultimate fate in the environment. These compounds are not expected to cause direct ecotoxicological effects at the levels typically found in natural waters. However, they do contain nitrogen and have the capability to affect metal balance in aquatic ecosystems.

Ethylenediaminetetraacetic acid (EDTA) and diethylenetriaminepentaacetic acid (DTPA) are synthetic complexing agents that have been utilized extensively as metal sequestrants by a wide variety of industries. EDTA has been used mainly as a chelating agent, e.g., in the metal, rubber, leather, photographic, textile, pulp and paper, pharmaceutic, cosmetic, and food industries. Applications of DTPA include the inaction of metal ions in the pulp and paper industry and use as a drug in heavy metal poisoning. The huge amounts of these compounds used in industrial applications as well as in households and agriculture have raised concern about their ultimate release to the aquatic environment. The production of EDTA in Europe was about 32,740 metric tons in 1998 (1). The sales quantity of DTPA was 14,000 tons in 1997 (2). EDTA and DTPA are expected to be released to aquatic environments due to their relatively low degradability. This chapter discusses the possible environmental impacts that these compounds may have in receiving aquatic environments.

Environmental Occurrence

EDTA is commonly found in natural waters, typical concentrations range from few μg/L up to 100 μg/L (3-5). Thus, EDTA is classified as one of the major organic anthropogenic pollutants in Central Europe. DTPA has received much less attention. However, DTPA is found at concentrations between 2 and 15 μg/L in the river Rhine, Germany (6) and between 9 and 18 μg/L in a Finnish lake near a pulp mill (7).

Chemical equilibrium calculations are complicated by the presence of competing natural and anthropogenic ligands, concentrations of different cations, and suspended particles and sediments. The initial input of EDTA and DTPA also plays a significant role. EDTA is known to react with metal cations at a molar ration of 1:1. Because of slow dissociation, the proportion of Fe(III)-EDTA cannot be assessed by means of calculated equilibrium speciation (8). It has been proven that exchange reactions do occur, in the presence of excess of Ca over trace metal, under natural aquatic conditions (9). Fe-EDTA also converts slowly into Zn thermodynamically favorable species under investigated conditions (8). However, it was also reported that, under anoxic conditions, the reduction of Fe could lead to increased exchanged reaction. Ca and Mg chelates were found to be the dominant EDTA species in sea water (10). Metal-DTPA interactions in natural waters has recently been studied (11). The detailed study on the speciation of complexing agents is presented elsewhere in this book.

Adsorption of free EDTA onto metal oxides has been demonstrated (12,13). EDTA was negligibly sorbed to clay minerals and sediments. Adsorption of various metal-EDTA complexes onto aluminum oxides and crystalline and amorphous iron oxides as well as onto goethite has also been investigated (14,15). It was reported that the maximum adsorption capacity of Fe-EDTA chelated and that several adsorption mechanisms were revealed, depending on the structure of divalent and trivalent metal-EDTA chelates (14). The adsorption behavior of DTPA and its metal complexes has received little research attention.

Degradation

Knowledge of the rate of degradation of pollutants by both biotic and abiotic processes when released to natural water is a key issue when estimating their fate. From the ecological point of view, one of the most important properties of environmentally compatible organic chemicals is their biological degradability. Biodegradation is a primary means of organic compound removal in the environment (16). The biological degradation has received recent attention and is discussed extensively elsewhere in this book. However, it can be concluded here that EDTA was long considered non-biodegradable, but recent studies show its ability to be oxidized by microorganisms (2).

Photochemical transformation is strongly dependent on natural conditions. Thus, the results obtained at maximum conditions in the laboratory cannot directly be applied to natural waters, where several factors may have an impact, when estimating the half-lives of pollutants. In natural waters, sunlight intensity is attenuated through adsorption and scattering. The photolysis rates of pollutants in pure water, seawater, and inland water and their dependence on conditions such as season, latitude, time of day, depth, ozone layer thickness and light attenuation are convincingly discussed in Reference (17).

There exsists wide-spread unanimity concerning the photolability of the Fe-EDTA complex (18-22). Also, Fe-DTPA was found to be photolabile (). Free DTPA is shown to be much more photodegradable than free EDTA (23). The possible interactions of EDTA, DTPA, and their metal complexes with sediments and, perhaps more importantly, suspended matter were not considered in the foregoing studies; therefore, the results apply only when in solution.

It has been reported that Fe-EDTA is the only environmentally relevant EDTA species that undergoes direct photolysis (24,25). It was concluded that in shallow rivers Fe-EDTA is rapidly photodegraded in summer, but the rate may be decreased in lakes because of additional light attenuation. It was also sugguested that the EDTA found in the North and Black Seas consists of species that are resistant to photochemical and biological decomposition.

As indicated, photolytic degradation may be a remarkable pathway to prevent the environmental accumulation of EDTA and DTPA. These compounds are already released to receiving waters as iron chelates to some extent. Also natural waters contain ferric ions and thus other metal complexes might be converted into Fe complexes in due course. On the other hand, iron in natural waters is present in its colloidal and amorphous iron hydroxides and thus may not be complexed by EDTA (18). Also the slow dissociation of Fe-EDTA in natural waters has been convincingly reported (8). Laboratory conditions do not directly apply to natural aquatic environment. At a minimum, the absorption of UV light by solids, plants, and macrophytes, and the daily and annual periodicity of light, cloudiness, ice cover, and shadowing effects, should be taken into consideration. Also, the intensity of sunlight decreases with angular height of the sun, from the tropics to higher latitudes. Thus, it is extremely difficult to estimate the amount of available light within the waters column. Actually, photodegradation might be a significant factor in EDTA decomposition, because, in the case of water environments, little or no light is received except for immediate surfaces (22,26).

Other abiotic processes than photodegradation are not suggested to contribute significantly to EDTA degradation (18). EDTA and DTPA are rather stable in industrial processes (27). Environmentally relevant degradation processes of EDTA have been discussed (24). It was concluded that hydrolysis, reaction with solvated electrons, organic peroxyradicals, singlet oxygen, and hydroxyl radicals are irrelevant as transformation pathways for EDTA and its metal complexes in river water that have short residence time. However, they reported that chemical degradation, excluding direct photolysis of the Fe-EDTA, may have some significance in the long-term behavior of EDTA and its complexes. It is expected that these conclusions apply to DTPA as well.

Ecological Risks

Eutrophic Effects

Algal growth in natural waters is dependent on the availability of nitrogen and phosphorus. Nitrogen is often the limiting factor in the marine environment, while phosphorus usually limits growth in limnic systems (28). When these nutrients are present in excess, algal growth can be inhibited also by the lack of essential trace elements such as Cu, Zn, and Fe that are indispensable for photosynthetic production. The importance of iron as a growth-limiting factor in offshore areas has been emphasized (29).

EDTA and DTPA contain about 10% nitrogen, and thus have, particularly once mineralized, the potential to contribute to eutrophication. As significant amounts of these compounds are used in several industries as well as in agriculture and households and cannot effectively be removed in waste water treatment plants, the role of EDTA and DTPA as a nitrogen source for algae growth is one of the main concerns in assessing their environmental impact. For example, EDTA has been found to be necessary for the growth of *Microcystis aeuginosa* (30). The addition of EDTA also enhanced the growth of *Dunadiella* and *Amphidinium* (28).The nitrogen in EDTA and DTPA cannot likely be used directly by algae; I, e., degradation of the molecule is required. As indicated, under certain circumstances either biological or photochemical degradation might occur, causing nitrogen to be present in bioavailable form.

Inorganic phosphates are present in sediments mainly in their Fe- and Ca- bound forms (31). EDTA and DTPA are known to form the most stable complexes with trivatent iron. Therefore, EDTA and DTPA may be able to desorb Fe form sediments, remobilizing phosphates into natural waters and thus, indirectly, enhancing eutrophication. In summary, data are scarce and the knowledge of the direct influence of EDTA and DTPA on algal growth is rather limited.

Direct Ecotoxicological Effects

Daphnia, being an important food source for vertebrate and invertebrate predators, was used in many of the toxic studies. EDTA and DTPA seem to be relatively nontoxic, as shown by acute exposure (32-37). However, in general, little is known of their chronic toxicity, which is of importance because these agents are likely to be rather persistent in the aquatic environment and also because acute toxicity does not take into account the impact of sustained exposure during the entire life cycle. The toxicity of EDTA and DTPA is strongly dependent on their chemical speciation, by many orders of magnitude in some cases (34). Thus the cations present and their concentrations might partially explain the differences in toxicological response. Naturally, species also have different sensitivities toward chemicals.

It was reported that mortality of *Daphnia carinata* increased significantly at 100 mg/L of DTPA in 6 d, while significant decreases were observed of 50 mg/ L and 10 mg/L, respectively (36). On the other hand, it is reported that deinking plant effluent containing 113 mg/L of DTPA caused no mortality in *Melanotaenia fluviatilis* and had no impact on egg hatchability, growth or larval mortality after 14 d (38,39). The mechanism of action for toxicity may be the depletion of essential trace metals by DTPA, thus possibly leading to trace metal imbalances and nutritional deficiencies (36). It was found that chronic exposure to 10 mg/L DTPA significantly decreased brood size and the cumulative number of offsping per adult in *Daphnia carinata* (37). It was also observed that the toxicity of Fe-DTPA was noticeably lower than that of DTPA and concluded that direct toxicity of DTPA can occur only under extreme conditions.

It can be summarized that the concentrations of EDTA and DTPA in receiving waters are generally at least two orders of magnitude below the levels for direct ecotoxic effects. Therefore, it is not expected that direct, acute toxic effects occur in aquatic organisms. However, the possible long-term ecotoxicity remains unknown.

Desorption of Heavy Metals

The ultimate release of EDTA and DTPA, being hydrophilic and extremely strong chelating agents for metals, to the aquatic environment raises concern about their capability to remobilize heavy metal from soils and sediments, thus causing secondary pollution. There are several studies concerning the influence of EDTA on heavy metal balance in natural waters (40-46).

DTPA has received less attention than EDTA as a metal extractant. However, it is reported that both EDTA and DTPA were able to desorb all metals investigated from diverse soil materials (46). DTPA was found to have more effective remobilizing capacity than EDTA, which is logical as DTPA forms stronger complexes with metals.

Heavy metal remobilization capacity of complexing agents is discussed elsewhere in this book, but EDTA and DTPA are expected to be able to remobilize heavy metals in soils and sediments to some extent, depending on metal and chelate concentrations, pH, and the presence of other competing complexing agents and cations, whether anthropogenic or natural.

Heavy Metal Toxicity Effects

The impact of EDTA and DTPA on heavy metal toxicity has received considerable attention (34, 47-51). According to an early hypothesis, it has been suggested that by increasing metal solubility EDTA also enhances the availability of metals to phytoplankton (52). However, later articles reveal that it is mainly the concentration of free metal ions and not the total metal concentration that determines metal toxicity (49,53). The effect of copper on the photosynthetic activity of *Selenastrum capricornutum* in the presence and absence of different concentrations of EDTA has been investigated (51). It was found that inhibition of bacterial growth could not be explained by free Cu ions only; thus, the complexed copper is also bioavailable to a significant extent.

In all, there are controversial data concerning the effect of chelation on the toxicity of trace metals. Both decreasing and negligible effects of EDTA and DTPA have been reported on the toxicity of Cu, Cd, Zn, and EDTA and DTPA enhanced, decreased, and or were insignificant to the toxicity of Fe, depending on the organism. On the other hand, EDTA and DTPA have consistently decreased the toxicities of Mn and Pb and, in the cases of Cr and As, EDTA has no effect. It is inferred that the results depend on the organism, intial speciation, exposure time, chelate and metal concentrations, and aquatic media, amoung other factors. Thus, definitive conclusions cannot be drown, but generally the metal complexes of EDTA and DTPA are expected to be less toxic than free metals.

Conclusions

EDTA, and in some cases also DTPA, are generally found in the receiving waters of many industial areas, thus being classified as one of the major organic pollutant discharged in waters. The photochemical degredation of Fe complexes of these compounds is documented, but the extent to which these results can be applied to natural waters is not clear. There exist still some uncertainties in the chemical speciation, adsorption, overall degradation, and ultimately the eutrophication effect of EDTA and especially those of DTPA.

It can be inferred that EDTA can effect the essential and nonessential metal balance in natural waters as well as in aquatic organisms. The estimation of the chemical speciation of EDTA and DTPA is a challenging task because of the complexicity of the system and should be based not only on equilibrium calculations but also on direct analytical determinations of diverse metal species.

EDTA and DTPA are not expected to be acutely toxic to aquatic organisms. On the other hand, in natural waters, several compounds effect organisms simultaneously. Therefore, EDTA and DTPA may contribute to the aquatic toxicity at significantly lower concentrations than those determined by short-term toxicity tests. Also, more studies should be directed to estimating chronic effects, including the possible imbalance of body calcium in animals and other organisms. EDTA and DTPA may desorb heavy metals bound to sediments and also prevent heavy metal sedimentation, thus increasing their cycle in water. However, these metal complexes are not expected to be as bioavailable as free metal ions.

References

1. Fuerhacker, M.; Lorbeer, G.; Haberl, L. *Chemosphere* **2003,** 52, 253.
2. Nörtemann, B. *Appl.Microbiol. Biotechnol.* **1999,** 51, 751.
3. Trapp, S.; Brüggemann, K.; Frey, S. *Wasser – Abwasser* **1992,** 133, 495.
4. Frimmel, F. *Wasser-Abwasser* **1989,** 130, 106.
5. Pietsch, J.; Schmidt, W.; Sacher, F.; Brauch, H-J. *Fresenius'J. Anal. Chem.* **1995,** 353, 75.
6. Wanke, V.; Eberle, S. *Acta Hydrochim. Hybrobiol.* **1992,** 20, 192.
7. Sillanpää, M. *Chemosphere* **1996,** 33, 293.
8. Xue, H.; Sigg, L.; Kari, F. *Environ. Sci. Technol.* **1995,** 29, 59.
9. Hering, J.; Morel, F. *Environ. Sci. Technol.* **1988,** 22, 1469.
10. Hudson, R.; Covault, D.; Morel, F. *Mar. Chem.* **1992,** 38, 209.
11. De Stefano, C.; Gianguzza, A.; Piazzese, D.; Sammartano, S. *Anal. Bioanal. Chem.* 2003, 375, 956.
12. Bowers, A.; Huang, C. *J. Colloid Interface Sci.* **1986,** 11, 575.
13. Chang, H-C.; Healy, T.; Matijevic, E. *J. Colloid. Interface. Sci.* **1983,** 92, 469.
14. Nowack, B.; Sigg, L. *J. Colloid. Interface Sci.* **1996,** 177, 106.
15. Nowack, B.; Lützenkirchen, J.; Behra, P.; Sigg, L. *Environ. Sci. Technol.* **1996,** 30, 2397.
16. Alexander, M. *Science* **1981,** 211, 132.
17. Zepp, J.; Cline, D. *Environ. Sci. Technol.* **1977,** 11, 359.
18. Frank, R.; Rau, H. *Ecotoxicol. Environ. Saf.* **1990,** 19, 55.
19. Lockhart, H.; Blakeley, R. *Environ. Lett.* **1975,** 9, 19.
20. Lockhart, H.; Blakeley, R. *Environ. Sci. Technol.* **1975,** 9, 1035.
21. Svenson, A.; Kaj, L.; Björndal, H. *Chemoshere* **1989,** 18, 1805.
22. Tiedje, J. *J. Environ. Qual.* **1977,** 6, 21.
23. Means, J.; Kucak, T.; Crerar, D. *Environ. Pollut. Ser. B Chem. Phys.* **1980,** 1, 45.
24. Kari, F.; Giger, W. *Environ. Sci. Technol.* **1995,** 29, 2814.
25. Kari, F.; Hilger, S.; Canonica, S. *Environ. Sci. Technol.* **1995,** 29, 1008.
26. Tiedje, J. *Appl. Microbiol.* **1975,** 30, 327.
27. Sillanpää, M.; Rämö, J. *Environ. Sci. Technol.* **2001,** 35,1379.
28. Horstmann, U.; Gelpke, N. *Revue. Internat. Oceanogr. Med.* **1991,** 104, 260.
29. Davies, V. *Nature* **1990,** 345, 114.
30. Sudo, R.; Okada, M. *The Contribution of Sediment to Lake Eutrofication as determined by algal assay;* Environmental Protection Agency, Washington, DC, 1979; p 161.
31. Golterman, H.; Booman, A. *Verh. Internat. Verein. Limnol.* **1988,** 23, 904.
32. Batchelder, T.; Alexander, H.; McCarty W. *Bull. Environ. Contam. Toxicol.* **1980,** 24, 253.
33. Bringmann, G.; Kühn, R. *Z. Wasser-Abwasser-Forsch.* **1982,** 15, 1.
34. Sillanpää, M.; Oikari, A. *Chemosphere* **1996,** 32, 1485.
35. Zuiderveen, J.; Birge, W. *12th Ann. Meeting Soc. Env. Toxicol. Chem.* **1991.**

36. van Dam, R.; Barry, M.; Ahokas, J.; Holdway, D. *Ecotoxicol. Environ. Saf.* **1995,** 31, 117.
37. van Dam, R.; Barry, M.; Ahokas, J.; Holdway, D. *Arch. Environ. Contam. Toxicol.* **1996,** 31, 433.
38. Holdway, D. *Proc. 47th Appita Ann. Gen. Conf.* **1993,** 2, 803.
39. Holdway, D. Aust. *J. Ecotoxicol.* **1996,** 2, 17.
40. Barica, J.; Stainton, M.; Hamilton, A. *Water Res.* **1973,** 7, 1791.
41. Dehnad, V.; Föstner, U. *Z. Wasser- Abwasser- Forsch.* **1988,** 21, 46.
42. Erel, Y.; Morgan, J. *Geochim. Cosmochim. Acta* **1992, 56,** 4157.
43. Li, Z.; Shuman, L. *Soil Sci.* **1996,** 161, 226.
44. Means, J.; Crerer, D.; Duguid, J. *Science* **1978,** 200, 1477.
45. Müller, G.; Föstner, U. *Z. Wasser- Abwasser- Forsch.* **1976,** 9, 150.
46. Norvell,W. *Soil Sci. Soc. Am.* **1984,** 48, 1285.
47. Castille, F.; Lawrence, A. *J. World Maricult. Soc.* **1981,**12, 292.
48. Chaudri, A.; McGrath, S.; Giller, K.; Angle, J.; Chaney, R. *Environ. Toxicol Chem.* **1993,** 12, 1643.
49. Huebert, D.; Shay, J. *Aquat. Toxicol.* **1992,** 22, 1469.
50. Muramoto, S. *Bull. Environ. Contam. Toxicol.* **1981,** 26, 641.
51. Rijstenbil, J.; Poortvliet, T. *Environ. Toxicol. Chem.* **1992,** 11, 1615.
52. Johnston, R. *J. Mar. Biol. Assoc.* **1964,** 44, 87.
53. Anderdon, M.; Morel, M. *Limnol. Oceanogr.***1982,** 27, 789.

Chapter 14

Reactions of Phosphonic Acids at the Solid–Water Interface

Klaus Fischer

Department of Analytical and Ecological Chemistry, FB VI-Faculty of Geography and Geosciences, University of Trier, Trier, Germany

Phosphonic acids contain one or more C-$PO(OH)_2$ groups and often additional functional groups, i.e. amino or hydroxy groups. Compounds with at least two functional groups have chelating properties, and they are produced for numerous technical and industrial applications, e.g. scale and corrosion inhibitors. Aminopolyphosphonates are often structurally analogous to aminopolycarboxylates, being formally derived from the latter by substitution of carboxylate groups. Reactions at the solid/water-interface are critical for the fate of phosphonates in aquatic environments as well as for their technical use. This review focuses on adsorption–desorption reactions, reflects effects on scale formation, crystal growth and crystal morphology, and considers metal remobilization from sediments. Conclusions concerning the expected environmental behavior of phosphonates were drawn, but the considerable lack of information (e.g. monitoring data, speciation, remobilization of phosphonates from natural sorbents) limits the reliability of such predictions.

Introduction

Surface reactions are critical for studying the environmental behavior of phosphonates. These reactiosn are also of fundamental importance for modeling predictable environmental concentrations and for the elaboration of an environmental risk assessment for this class of synthetic chelating agents. Several natural phosphonates exist, but so far as known, they do not function as "biochelates" (1).

Due to their physico-chemical properties (high water solubility, negative log K_{ow} values, very low vapor pressure) and their practical application, most of the released phosphonates, if not eliminated during wastewater treatment, enter aquatic ecosystems. Thus reactive components and physico-chemical conditions that are constitutive for our rivers, lakes and estuarine ecosystems, should be applied in phase transfer experiments. Relevant sorbents are living and dead microorganisms, geochemically generated organic particles, suspended matter, complete sediments and the main sediment constituents, i.e. metal (hydr)oxides, clay minerals, carbonates, silicates and detritus. Viewed by this standard, the available experimental data are fragmentary and not sufficient to draw sound conclusions. On the other hand, investigations of the application properties of phosphonic acids have built a broad knowledge base which can be exploited for environmental research purposes (2, 3). This applies especially to such effects that belong to reactions at solid/water-interfaces (e.g. inhibition of scale formation (4-6), modification of crystal growth and morphology (7-10), retardation of cement hardening and gypsum settling (11, 12), reduction of the viscosity of clay suspensions and ceramic pastes (2), support of biomineralisation processes (13, 14) and action as dispersants (15)). Thus, this review makes no difference between studies focused on environmental behavior of phosphonates and those engaged in fundamental physico-chemical or technical aspects. Nevertheless it is an open question whether the results of some of the studies attributable to the latter category, e.g. testing the effectiveness of specific phosphonate formulations to inhibit carbonate precipitation (16), are of any significance for chemical reactions in the environment.

Adsorption

General Aspects

The majority of the adsorption studies aimed to compare the adsorption behavior of differently structured phosphonic acids and to rank the adsorption capacity of various solids on a rather phenomenological level. Usually, the pH value, the electrolyte concentration, the cation coverage of the sorbent surface, the solid-to-liquid ratio and the reaction time are important reaction parameters.

Considering heterogenous solids such as soils or sediments, the different adsorption capacities found were related to their physical properties (specific surface area, grain size distribution, etc.), to their mineralogical composition, and to their organic matter content.

Several investigations dealt with mechanistic aspects of the adsorption process. Central questions were the bonding mode and its modification by the reaction conditions (bonding via surface complex formation, electrostatic forces, or hydrogen interactions), the structure of surface complexes formed, and thermodynamic aspects, such as stability of surface complexes, bonding energies and reversibility of the adsorption reaction.

Although not explicitly formulated, a few studies (17, 18) intended to formulate qualitative structure-activity relations for the adsorption properties of phosphonates, based on a stringent application of coordination chemistry principles, combined with stereochemical aspects. In this context, the following properties of the ligand functional groups are of high importance: chemical hardness, Lewis basicity, pk_a value(s), size, and spatial arrangement. At the level of the entire molecule, the maximum denticity of the ligand as well as its dimensions and possible coordination geometries are decisive factors. Consequently the speciation of the solved ligand (i.e. the nature of the coordinated metal ion and the structure of the formed complex), is also regarded as a determinant for the phase distribution of phosphonates (17-21).

To elucidate the structure of the formed surface complexes, several spectroscopic techniques were used. Specific questions were the differentiation between inner-sphere and outer-sphere surface complexation, the identification of the binding donor atoms or groups of the ligand, the distinction between mono-, bi- and polydentate surface binding, the distinguishing between mono- and binuclear complexation, the discrimination between "metal-linked" and "ligand-linked" ternary surface complexes, and the characterization of the surface conformation of the adsorbed phosphonates. For this purpose, various FT-IR techniques, i.e. diffuse reflectance spectroscopy, attenuated total reflectance (ATR) spectroscopy, and cylindrical internal reflectance (CIR) spectroscopy, were predominately used (15, 22-29). Other analytical techniques applied are extended X-ray absorption fine structure (EXAFS) spectroscopy (29, 30), X-ray photoelectron spectroscopy (XPS) (28, 31, 32), ^{31}P and ^{13}C magic angle spinning (MAS) NMR (22, 33), FT-Raman spectroscopy (23), Auger electron spectroscopy (15) and atomic force microscopy (AFM) (32).

Energetic aspects of the interaction between phosphonates and solid surfaces were examined, and corresponding data are completely missing for heterogeneous environmental solids. The reactive chemisorption of methylphosphonic acid at clean and oxidized Al (111) surfaces has been investigated and core binding energies for the molecule units were reported (31). Research on the inhibition of crystal growth and the modification of crystal morphology by phosphonates has paid more attention to energetic reaction terms. For instance binding energies for the docking of several diphosphonates at specific sites, defined for individual crystal faces of barite were calculated using various molecular modeling techniques (34). A similar approach was pursued to compute the energetics of docking of phosphonates on ettringite (11).

Even the reversibility of phosphonate adsorption is rarely investigated. In two studies, the reversibility of the adsorption onto goethite and onto an immobilized humic acid was measured after a rise of the suspension pH (17, 20). In another study desorption of HEDP from several river sediments by differently concentrated salt solutions, diluted nitric acid and natural river water was measured and Freundlich desorption isotherms were calculated (35).

In a few examinations differences between the adsorption capacities of aged ("well-developed") and newly generated rough surfaces were highlighted and the distinct adsorption properties of specific crystal faces were described (34, 36). The latter aspect is more recognized in the context of investigations concerned with scale-inhibition reactions and modification of crystal growth and morphology.

A synopsis of adsorption studies with aliphatic phosphonic acids is provided in Table I.

The elucidation of the adsorption mechanism on a molecular scale creates the prerequisites for the interpretation of resulting phase distribution equilibria and for the rationalization of effects exerted on the sorbents by the adsorbates. Research on the interactions of phosphonates with surfaces of crystalline materials is directed toward the recognition of the critical factors responsible for a specific spatial and electronic matching between adsorbate and substrate. Amongst others, the goodness of this match depends on the correspondence between the dimensions and geometry of the lattice anions and of the binding groups of the adsorbate. A further criterion is the adequacy of the interatomic distances between surface binding sites and between the active functional groups of the adsorbate.

Metal(hydr)oxides

Phosphonates adsorb very strongly to almost all mineral surfaces. This behavior is pronounced in the case of metal (hydr)oxides as sorbents.

Goethite (α-FeOOH)

Nowack and Stone (17, 18) studied the adsorption of one monophosphonate (MPA), two hydroxyphosphonates (HMP and HEDP) and five aminophosphonates (AMP, IDMP, NTMP, EDTMP and DTPMP) onto goethite as a function of pH. At phosphonate concentrations significantly lower than the total number of available surface sites, nearly 100 % adsorption was observed below pH 8.0. The adsorption approximated to zero percent between pH 9.5 and 12.0, depending on the number of phosphonate groups. This confirms that a certain adsorption capacity is present at pH values above the point of zero charge (8.5). At an excess of surface sites, the adsorption of NTMP as a function of pH was nearly independent from the ionic strength. After a rise of the pH from 7.0 to 12.2, NTMP was completely desorbed within 5 h. At phosphonate concentrations close to the total number of available surface sites, adsorption was highest at lowest pH tested (pH 3.0) and decreased over a broader range in pH. At pH 7.2,

Table I: Synopsis of adsorption studies with aliphatic phosphonic acids

compound	formula	sorbents[f]	references
methylphosphonic acid (MP)	CH_5O_3P	goe	17, 25
hydroxymethylphosphonic acid (HMP)	CH_5O_4P	goe	17
aminomethylphosphonic acid (AMP)	CH_6NO_3P	goe	17, 18
N-(phosphonomethyl)glycine[a] („glyphosate")	$C_3H_9NO_5P$	cm, goe, gi, ht, mo, soil,	27-32, 39-41, 51-56
2-phosphonobutane-1,2,4-tricarboxylic acid (PBTCA)	$C_7H_{11}O_9P$	al	42
methylenediphosphonic acid (MDP)	$CH_6O_6P_2$	hap	59
aminomethanediphosphonic acid (AMDP)	$CH_7NO_6P_2$	cal, gyp	7, 48
iminodi(methylenephosphonic acid) (IDMP)	$C_2H_{10}NO_6P_2$	goe, MnOOH	17, 19, 57
1-hydroxyethane-1,1-diphosphonic acid[b] (HEDP)	$C_2H_8O_7P_2$	al, bar, cal, cm, goe, gyp, hap, sed, sl, soil	7, 9, 17, 19, 35, 37, 43-45, 47-50, 59
1,2-dihydroxyethane-1,2-diphosphonic acid (DHEDP)	$C_2H_8O_8P_2$	cal	7, 9
nitrilotris(methylenephosphonic acid)[c] (NTMP)	$C_3H_{12}NO_9P_3$	al, bar, cal, goe, hap, iha, hfo, hzo, SiO_2, sl, soil	6, 17, 19, 20, 37, 46, 49, 50
ethylendiaminetetra(methylenephosphonic acid)[d] (EDTMP)	$C_6H_{20}N_2O_{12}P_4$	cal, goe, hap	17, 19, 36, 38, 49, 50, 58
hexyldiaminetetra(methylenephosphonic acid) (HDTMP)	$C_{10}H_{28}N_2O_{12}P_4$	hap	50
diaminoethoxytetra(methylenephosphonic acid) (DETMP)	$C_8H_{24}N_2O_{13}P_4$	hap	50
diethylenetriaminepenta(methylene-phosphonic acid)[e] (DTPMP)	$C_9H_{28}N_3O_{15}P_5$	al, bar, goe, hap, sl, soil	6, 17, 19, 37, 50
dihexyltriaminepenta(methylene-phosphonic acid) (DHTPMP)	$C_{17}H_{44}N_3O_{15}P_5$	bar, hap	6, 50

a: not systematically registered, **b**: CAS: 2809-21-4, **c**: CAS: 6419-19-8, **d**: CAS: 1492-50-1, **e**: CAS: 15827-60-8, **f**: sorbents: al: α-Al_2O_3, bar: barite, cal: calcite, cm: clay minerals, gi: gibbsite, goe: goethite, gyp: gypsum, hap: hydroxyapatite, hfo: hydrous ferric oxide, ht: hydrotalcite, hzo: hydrous zinc oxide, iha: immobilized humic acids, mo: metal(hydr)oxides, sed: sediments, sl: sewage sludge.

the maximum surface concentration of the phosphonate molecules increased with decreasing number of phosphonic acid functional groups, presumably explainable by the concomitant decrease of the molecule size.

The adsorption as a function of phosphonate concentration and pH was modeled using a 2-pK constant capacitance model (hypothesis: formation of a 1:1 surface complex involving one surface site and one phosphonate functional group). For every specific protonation level "n" of the formed surface complex, the corresponding overall complex formation constant $\log\beta_{n,surf}$ was calculated. A linear relationship between $\log \beta_{n,surf}$ and the surface complex charge was found. The formation constants log K of the fully deprotonated surface complexes ranged from 13.3 to 41.0 and increased with increasing number of phosphonate groups.

The examination of the pH-depending MPA adsorption onto goethite by Barja et al. (25) parallels the above summarized findings. The maximum adsorption densities spanned between 182 (pH 3.5) and 84 (pH 9.0) μmoles g^{-1}. ATR-FTIR spectroscopy and measurement of electrophoretic mobilities gave indications for different surface complex structures, depending on sorption pH and surface coverage. At low pH and high sorption density, MPA was bound predominantly as a monodentate protonated species, while at high pH and low coverage, the prevailing structure was that of a bridging bidentate complex.

According to Nowack and Stone (19) the influence of equimolar concentrations of Ca^{2+}-, Cu^{2+}-, Zn^{2+}-, and Fe^{3+}-ions on the adsorption of AMP, IDMP, HEDP, NTMP, EDTMP, and DTPMP onto goethite is negligible or slight. The authors explain this result by dissociation of the metal-phosphonate complex formed in solution and separate adsorption of the metal ion and the ligand onto different surface sites. The addition of a 100-fold excess of Ca^{2+} to the adsorbate solutions did not change the phase distribution of AMP, the smallest phosphonate tested, but increased the adsorption of the other components by 38 % to 84 %, having the strongest effect on NTMP. Increasing Ca^{2+} concentrations shifted the NTMP adsorption edge toward increasing pH values. In the case of excess Ca^{2+} or Zn^{2+} concentrations it is assumed that ternary surface complexes are formed.

Other Metal(hydr)oxides

At pH 7.2 the adsorption capacity of hydrous ferric oxide (HFO) was 15 times (for NTMP) and 19 times (for DTPMP) that of goethite, reflecting the correspondingly higher surface area of HFO (20). As in the case of goethite, the adsorbed NTMP amount nearly doubled in the presence of an excess of Ca^{2+} ions.

Laboratory experiments and a field trial in a municipal sewage treatment plant revealed that HEDP is synchronously eliminated with phosphate from the water phase by adsorption onto in-situ forming iron(hydr)oxides after addition of $FeCl_3$ solutions (47). Further evidence of a strong HEDP fixation on this substrate was provided by another study concerning the influence of HEDP on the phosphate elimination by $FeCl_3$ solutions (60).

The adsorption of IDMP onto MnOOH (pH 7.0) surpassed that onto goethite (pH 7.2) by a factor of about 6.5 (57). The maximum surface coverage of MnOOH by IDMP and the calculated Langmuir sorption constant were in the same range as reported for the sorption of NTMP and DTPMP onto HFO. It was shown that zinc(hydr)oxide is also very efficient at adsorbing NTMP at pH values around or below 9.0, whereas silica did not adsorb this aminopolyphosphonate between pH 5.0 and 10.5 in the absence of divalent metal ions (19).

Adsorption of PBTCA onto α-Al_2O_3 resembled the high affinity Langmuir type (42). The saturation adsorption amount increased with increasing PBTCA concentration until a pH-dependent plateau region was obtained. The interaction of the carboxylate groups with the alumina surface was studied by FTIR spectroscopy. Due to the adsorption of the multi-charged phosphonocarboxylate, the isoelectric point of α-Al_2O_3 is strongly shifted toward the acidic pH region and the absolute zeta potential is remarkably increased.

Clay Minerals

The adsorption of HEDP onto kaolinite, bentonite (main constituent: Ca-montmorillonit) and two illitic clays (Fithian illite and engobe clay) was studied at initial phosphonate concentrations between 2.5 and 50.0 mg L^{-1} (43). At near neutral pH the Freundlich constants spanned from 293 (kaolinite) to 2378 (engobe clay). The sorption densities ranked from 26.1 $\mu g\ m^{-2}$ (bentonite) to 41.9 $\mu g\ m^{-2}$ (kaolinite). Pretreatment of bentonite with hydrochloric acid or addition of HCl to maintain a sorption pH of about 4.0 reduced the surface binding of HEDP at low and medium initial concentrations significantly. The high adsorption strength of the illitic clays is presumably caused by their high iron(hydr)oxide content.

The exposition of bentonite samples to a 10 % (w/v) Na_2HEDP solution provoked an intercalation of the phosphonate into the interlamellar spaces of the expandable clay mineral montmorillonite. In contrast to the usually effected widening of the layer distance a shortening occurred presumably caused by a simultaneous interaction of the phosphonate molecules with both interlamellar surfaces which requires an opposite orientation of the two phosphonic acid groups.

The adsorption of glyphosate onto two smectites (montmorillonite and nontronite, the latter characterized by isomorphic substitution of Si^{4+} ions through Fe^{3+} ions) and kaolinite was found to depend on sorption pH, cation coverage of the clays and on their specific surface areas (41). With a few exceptions the adsorption decreased with pH increasing from 4.5 to 11.0. At pH 7 and Na^+ coverage of the sorbents, the Freundlich constant increased following the sequence kaolinite < montmorillonite < nontronite, reflecting the order of specific surface areas. At acidic pH the Al^{3+} covered clays were the strongest adsorbents, but at neutral pH, adsorption onto Na^+ or Ca^{2+} covered surfaces was favored. Morillo et al. (40) found that the adsorption of glyphosate onto montmorillonite is partially reversible. Cu^{2+} ions enhanced desorption of glyphosate. The depression of the glyphosate surface binding in the presence of Cu^{2+} ions was higher at pH 6.8 than at pH 4.2. Cu speciation calculations showed a very high proportion of a solved Cu(II)-glyphosate complex, which had a lower tendency to be adsorbed on the clay mineral than the free ligand.

Calcite, Hydroxyapatite and Gypsum

A "high affinity type" adsorption of HEDP, DHEDP and EDTMP onto calcite at slightly alkaline pH was examined by various authors (7, 9, 36). The surface of aged calcite was saturated with phosphonates at a calculated surface coverage of about 4.5 % (EDTMP) and 9.0 % (DHEDP), respectively. The surface of freshly crushed calcite was almost fully covered with EDTMP (36). The HEDP sorption was modeled with the Freundlich isotherm; no surface saturation was observed.

All phosphonates tested inhibited the calcite growth strongly. The Langmuir adsorption model could be used to describe empirically the reduction in crystal growth rates. In this transformed isotherm ("kinetic adsorption isotherm") a term belonging to the relative inhibition efficiency of the phosphonates is correlated with their reciprocal equilibrium concentration in solution. The slope of the straight line, generated by a plot of this term against the reciprocal phosphonate concentration, gives the ratio between the rate constants of the adsorption and desorption reaction ("affinity constant"). It was found that the slopes of the conventional and of the kinetic Langmuir isotherms are very similar thereby demonstrating that the adsorption of the phosphonate and its inhibition activity are closely related. The fact that crystal growth is reduced at low surface concentration of phosphonates provides additional evidence that these inhibitors adsorb preferentially or exclusively at the active growth sites (kink and step sites) of the crystals.

Also motivated by the aim to elucidate mechanisms of crystal growth inhibition, Amjad (49) and Zieba et al. (50) studied the interaction of phosphonates with hydroxyapatite, and Weijnen and van Rosmalen (48) the interaction with gypsum. According to Amjad, the affinity of several phosphonates to hydroxyapatite and likewise their inhibition efficiency increased following the sequence NTMP < HEDP < HDTMP < EDTMP. The affinity constants ranged between $0.62 \cdot 10^6$ and $1.8 \cdot 10^6$ L mol^{-1}. The inhibition sequence did not agree with the order of formation constants of the corresponding Ca phosphonate complexes in solution. The affinity sequence for phosphonates was NTMP < HDTMP < DHTPMP ≤ HEDP < ETPMP < EDTMP < DETMP (50). The findings that some of the affinity constants were higher as measured by Amjad and the reverse position of HDTMP and HEDP might be caused by the lower reaction temperature (25°C vs. 37°C).

The isotherm of the irreversible HEDP adsorption at pH 5 onto gypsum, suspended in supersaturated $CaSO_4$ solutions, exhibited two plateaus at less than 1 % surface coverage and at about 5 % surface coverage, indicating different kinds of binding surface sites (48). The adsorption of AMDP at pH 5 and of HEDP at pH 7 was even stronger, reaching coverage degrees of about 58 % and 43 %, respectively. The high adsorption capability of AMDP correlated with its higher inhibition efficiency.

Sewage Sludges, Sediments and Soils

Applying sewage sludges from different municipal waste water treatment plants at a solid: liquid ratio of about 1: 400, removal degrees of greater than 90 % were determined for HEDP and NTMP at initial concentrations of $\leq$ 10 mg L^{-1} (45, 46). The pH was not controlled in these experiments. Calculated for initial phosphonate concentrations of 0.6 to 6.4 mg L^{-1}, the Freundlich adsorption constants ranged from 1000 to 4000 (NTMP) and from 2600 to 12700 (HEDP). With less concentrated sewage sludge suspensions, elimination degrees of about 87 % (initial HEDP concentration: 1.25 mg L^{-1}) and of 70 % (HEDP: 2.4 mg L^{-1}) were found (44, 47). Even at an HEDP initial concentration of 50 mg L^{-1}, the adsorption degree remained above 50 %. A Freundlich constant of 2718 yielded from a laboratory test series with HEDP starting concentrations between 1.25 and 50.0 mg L^{-1} (44).

Nowack studied the adsorption of NTMP and its Cu(II)-, Fe(III)- and Ca(II) complexes at different pH values onto activated sludge having a low iron content (20). The sludge binding of NTMP decreased with increasing pH from 3 to 8 and increased again at pH above 8. A clear influence of the coordinated metal ion on adsorption was not discernible. A great excess of Ca^{2+} ions strongly favored the phosphonate adsorption at pH 6.5 and at higher NTMP concentrations. It is argued that iron(hydr)oxides may have significantly contributed to the sludge adsorption capacity despite of their low solid content. Humic acids, immobilized on silica, were selected to determine whether NTMP can be adsorbed by high molecular weight organic matter. The ineffectiveness of silica to bind NTMP within the relevant pH range was proved before. It was found that NTMP transfers to the organic matter in a completely reversible manner. Below pH 5 adsorption was higher than 90 % but declined sharply with increasing pH and approached zero percent at about pH 6.5. It is assumed that electrostatic interaction with protonated carboxylic functions and hydrogen bonding is responsible for NTMP adsorption onto humic acids.

The adsorption of HEDP onto several river sediments and a North Sea sediment was studied at initial phosphonate concentrations between 1.0 and 50.0 mg L^{-1} without pH adjustment (44). The Freundlich adsorption constants ranged between 407 and 2107, the slopes of the linearized isotherms spanned from 0.24 to 0.70. Suspension of the sediments in river water (Ca content between 130 and 180 mg L^{-1}) increased the phase transfer of HEDP significantly. The sorption strength, expressed by the sorption constant, correlated with the content of metal(hydr)oxides, especially iron(hydr)oxides, and with the illite content of the clay fraction. A high organic matter content favored the adsorption capacity. This conclusion is supported by a comparison of the adsorption properties of sediments and of clay minerals, generalized by the Freundlich model. Whereas the means of the sorption constants of the sediments (mean 1160) and of the clay minerals (mean 1380) did not differ much, the slopes of the sediment-related isotherms were significantly higher than the clay-related isotherms, indicating a higher sorption capacity of the former. In contrast to the almost pure clay minerals the river sediments contained between 2.8% and 4.9% organic carbon. Qualitatively similar results were found by a preceding study with seven sediments (47). Adsorption degrees between 48 % and 84 % were ascertained at an initial HEDP concentration of 2.4 mg L^{-1}.

Resuspension of HEDP loaded river sediments in river water, in differently concentrated solutions of potassium salts, and in diluted mineral acids (5 and 50 mM HNO_3 or HCl) verified the strong and partially irreversible sediment binding of the hydroxyphosphonate. Desorption decreased with decreasing HEDP load (from 9.1 mg kg^{-1} to 2.2 mg kg^{-1}). Desorption degrees above 25 % were only achieved with a high excess of mineral acids (35).

Adsorption of NTMP and HEDP onto soils is remarkably lower than onto sediments. Steber and Wierich calculated Freundlich constants for NTMP adsorption onto three soils in the range from 30 – 240. The interpretation of the results was complicated by the onset of NTMP degradation processes (46). According to the same authors, the K values for HEDP adsorption onto 6 differently composed soils ranked from 70 to 190 (45), which is comparable to a K value of 240, deduced from sorption tests with a sandy soil (44). It was not possible to trace back the adsorption behavior of the soils to specific soil properties (texture, organic matter content, and pH). Soil column leaching tests proved the low leachability of HEDP.

Effects on Scale Formation, Crystal Growth and Crystal Morphology

Many of the commercial and technical applications of polyphosphonic acids make use of their ability to prevent or at least retard scale formation, i.e. the precipitation of sparingly soluble salts and their adhesion on various carrier materials, at substoichiometric amounts ("threshold effect"). The chemical engineering of "scale inhibitors" is mainly directed toward the suppression of the formation of calcium and barium salts, e.g. carbonates, sulfates and phosphates. Important areas of application are cooling water technology, desalination, oil field applications, industrial cleaners, and house hold detergents (2, 61).

Despite of intense research, a generally accepted unifying model for crystallization inhibition at the molecular or process level does not exist. Models for poisoning of crystal growth include inhibition of nucleation, adsorption onto growth sites (kink and step sites), distortion of the crystal lattice, changes in surface charge and association with precursors of crystal formation (62, 63). The rationalization of the inhibition effectiveness has to be done for each crystal phosphonate combination specifically.

As mentioned earlier, the search for molecular properties, which enable phosphonates to act as scale inhibitors and as crystal growth retarders, is focused on the following topics:

- molecular structure:
  number of and distance between phosphonate groups, type and position of other functional groups, size of hydrophobic molecule units, conformational flexibility and energy barriers between different conformational arrangements

- charge properties:
  effective charges and charge densities, pK_a values

- binding properties:
  strength of coordinative interactions with the lattice cations, additional interactions with the lattice anions, e.g. via hydrogen bonding.

Several studies confirmed that the lowest phosphonate concentration needed to initiate inhibition effects ("threshold limit") is between 10^{-8} and 10^{-6} M (7, 9, 48). Concentrations between 10^{-6} and 10^{-4} M are usually required to achieve surface coverage of a few percent which is enough to block most of the active growth sites of various calcium minerals. Using atomic force microscopy the poisoning of calcite growth by HEDP was monitored in real-time (62). The AFM images gave hints for a selective adsorption of HEDP onto step sites, combined with a characteristic change of step morphology.

Alterations of the morphology of crystals precipitated in the presence of phosphonates are described several times, e.g. for calcium carbonates (7, 58), gypsum (64) and barium sulfate (8, 65, 66). For instance the rhombohedral shape of calcite crystals changed into a spherulitic crystal form when precipitated in the presence of HEDP (7). Additionally the size of the formed crystals is reduced. Furthermore phosphonates retard or inhibit the transformation of thermodynamically instable crystal modifications or minerals into the stable forms, e.g. vaterite into calcite (9, 58) or brushite into hydroxyl apatite (2).

According to the low solubility of some Ca HEDP salts (solubility of Ca_2 HEDP · 9 H_2O at pH 7: 16 mg L^{-1}), pure and sedimentary calcium carbonates (calcite, Mg calcite, aragonite, but not dolomite) were transformed into Ca HEDP salts after suspension in Na_2 HEDP solutions (67).

Metal Remobilization from Sediments

The possible remobilization of toxic heavy metals from sediments by synthetic chelating agents is a central point within risk assessments of their behavior in aquatic ecosystems. This aspect is subject of many studies about the environmental effects of aminopolycarboxylic acids, e.g. NTA and EDTA (68-70).

Less attention was paid to the interaction of phosphonates with sediment bound metals. Müller et al. (47) exposed several polluted sediments from German rivers in batch tests to HEDP solutions, containing 2.4 mg L^{-1} of phosphonate. Only the solubilization of Fe was promoted by the HEDP solution compared to the blank sample. No effect on the phase distribution of Zn, Cr, Ni, Cu, Pb, and Cd was recognized. According to another study only NTMP concentrations above 0.1 mM were able to mobilize Cu, Cd and Pb from a river sediment at pH 7.2 (71).

With respect to predicted environmental concentrations (72) it can be assumed that phosphonates will not enhance the solubility and mobility of heavy metals in the aquatic environment except areas which are subjected to high volume point sources.

Conclusions

Interactions of phosphonates at solid/water interfaces are of high importance for their technical applications and for their behavior in aquatic ecosystems. At circumneutral pH, phosphonates strongly adsorb onto river sediments, typical sedimentary minerals and sewage sludges. The adsorption onto sediments is partially irreversible and correlates with the iron(hydr)oxide content of the sorbents. Usually the affinity of phosphonic acids for mineral surfaces increased with increasing number of phosphonate groups.

The adsorption of polyphosphonates onto metal(hydr)oxides is the result of specific interactions at the particle surface mainly. Predominantly inner-sphere complexes are formed. This process can be described applying various surface complex formation models, e. g. the 2-pk constant capacitance model. Indications exist that the structure of the formed surface complexes depends on the extent of surface coverage and on the pH value. As it was proven for goethite at pH 7.2, the maximum adsorption extent decreases with increasing number of phosphonate groups, presumably caused by a parallel increase of the surface area occupied by each molecule. In contrast to weak effects exerted by transition metal ions, higher concentrations of Ca^{2+}-ions favor the adsorption strongly. It is assumed that a "ligand-like" ternary surface complex is formed under these conditions.

Phosphonic acids retard the precipitation of Ca and Ba carbonates, sulfates, and phosphates, and they inhibit crystal growth by preferred adsorption onto active growth sites, leading to a change of the crystal morphology also. Despite the high technical importance of these processes, their mechanistic principles are still controversial, and a unifying model for the prediction of inhibitor effectiveness does not exist.

Studies do not indicate phophonic acids enhance removal of toxic heavy metals from polluted sediments under environmentally relevant conditions. Although fundamental aspects of the distribution properties of phosphonates are well examined, the small number of tests performed with environmental substrates, the insufficient data concerning the reversibility of adsorption, and the lack of information about the speciation of phosphonates in sewages and natural waters limits the conclusions that can be drawn.

References

1. Hilderbrand, R.L.; Henderson, T.O. In *The Role of Phosphonates in Living Systems*; Hilderbrand, R.L., Ed.; CRC Press: Boca Raton, FL, 1983, pp 5-30.
2. Heins, A.; Plöger, W. *Henkel-Referate* **1980**, *16*, 19-26.
3. Gledhill, W.E.; Feijtel, T.C.J. In *The Handbook of Environmental Chemistry*; Huntzinger, O., Ed.; Springer: Berlin, GE, 1992, Vol 3, part F, pp 260-285.
4. Graham, G.M.; Dyer, S.J.; Shone, P. *SPE Production & Facilities* **2002**, *17,* 212-220.
5. Ojo, S.A.; Slater, B.; Catlow, C.R.A. *Molecular Simulation* **2002**, *28,* 591-606.

6. Tomson, M.B.; Fu, G., Watson; M.A., Kan, A.T. *SPE Production & Facilities* **2003**, *18*, 192-199.
7. Xyla, A.G.; Koutsoukos, P.G. *J. Chem. Soc. Faraday Trans. I* **1987**, *83*, 1477-1484.
8. Black, S.N.; Bromley, L.A.; Cottler, D.; Davey, R.J.; Dobbs, B.; Rout, J.E. *J. Chem. Soc. Faraday Trans.* **1991**, *87*, 3409-3414.
9. Xyla, A.G.; Mikroyannidis, J.; Koutsoukos, P.G. *J. Colloid Interface Sci.* **1992**, *153*, 537-551.
10. Mutin, P.H.; Guerrero, G.; Vioux, A. *Comptes Rendus Chimie* **2003**, *6*, 1153-1164.
11. Conveney, P.V.; Humphries, W. *J. Chem. Soc. Faraday Trans.* **1996**, *92*, 831-841.
12. Bishop, M.; Bott, S.G.; Barron, A.R. *Chem. Materials* **2003**, *15*, 3074-3088.
13. Francis, M.D.; Centner, R.L. *J. Chem. Educ.* **1978**, *55*, 760-766.
14. Fleisch, H. *Recent Results Cancer Res.* **1989**, *116*, 1-28.
15. Liu, Y.Q.; Gao, L.; Guo, J.K. *J. Inorg. Mat.* ***2000***, *15*, 855-861.
16. Graham, G.M.; Deyer, S.J.; Shone, P. *SPE Production & Facilities* **2002**, *17*, 212-200.
17. Nowack, B.; Stone, A.T. *J. Colloid Interface Sci* **1999**, *214*, 20-30.
18. Stone, A.T.; Knight, M.A.; Nowack, B. In *Chemicals in the Environment – Fate, Impacts and Remediation*, Lipnick, R.L.; Mason, R.P.; Philliops, M.L.; Pittman, C.U., Eds.; *ACS Symposium Series 806*; ACS: Washington, DC, 2002, pp 59-94.
19. Nowack, B.; Stone, A.T. *Environ. Sci. Technol.* **1999**, *33*, 3627-3633.
20. Nowack, B. *Water Res.* **2002**, *36*, 4636-4642.
21. Stone, A.T.; Morgan, J.J. In *Aquatic Chemical Kinetics*, Stumm, W., Ed.; Wiley-Interscience: New York, 1990, Chapter 1.
22. Gao, W; Dickinson, L.; Grozinger, C.; Morin, F.G.; Reven, L. *Langmuir* **1996**, *12*, 6429-6435.
23. Persson, P.; Laiti, E.; Öhman, L.-O. *J. Colloid Interface Sci.* **1997**, *190*, 341-349.
24. Laiti, E.; Persson, P.; Öhman, L.-O. *Langmuir* **1998**, *14*, 825-831.
25. Barja, B.C.; Tejedor-Tejedor, M.I.; Anderson, M.A. *Langmuir* **1999**, *15*, 2316-2321.
26. Kim, C.S.; Lad, R.J.; Tripp, C.P. *Sensors Actuators B* **2001**, *76*, 442-448.
27. Martín, M.J.S.; Villa, M.V.; Sánchez-Camazano, M. *Clays Clay Min.* **1999**, *47*, 777-783.
28. Sheals, J.; Sjöberg, S.; Persson, P. *Environ. Sci. Technol.* **2002**, *36*, 3090-3095.
29. Sheals, J.; Granström, M.; Sjöberg, S.; Persson, P. *J. Colloid Interface Sci.* **2003**, *262*, 38-47.
30. Dubbin, W.E.; Sposito, G.; Zavarin, M. *Soil Sci.* **2000**, *165*, 699-707.
31. Davies, P.R.; Newton, R.G. *Appl. Surface Sci.* **2001**, *181*, 296-306.
32. Dideriksen, K.; Stipps, S.L. *Geochim. Cosmochim. Acta* **2003**, *67*, 3313-3327.
33. Grossmann, G.; Grossmann, A.; Ohms, G.; Breuer, E.; Chen, R.; Golomb, C.; Cohen, H.; Hagele, G.; Classen, R. *Magn. Reson. Chem.* **2000**, *38*, 11-16.

34. Rohl, A.L.; Gay, D.H.; Davey, R.J.; Catlow, C.R.A. *J. Am. Chem. Soc.* **1996,** *118*, 642-648.
35. Fischer, K. *Wat. Res.* **1993,** *27,* 485-493.
36. Sawada, K.; Abdel-Aal, N.; Sekino, H.; Satoh, K. *J. Chem. Soc. Dalton Trans.*, **2003**, 342-347.
37. Held, S. *Textilveredlung* **1989,** *24*, 394-398.
38. Chirby, D.; Franck, S. *Int. J. Radiat. Appl. Instrum. Part A* **1988,** *39,* 495-499.
39. Day, G.M.; Hart, B.T. *Environ. Technol.* **1997,** *18*, 781-794.
40. Morillo, E.; Undabeytia, T.; Maqueda, C. *Environ. Sci. Technol.* **1997,** *31*, 3588-3592.
41. McConnell J.S.; Hossner, L.R. *J. Agric. Food Chem.* **1985,** *33*, 1075-1078.
42. Liu, Y.; Gao, L.; Yu, L.; Guo, J. *J. Colloid Interface Sci.* **2000**, *227*, 164-170.
43. Fischer, K. *Chemosphere* **1991,** *22*, 15-27.
44. Fischer K. *Chemosphere* **1992,** *24*, 51-62.
45. Steber, J.; Wierich, P. *Chemosphere* **1986,** *15*, 929-945.
46. Steber, J.; Wierich, P. *Chemosphere* **1987,** *16*, 1323-1337.
47. Müller, G.; Steber, J.; Waldhoff, H. *Vom Wasser* **1984,** *63*, 63-78.
48. Weijnen, M.P.C.; van Rosmalen, G.M. *J. Crystal Growth* **1986,** *79*, 157-168.
49. Amjad, Z. *Langmuir* **1987,** *3*, 1063-1069.
50. Zieba, A.; Sethuraman, G.; Perz, F.; Nancollas, G.H.; Cameron, D. *Langmuir* **1996,** *12*, 2853-2858.
51. Sprankle, P.; Meggitt, W.F.; Penner, D. *Weed Sci.* **1975,** *23*, 229-234.
52. Glass, R.L. *J. Agric. Food Chem.* **1987,** *35*, 497-500.
53. Shoval, S.; Yariv, S. *Clays Clay Min.* **1979,** *27*, 9-28.
54. Maqueda, C.; Morillo, E.; Undabeytia, T. *Soil Sci.* **2002,** *167,* 659-665.
55. Gimsing, A.L.; Borggaard, O.K. *Clays Clay Min.* **2001,** *49,* 270-275.
56. Gimsing, A.L.; Borggaard, O.K. *Intern. J. Environ. Anal. Chem.* **2002,** *82,* 545-552.
57. Nowack, B.; Stone, A.T. *J. Phys. Chem. B* **2002,** *106*, 6227-6233.
58. Abdel-Aal, N.; Sawada, K. *J. Crystal Growth* **2003,** *256*, 188-200
59. Claessen, R.A.M.J.; Kolar, Z.I. *Langmuir* **2000,** *16*, 1360-1367.
60. Horstmann, B.; Grohmann, A. *Z. Wasser Abwasser-Forsch.* **1984,** *17*, 177.
61. Knepper, T.P.; Weil, H. *Vom Wasser* **2001,** *97*, 193-232.
62. Gratz, A.J.; Hillner, P.E. *J. Crystal Growth* **1993,** *129*, 789-793.
63. Gal, J.-Y.; Bollinger, J.C.; Tolosa, H.; Gache, N. *Talanta* **1996,** *43*, 1497.
64. Weijnen, M.P.C.; Rosmalen, G.M. In *Industrial Crystallization 84*; Jancic, S.J., DeJong, E.J., Eds.; Elsevier: Amsterdam, NL, 1984, p 61.
65. van der Leeden, M.C.; Rosmalen, G.M. In *Industrial Crystallization 84*; Jancic, S.J., DeJong, E.J., Eds.; Elsevier: Amsterdam, NL, 1984, p 325.
66. Benton, W.J.; Collins, I.R.; Grimsey, I.M.; Parkinson, G.M.; Rodger, S.A. *Faraday Discuss.* **1993,** *95*, 281-297.
67. Fischer, K.; Müller, G. *Naturwissenschaften* **1990,** *77*, 533-534.
68. Wolf, K.; Gilbert, P.A. In *The Handbook of Environmental Chemistry*; Hutzinger, O., Ed.; Springer: Berlin, GE, 1992, Vol. 3, Part F, pp 243-259.
69. Bucheli-Witschel, M.; Egli, T. *FEMS Microbiol. Rev.* **2001,** *25*, 69-106.
70. Nowack, B. *Environ. Sci. Technol.* **2002,** *36*, 4009-4016.
71. Bordas, F.; Bourg, A.C.M. *Aquat. Geochem.* **1998,** *4*, 201-214.
72. Jaworska, J.; Van Gendersen-Takken, H.; Hanstveit, A.; van de Plassche, E.; Feijtel, T. *Chemosphere* **2002,** *47*, 655-665.

Chapter 15

Reaction of Phosphonates and Phosphinopolycarboxylate in the Subsurface

Amy T. Kan, Gongmin Fu, and Mason B. Tomson

Department of Civil and Environmental Engineering, Brine Chemistry Consortium. Rice University, MS–519, Houston, TX 77005

Chemical scale inhibitors, e.g. phosphonates and polyacrylic acids, are used in nearly all industrial waters. However, there is little agreement regarding the primary mechanism by which the threshold scale inhibitors are adsorbed and later released. In this paper, the reactions of three phosphonates and one phosphinopolycarboxylic acid are compared. From these large numbers of studies and observations, the following conclusions are made: (1) Although formation mineralogy can be a factor, for many formation types, the inhibitor chemistry is also a primary factor. Adsorption or precipitation of phosphonate to calcite rich formation rock is essentially identical to the adsorption/precipitation of phosphonate to pure calcite. (2) When the inhibitor is injected into formations, a complex, well defined series of reactions take place: acid dissolution of calcite, phase separation, formation of a crystalline calcium phosphonate layer at the solid-liquid interface, slow dissolution of calcite; and formation of several different calcium phosphonate solid phases.

Introduction

Phosphonates and short chain polyacrylic acids are widely used in a broad variety of applications as scale inhibitors. Their ability to prevent precipitation of calcium and barium salts at substoichiometric concentrations (threshold inhibition) finds wide application in water treatment, water conditioning, cooling tower and oil and gas wells. These materials are also used as corrosion inhibitors, in industrial cleaning, metal finishing, and agrochemical applications. Three phosphonic acids [nitrilotris(methylene phosphonic acid) or NTMP, diethylenetriamine penta(methylene phosphonic acid) or DTPMP, and bis-hexamethylenetriamine penta(methylene phosphonic acid) or BHPMP) and one polyacrylic acid (phosphinopolycarboxylic acid or PPCA, 3600 MW) are commonly used as scale inhibitors in oil and gas production (See Figure 1), hence most produced waters contain inhibitors. Inhibitor is typically applied to the oil and gas well via inhibitor squeeze, where inhibitor solution (0.5-20% wt/vol.) of various acidity is pumped into a producing formation and allowed to react. A large portion of inhibitor is retained in the formation. There is little agreement regarding the primary mechanism by which threshold scale inhibitors are retained in the producing oil or gas well formation as a result of the squeeze procedure. In this paper, the interaction of scale inhibitors with calcite and calcite rich core material and how it relates to their retention and release with subsurface materials are discussed.

Solution Speciation And Metal Salt Solubilities

To predict the fate of inhibitors in diverse fluid compositions, it is necessary either to know or to be able to predict the simple acid-base and complexation equilibria of inhibitors and divalent metals, such as calcium, magnesium, barium, and iron. It is also essential to understand the effects of temperature, pH and ionic strength on these equilibria. The water or brine pH can be deduced from surface measurement of brine composition, T, P and production volumes (*1,2*). Theory and experience suggest that the effect of pressure on acid-base and complex reactions is both small and reasonably predictable to the accuracy needed from known and estimated partial molar volumes. Previously, several of these equilibrium have been described empirically to within experimental error by an electrostatic-type equation (*3-5*).

Xiao of this research group (*6*) has systematically developed the parameters using electrostatic theory (*7*) to address the solution speciation of polymers, e.g., phosphinopoly carboxylic acid (PPCA). A similar approach has been used to study protein ionizations and environmental chemistry of organic matter (*8*).

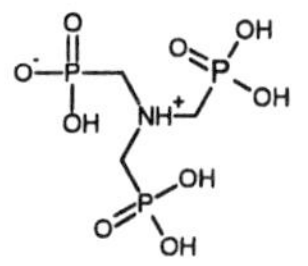

NTMP

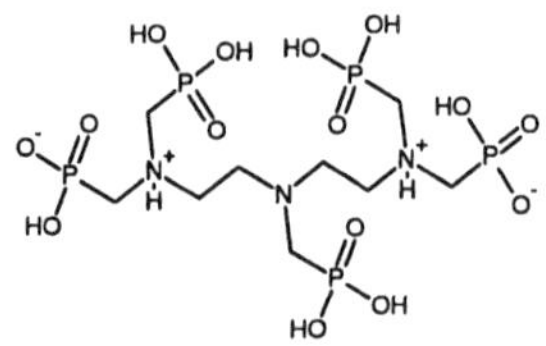

DTPMP

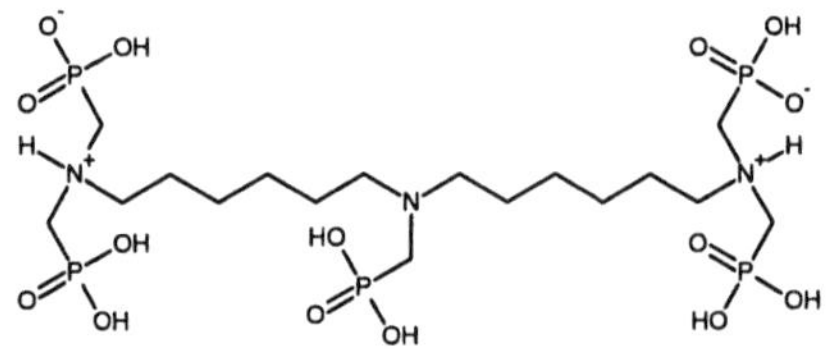

BHPMP

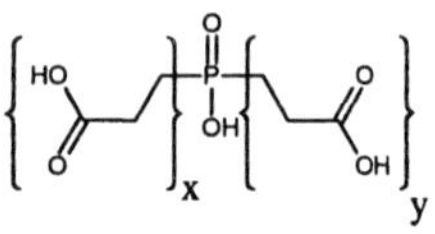

x + y =50

PPCA

Figure 1. Structure of common phosphonats and phosphinocarboxylate scale inhibitors.

It is proposed that the equilibrium constant for the ionization of any functional group can be considered as consisting of two parts. The first is a constant related to the breakage of the covalent bond between the oxygen and the hydrogen of a particular acid group, RO-H. This is represented as $K_{intrinsic}$. The second is variable that related to the fact that as a polyprotic molecule becomes progressively ionized there is increasing electrostatic energy (a_{elec}) required to separate a proton from the increasing negative charge of the molecule. This is treated as an electrostatic part and should be readily calculated from electrostatics theory and the dissociation fraction (θ_u). Therefore, the overall equilibrium constant, $K = K_{intrinsic} \cdot e^{a_{elec} \cdot \theta_u}$. The unit of polymer concentration used in calculating PPCA speciation is expressed as the acrylic acid monomer (A), which may be extended to polymers of different sizes. Similarly, the stoichiometry of the calcium salt is expressed as three Ca and two acrylic acid trimers. To our knowledge, this is the first attempt to establish a simple set of substituent constants for inhibitor acids at realistic oilfield conditions.

At 1.0 M ionic strength and 70°C, the stability constants for the complexation of Ca^{2+} with a dinegative phosphonate ion are $10^{1.12}$, $10^{1.24}$, and $10^{0.76}$ in the order of NTMP, DTPMP and BHPMP, respectively. This is comparable to the stability constants of Ca/methyl and aminomethyl phosphonates (see Smith and Martell: Critical stability constants). The average stability constant of Ca with twelve common monophosphonates is $10^{1.55 \pm 0.14}$ at 0.1 M I and 25°C. It should be noted that the stability constant for

$Ca^{2+} + HPO_4^{2-} \rightarrow CaHPO_4^0$ is about $10^{1.3}$ at 1 m I and 25 C. The difference between the stability constants of the polyphosphonates measured by this research group and those reported in the literature may be due to electrostatic effect of these molecules, or due to an ionic strength effect, or related to more complicated relations of structure, but clearly the complexation is mostly electrostatic in origin.

Using these speciation models, we are able to establish the solubility products of various metal phosphonate and metal-polymer salts and their ionic strength and temperature dependence by correlating the laboratory-measured solubilities to the equations (*4,5,9,10*). In Table 1 is listed the representative solubilities of various phosphonate salts at 1 M ionic strength and 70 °C. For all three phosphonate salts and the Ca-PPCA salt, we typically observed the formation of a high solubility amorphous phase of metal inhibitor salt when mixing the inhibitor with calcium at high concentrations (*4,5,9-11*). The amorphous high solubility material will eventually develop into a crystalline phase with much lower solubility, often by flowing brine over the metal inhibitor salt via a membrane filter to remove the readily dissolved amorphous Ca-Phn salt. The crystalinity of both Ca-NTMP and Ca-DTPMP solid phases have been confirmed by XRD analyses. The solubility of Ca-NTMP and Ca-DTPMP are very similar, while Ca-BHPMP is significantly more soluble than that of Ca-NTMP (approximately seven times higher than that of Ca-NTMP at 70 °C, 1 M ionic strength, 4000 mg/l Ca and 5.5 pH). The solubility of Ca-PPCA is lower than Ca-BHPMP and higher than Ca-NTMP. Note that the solubility product of Fe(II)-NTMP is many orders of magnitude lower than Ca-NTMP. Therefore, the solubility of Fe(II)-NTMP may also play a significant role in controlling the fate of phosphonate during production, even though the iron concentration is typically much lower than the calcium concentration in brine (*5*). Al Thubaiti et al. (*12*) observed an acidic calcium phosphonate phase at pH < 2.

Phosphonate/Calcite Interaction

The chemical/physical properties of phosphonates bear many similarities to those of orthophosphate (*3*). In calcareous soil, the predominant sorbent for phosphate is calcite. At low phosphate concentrations, the phosphate adsorption is strongly favored (*13-15*). Phosphonate and phosphate are adsorbed to the growth sites (kink site, step edges, and terraces) and further growth of calcite is therefore inhibited. In the presence of Ca^{2+}, the adsorption of phosphate and

Table 1. Comparison of different phosphonate and phosphinopolycarboxylate salt solubility at 1 M ionic strength and 70 C.

Inhibitor	Type	Ca/Fe (mg/L)	pH	Solubility (mg/L)
NTMP	Am., $Ca_{2.5}HNTMP$	4,000	5.5	174
	Cr., $Ca_{2.5}HNTMP$	4,000	5.5	0.92
	Cr., $Fe_{2.5}HNTMP$	40	5.5	0.095[3]
	Acidic, $Ca_{1.25}H_{3.5}NTMP$	4,000	3.0	930[4]
DTPMP	Am., Ca_3H_4DTPMP	4,000	5.5	250
	Cr., Ca_3H_4DTPMP	4,000	5.5	1.05
BHPMP	Am., Ca_4H_2BHPMP	4,000	5.5	385
	Cr. Ca_4H_2BHPMP	4,000	5.5	7.0
PPCA	Aged, $Ca_3(AAA)_2$	4,000	5.5	1.45

phosphonate to vaterite, calcite, barite, and goethite is much more strongly favored than in the absence of Ca^{2+} (*16-21*). Tomson et al. (*20*) proposed that the primary driving force for phosphonate adsorption is related to simple hydrophobic repulsion from solution of a macroneutral molecule and not, as is generally presumed, some specific phosphonate-surface interaction. At high concentration phosphate sorption occurs in two steps, the fast chemisorption of phosphate to a limited number of specific surface sites followed by the spontaneous crystal growth on the cluster of surface phosphate ions (heteronuclei) (*16,22*). Calcium phosphate crystals, e.g., amorphous calcium phosphate (*23*), dicalcium phosphate (*15*) or octacalcium phosphate (*22*) and/or calcium-carbonate-phosphate surface complexes (*24*) have been identified on calcite surfaces. Several authors have proposed the phosphate calcite reaction to be a surface precipitation of brushite that slowly transforms to the more stable octa-calcium phosphate (*25*). Stumm and Leckie (*26*) postulate that the reaction occurs in three steps: (1) chemisorption of P to form amorphous calcium phosphate; (2) slow transformation of the amorphous calcium phosphate to hydroxyapatite; and (3) crystal growth of apatite.

It is proposed that the retention of phosphonate by calcite also involves three steps, i.e., (1) the formation of an adsorbed layer of phosphonate on the calcite surface, (2) crystal growth of four to five layer thick of a calcium phosphonate solid phase with a stoichiometry of $Ca_{2.5}HNTMP$, and (3) growth of mixed calcium phosphonate solid phases (Figure 2). The Langmuir adsorption parameters are similar to that of phosphate/calcite reactions reported in the literature (*27*). Only approximately 8% of the surface is covered with NTMP before reaching adsorption maximum. Apparently, the adsorption is at the kink site or step edges. The crystal growth phase yields a constant solubility product of 24.11 ± 0.22 by assuming a stoichiometry of $Ca_{2.5}HNTMP$. The

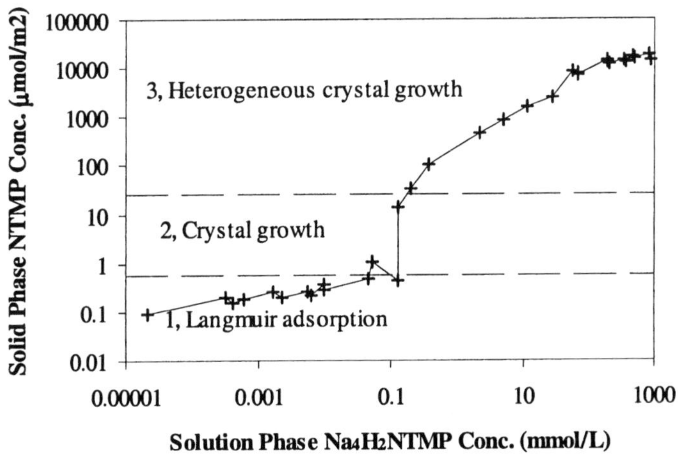

Figure 2. Adsorption and precipitation of Na_4H_2NTMP to calcite in 1 M NaCl solution at 70 C.

measured solid phase Ca/Phosphonate ratio is 2.31 ± 0.42. Note that this solid phase has never been identified before. The solubility product of this solid phase is different from the other three calcium phosphonate solubility products listed in Table 1. Interestingly, the 4.5^{th} root (i.e., per ion) of the solubility product of this new calcium phosphonate salt is equal to 5.34. The 9th root of the solubility product of hydroxyapatite is 5.42, which is remarkably similar to that of the calcium phosphonate phase. Since the other three calcium phosphonate salts of Table 1 (the acidic, amorphous and crystalline calcium phosphonate salts) were formed at much lower pHs (pH 1 - 6), they could be different from the solubility product measured in this study (pH 9). We expected the proposed solid phase to be the most thermodynamic stable phase for NTMP since the most basic form of calcium phosphate (hydroxyapatite) is also the most thermodynamic stable phase under many environmental conditions. It has been shown that this solid phase apparently controls the long term slow release of phosphonate following an inhibitor squeeze in numerous oil and gas wells (*28*) and similar results are expected in other subsurface calcareous deposits. In region 3, calcite dissolution was inhibited and the solutions were at lower pH, a condition that would allow the formation of the more acidic/amorphous calcium phosphonate phases. Similar phenomenon is well know with the different calcium phosphate salts. For example, brushite is the more stable solid phase than hydroxyapatite at pH 4.3, while hydroxyapatite is the least soluble phase at neutral pH (*29*). More detailed research is needed to further understand the interrelation of calcite dissolution and phosphonate precipitation at high concentrations.

Calcite Dissolution Kinetics

The rate and extent of calcite dissolution is a significant factor that controls the fate of scale inhibitor retention in the subsurface. Numerous papers have shown that calcite surface is poisoned when low concentrations of phosphonate is adsorbed onto calcite (*30*). The interrelation between calcite surface poisoning and inhibitor retention is not yet understood.

According to Plummer et al (*31*), the dissolution of calcite can be expressed as simultaneous reactions of calcite with H^+, $CO_{2,aq}$, and H_2O.

$$CaCO_3 + H^+ \xrightarrow{k_1} Ca^{2+} + HCO_3^-$$
$$CaCO_3 + CO_{2(aq)} \xrightarrow{k_2} Ca^{2+} + 2HCO_3^-$$
$$CaCO_3 + H_2O \xrightarrow{k_3} Ca^{2+} + HCO_3^- + OH^-$$

Below a pH of about 3.5, the rate controlling step is the first reaction. The dissolution rate of Iceland spar calcite was measured with both hydrochloric acid and with phosphonic acids, at the same concentrations and pHs, from 1.0 to about 2.75 pH. It was expected that the phosphonic acid would adsorb onto the calcite and inhibit dissolution. In fact, this did not occur over this pH range and the dissolution rate seemed to be essentially identical to that of hydrochloric acid at the same pHs and temperatures. In Figure 3 are plotted the logarithm of calcite dissolution rate versus pH in the presence of HCl and NTMP at 70°C of this study. By fitting data of Figure 3 to Eq. 1, a rate coefficient (k, cm/sec) of 0.0181 cm/sec for HCl (σ=0.0005 cm/sec, r=0.992) was observed in this study.

$$\log r = \log k - pH \qquad (1)$$

By assuming that the apparent rate coefficient is primarily diffusion limited, the diffusion layer thickness corresponds to 27 μm in this experiment. This dissolution rate coefficient compares reasonably to that reported in literature, e.g., that of Alkattan et al. (0.016 cm/sec at 50 °C and 0.025 cm/sec at 80 °C using a rotating disk experiment). The rate coefficient for NTMP/calcite reaction is 0.0126 cm/sec for NTMP (70 °C, σ=0.0005 cm/sec, r=0.965), which is only slightly slower than that of HCl (0.018 cm/sec). Similarly, Alkattan et al. reported a smaller calcite dissolution rate coefficient in the presence of phosphoric acid (0.00163 cm/sec, σ=0.00014, r=0.983 at 25 °C) than that of HCl (0.007 cm/sec at 25 °C). The above kinetics study would suggest that calcite dissolution kinetics should not be a limiting factor in inhibitor/calcite reaction. This contradicts previous observations that inhibitors inhibit calcite dissolution. Establishing this baseline of "no inhibition" of dissolution is important to any model of an inhibitor-solid reaction. In Figure 4 is shown the

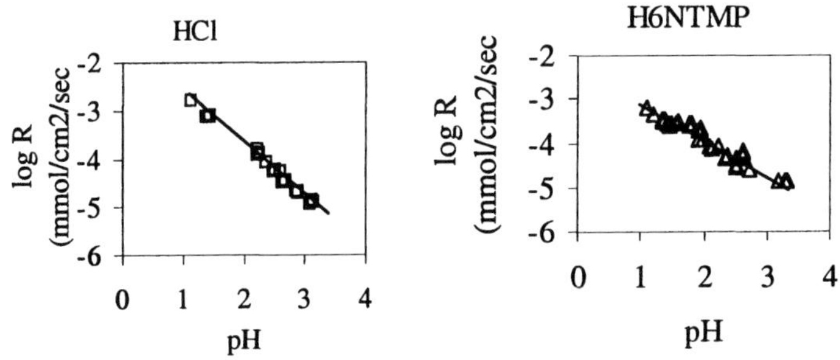

Figure 3. Plot of the logarithmic rate of calcite dissolution versus solution pH at 1 M ionic strength and 70 C.

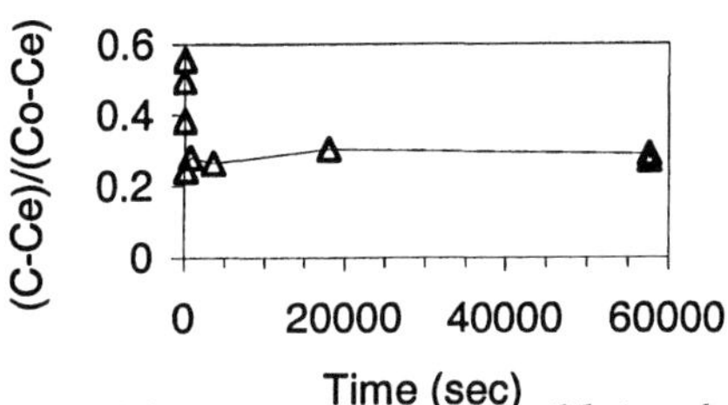

Figure 4. Plot of the progress toward equilibrium { $(C-Ce)/(C_0-Ce)$ } *versus reaction time for the calcite/phosphonate reactions.*

progress of calcite dissolution toward equilibrium versus reaction time, where the progress toward equilibrium is defined as { $(C-Ce)/(C_0-Ce)$ } and C is the solution phase dissolved carbonate concentration, Ce is the carbonate concentration at equilibrium with calcite, and C_0 is the initial carbonate concentration. Calcite dissolution was very fast between 0 - 200 seconds reaction time. The fast dissolution corresponds to the neutralization of the first two NTMP protons. However, nearly complete inhibition of calcite dissolution was observed after 200 second reaction time. This is presumably caused by surface poisoning of calcite with adsorbed phosphonate. Additional testing to identify the transition from diffusion controlled to surface inhibition controlled dissolution is ongoing and will be discussed in the future. These data imply that acidic phosphonate will dissolve calcite or other minerals to stoichiometric equivalents of at least 1 calcite per phosphonate during squeeze before the dissolution is inhibited by surface poisoning.

Comparison of Calcite to Calcium Rich Formation Rock

In calcareous soil, the predominant sorbent for phosphate is calcite. It is proposed that the predominant sorbent for phosphonate in calcite rich oil bearing formation is also calcite. The calcite rich oil bearing formation typically contains a large fraction of clay. The sorbent should be either clay or calcite. To definitely demonstrate that calcite is the primary sorbent, some electron microprobe studies were done on the solid that was adsorbed with phosphonate.

Formation material from a gas well in A. E. Guerra ranch, McAllen, TX was used. It contains similar fractions of clay and calcite. But, the specific surface area of clay should be ten times higher than the specific surface area of calcite. In Figure 5 is plotted an electron microprobe line scan of the A.E. Guerra core. The line scan is done over 20 μm distance. The EDAX spectroscopic intensity of Al, Ca, and P are used to trace the distribution of clay, calcite and phosphorus, respectively. The three granules are identified as two calcite particles and one clay particle by the spectrum of Al and Ca. The phosphorus signal is strongly associated with calcite and weakly associated with clay. Since the surface area of clay should be much higher than the surface area of calcite, the phosphorus signal should be strongly associated with clay if clay is the sorbent. The line scan clearly demonstrates that phosphonate is adsorbed on the calcite/solution interface.

In Figure 6 is plotted the adsorption/precipitation isotherm of phosphonate/calcite versus phosphonate/A.E.G. rock material. After normalization of the solid phase phosphonate concentration to the weight of calcite in AEG rock, it can be seen that the adsorption/precipitation reaction between phosphonate/calcite is very similar to that of phosphonate/AEG rock.

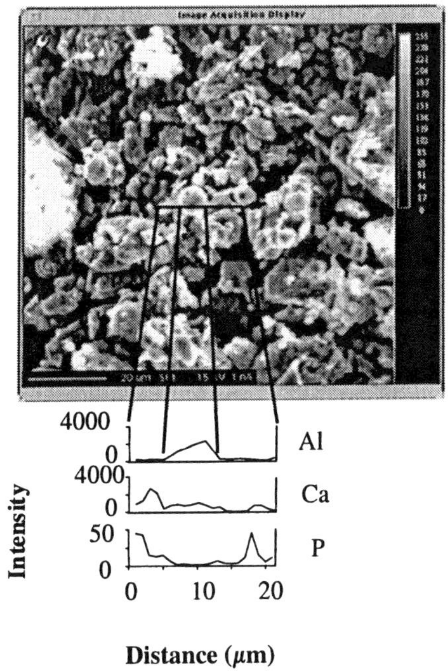

Figure 5. Electron microprobe analysis of phosphonate/ A.E.G. core material and EDAX line scan for Al, Ca, and P spectra over three particles.

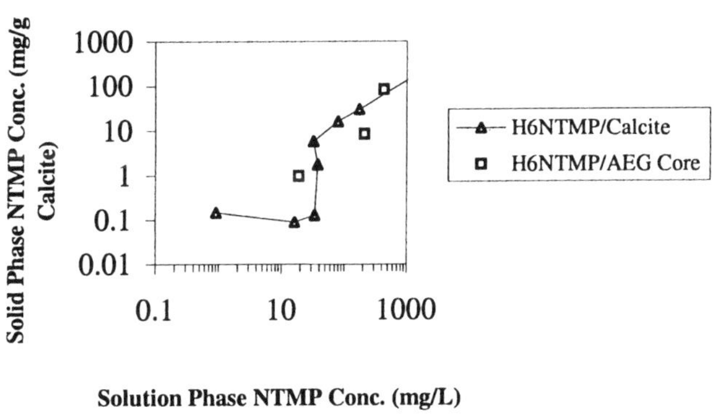

Figure 6. Plot of the adsorption/precipitation isotherm of H_6NTMP/calcite versus H_6NTMP/A.E.G. rock. The solid phase concentration is normalized to calcite weight.

Comparison of Inhibitors

The reactions between different scale inhibitors and A. E. Guerra core materials are compared to determine if the inhibitor/core reaction is primarily controlled by calcium phosphonate equilibruim dissolution. If the inhibitor core reaction is simple acid/base and precipitation/dissolution reactions with calcite, one would expect that the solution should be at equilibrium with respect to both calcite and calcium phosphonate phases after sufficient reaction time. In figure 7a is plotted the solution phase inhibitor concentrations of four inhibitor experiments at 0 and 48 hours reaction time. The initial inhibitor concentration of these four experiments are ~ 0.6% active (6,787, 6,057, 6,457, and 5,863 mg/L for NTMP, DTPMP, BHPMP, and PPCA, respectively). Even though the initial inhibitor concentrations are similar, the acidity of these four inhibitor solutions correspond to 141, 112, 112, and 62 meq/L, which is calculated from the equivalent weight of phosphonic acids and acrylic acid and the amount of mineral acids or base in the original inhibitor solution. The final inhibitor concentrations in these experiments varied considerably. The mass of inhibitor that precipitated are in the order of NTMP (81%) > DTPMP (72%) > BHPMP (42%) >PPCA (26%). In Figure 7b are comparisons of final solution pH, and Ca concentrations of these experiments. Since Ca and carbonate are produced from the acid/base reaction of the inhibitor with core, the solution phase carbonate concentration equals the dissolved Ca concentration prior to its precipitation with inhibitors. Therefore, the difference between the dissolved and solution phase Ca concentration is assumed to be equal to the amount of Ca that is precipitated with inhibitors. The amount of calcium dissolved follows the order of DTPMP>NTMP>BHPMP>PPCA. Even though the acidity of DTPMP and BHPMP solutions are similar, much less Ca is dissolved by BHPMP than by DTPMP. Consistent with the Ca data, the pH of the BHPMP solution is much lower than that of the other experiments and much less BHPMP is precipitated. The observed bulk solid phase Ca/Inh ratios are 1.45, 3, 2.4, and 0.54 for NTMP, DTPMP, BHPMP, and PPCA (in terms of monomers). The previously reported Ca/Inh ratios for these calcium-inhibitor salts are 2.5, 3, 4, and 0.5 (See Table 1). The Ca/Inh ratios for DTPMP and PPCA are similar to that reported previously, while the Ca/Inh ratios for NTMP and BHPMP precipitates are much less than that reported previously. These observations possibly imply that a lower Ca stoichiomtry and more soluble solid phases of NTMP and BHPMP salts are formed.

In Figure 7c is plotted the negative logarithm of the solution phase $CaCO_3$ ion concentration product (pIP). At 70 °C and 1 M ionic strength, the $CaCO_3$ conditional solubility product is $pK_{sp} = 6.96$ (dashed line). Note that the solution is undersaturated with respect to the specific solid phases when the data is above the dashed line and vice versa. Interestingly, only the PPCA solution appears to

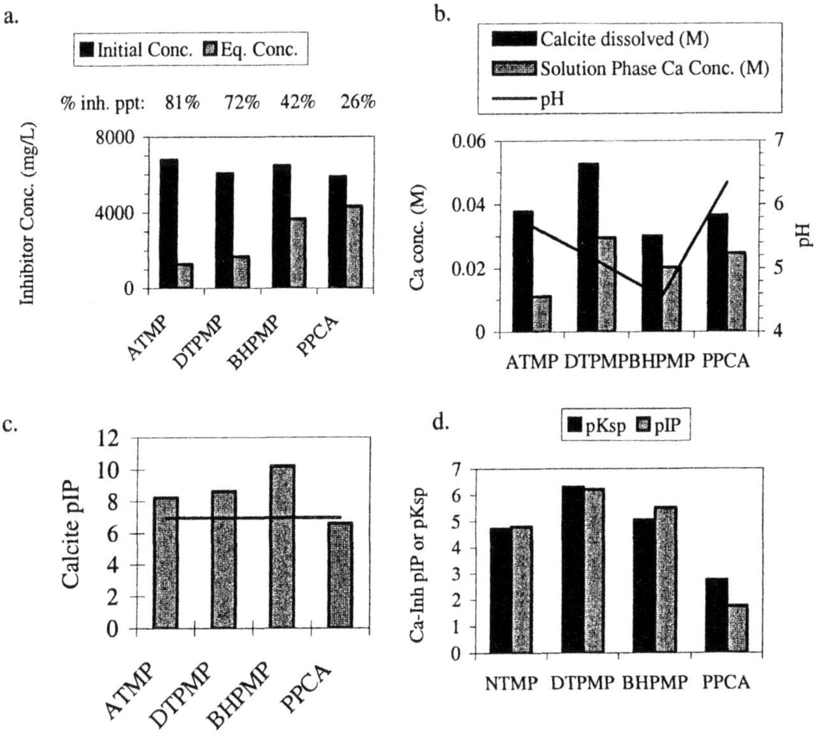

Figure. 7. Comparison of four inhibitor/A.E. Guerra core reactions. (a) Comparison of initial versus equilibrium inhibitor concentrations; (b) Comparison of the final solution pH, dissolved versus equilibrium solution phase Ca concentrations; (c) Comparison of the negative logarithm of the calcite ion concentration and calcite solubility products (solid line); and (d) Comparison of the root mean negative logarithm of the calcium inhibitor ion products versus solubility products.

be near saturation with respect to calcite. Calcite dissolution apparently is strongly inhibited by all three phosphonate inhibitors.

In Figure 7d is plotted the negative logarithm of the Ca inhibitor ion product versus the corresponding amorphous phase solubility product. In order to compare these solubility product and ion product results, negative logarithm of a root mean solubility product and ion product are plotted. The root mean solubility product or ion product is defined as the n-th root of the solubility product or ion product, where n is the sum of the number of stoichiometric ions (Eq. 2).

$$pIP_{\text{root mean}} = -\log\{([Ca^{2+}]^x[H^+]^y[Inh^{(2x+y)-}])^{1/n}\} \quad (2)$$
$$\text{where } n = x + y + 1$$

The calcium inhibitor ion products compare favorably, within expected error, to the amorphous phase solubility products for the three calcium phosphonate salts, indicating that the equilibrium phosphonate composition can be well represented by their amorphous solubility products. In Figure 5d, the calcium inhibitor ion product for the PPCA experiment is compared with the solubility product of the aged solid phase since the solubility product for the amorphous phase Ca-PPCA solid is not available. The data shows that there is a more soluble solid phase that exists and that the solubility product of the more soluble phase material is on the order of pKsp ~ 8.5, but more work is needed on this phase.

Conclusions

Phosphonate and phosphinopolycarboxylic acid interaction with calcite rich formation rock bear similarity with the corresponding reaction between orthophosphate with calcareous soil. At low inhibitor concentration, the reaction proceeds with a Langmuir type adsorption, followed by crystal growth of a low solubility calcium salt at the calcite water interface. At high inhibitor concentration, the inhibitor/calcite reaction is limited by solution phase Ca concentration due to surface inhibition controlled calcite dissolution. The high concentrations of phosphonate and dissolved calcium induce the formation of more soluble, less crystalline mixed solid phases of calcium inhibitor salts.

Acknowledgements

The financial support of the Center for Biological and Environmental Nanotechnology, the U.S. EPA Hazardous Substance Research Center/South & Southwest Region, and Rice University Brine Chemistry Consortium of companies: Aramco, B.J.-Unichem, Baker-Petrolite, Champion Technologies, Inc., Chevron-Texaco, Inc., ConocoPhillips, Inc., Marathon Oil, Nalco, Occidental Oil and Gas, is greatly appreciated.

References

(1) S. He; A. T. Kan; M. B. Tomson; Z. Amjad, Ed. (Eds.), *Water Soluble Polymers: Solution Properties and Applications*; Plenum Press: New York, 1997, pp 163-170.
(2) M. B. Tomson; A. T. Kan; G. Fu; L. Cong In *6th International Symposium on Oilfield Scale.*; SPE: Aberdeen, UK, 2004.
(3) M. B. Tomson; A. T. Kan; J. E. Oddo, *Langmuir 10*, (1994),1442-1449.
(4) L. M. Frostman; A. T. Kan; M. B. Tomson In *Calcium phosphates in biological and industrial systems*; Z. Amjad, Ed.; Kluwer Academic Publishers: Boston, MA, 1998, pp 493-506.
(5) S. Friedfeld "The Temperature and Ionic Strength Dependence of the Solubility Product Constant of Ferrous Phosphonate," Rice University, 1997.
(6) J. Xiao; A. T. Kan; M. B. Tomson, *Langmuir 17*, (2001),4661-4667.
(7) C. Tanford, *Physical Chemistry of Macromolecules*; John Wiley & Sons, Inc.: NY, 1967.
(8) E. Tipping, *Environ. Sci. Technol. 24*, (1990),1700.
(9) A. T. Kan; J. E. Oddo; M. B. Tomson, *Langmuir 10*, (1994),1450-1455.
(10) J. Xiao "Mineral Nucleation and Inhibition," Rice University, 2000.
(11) J. E. Oddo; M. B. Tomson, *App. Geochem. 5*, (1989),pp. 527-532.
(12) M. Al-Thubaiti; A. T. Kan; M. B. Tomson In *NACE 2004*; NACE: New Orlean, LA, 2004.
(13) J. C. Miltenburg; H. L. Golterman, *Hydrobiologia 364*, (1998),93-97.
(14) J. W. Bowden; S. Nagarajah; N. J. Barrow; A. M. Posner; J. P. Quirk, *Australian Journal of Soil Research 18*, (1980),49-60.
(15) C. V. Cole; S. R. Olsen; C. O. Scott, *Soil Sci. Soc. Am. Proc. 17*, (1953),352-356.
(16) K. Sawada; H. Ohtaki, Ed. (Eds.), *Crystallization Processes*; John Wiley & Sons: New York, NY, 1998.
(17) K. Sawada; S. Yoshida; T. Suzuki, *J. Chem. Soc. Faraday Trans. 88*, (1992),2227-2231.
(18) T. Suzuki; S. Inomata; K. Sawada, *J. Chem. Soc. Faraday Trans. 82*, (1986),1733-1743.
(19) F. Millero; F. Huang; X. Zheu; X.-H. Liu; J. Zhang, *Aquatic Geochemistry 7*, (2001),33-56.
(20) M. B. Tomson; G. Fu; M. A. Watson; A. T. Kan, *SPE Production and Facilities August 2003*, (2003),192-199.
(21) B. Nowack; A. T. Stone, *Environ. Sci. Technol. 33*, (1999),3627-3633.
(22) R. A. J. Griffin, J. J., *Soil Science Society of America Proceedings 37*, (1982),847-850.
(23) W. Stumm; J. O. Leckie In *Advan. Water Pollut. Res., Proc. Int. Conf., 5th*; S. H. Jenkins, Ed.; Pergamon: San Francisco, 1971.

(24) Y. Avnimelech, *Nature 288*, (1980),255-257.
(25) J. S. Freeman; D. L. Rowell, *J. Soil Sci. 32*, (1981),75-84.
(26) W. Stumm; J. O. Leckie In *Advances in Water Pollution Research*; Peregamon: Oxford, 1970, pp 1-16.
(27) A. T. Kan; G. Fu; M. B. Tomson, *J. Colloid Interface Sci (In press)*, (2004).
(28) M. B. Tomson; A. T. Kan; G. Fu In *SPE 6th International Symposium on Oilfield Scale*: Aberdeen, UK, 2004.
(29) W. E. Brown; E. J. Griffith, A. Beeton, J. M. Spencer, D. T. Mitchell, Ed. (Eds.), *Environmental Phosphorus Handbook*; John Wiley & Sons: New York, 1973, p 718.
(30) A. T. Kan; G. Fu; M. Al-Thubaiti; J. Xiao; M. B. Tomson In *SPE International Sympposium on Oilfield Chemsitry*: Houston, TX, 2003.
(31) L. N. Plummer; E. Busenberg, *Geochimica et Cosmochimica Acta 46*, (1982),1011-1040.

Chapter 16

Glyphosate

Anne Louise Gimsing[1] and Afonso Maria dos Santos[2]

[1]Department of Natural Sciences, The Royal Veterinary and Agricultural University, Thorvaldsensvej 40, 1871 Frederiksberg C., Denmark
[2] INQUIMAE and Departamento de Química Inorgánica, Analítica y Química Física, Facultad de Ciencias Exactas y Naturales, Universidad de Buenos Aires, Ciudad Universitaria Pabellŏn II 3er Piso, C1428EHA Buenos Aires, Argentina

The herbicide glyphosate (N-phosphonomethylglycine) interacts strongly with many soil components. It forms strong complexes with many metals in solution, and it is adsorbed through innersphere complexation to iron- and aluminium oxides. Glyphosate can also be adsorbed by clay minerals by forming complexes with interlayer cations. Because of these interactions, glyphosate is strongly adsorbed in soils. It is mainly the phosphonic acid moiety that participates in the adsorption, and therefore phosphate competes with glyphosate for adsorption sites.

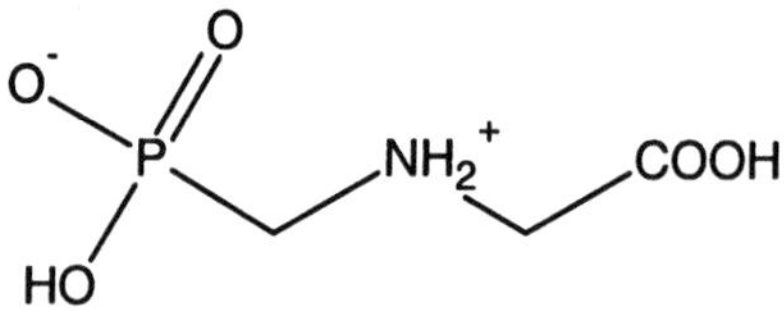

Scheme 1. Glyphosate chemical structure.

Physical and chemical properties

Glyphosate (*N*-phosphonomethylglycine, empirical molecular formula $C_3H_8NO_5P$), is a secondary aminomethylphosphonic acid. The structure is shown in scheme 1.

It is a white odourless solid, with a molecular weight of 169.1 g mol^{-1}. At 25 °C the solubility of glyphosate in water is 1.16 wt%. This relatively low solubility is due to strong intermolecular hydrogen bonds in the crystal lattice. Glyphosate is practically insoluble in organic solvents, and the log octanol/water partition coefficient (log K_{ow}) is -4.1. Glyphosate is very soluble in dilute bases and in strong acids, and it forms soluble salts. The monoanionic salts are used in the commercial herbicide formulations with glyphosate. The most common salts used in commercial products are the monoisopropylammonium (used in Roundup®) and the monotrimethylsulfonium (trimesium) salts (1).

The mechanism of action of glyphosate is the inhibition of the enzyme 5-enolpyruvylshikimate-3-phosphate (EPSP) synthase which produces EPSP from shikimate-3-phosphate and phosphoenolpyruvate in the shikimic acid pathway. EPSP synthase inhibition leads to a depletion of the aromatic aminoacids tryptophan, tyrosine and phenylalanine, all needed for protein synthesis or for biosynthetic pathways leading to growth.

The phosphonic acid group with its very stable C-P bond makes glyphosate extremely resistant to chemical hydrolysis, thermal decomposition and

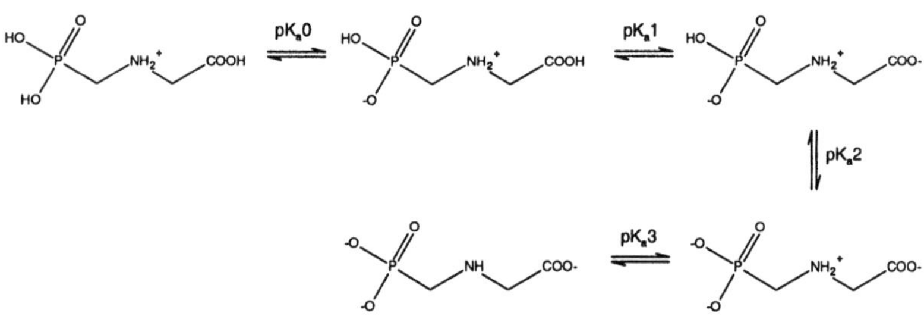

Scheme 2. Acid-base dissociation of glyphosate.

photolysis (2). All the three acid-base groups in glyphosate have the ability to dissociate in aqueous solution, and the overall dissociation of glyphosate is shown in scheme 2.

The pK_{a1}, pK_{a2} and pK_{a3} values (see Table 1) have been determined by different analytical methods: potentiometric titration (3-7) or NMR (8,9).

Table I. Acid dissociation constants for glyphosate. Blanks means that this information was not given.

pK_{a1}	pK_{a2}	pK_{a3}	Background electrolyte	Temperature	Reference
2.22	5.44	10.13	0.1 M NaCl	25 °C	(10)
2.53	5.68	10.25	0.1 M KCl		(11)
2.09	5.52	10.28	0.1 M NaCl	25 °C	(7)
2.21	5.46	10.15	0.1 M KNO_3	25 °C	(12)
2.13	5.37	10.03	0.2 M KCl	25 °C	(13)
1.88	5.37	10.03	0.2 M KCl	25 °C	(14)
2.24	5.53	10.30			(15)
2.23	5.46	10.14	0.1 M KNO_3	25°C	(9)
2.00	5.50	10.50		28°C	(7)
2.27	5.57	10.25	0.1 M KNO_3	25°C	(3)
2.32	5.86	10.86			(4)
2.60	5.60	10.60			(5)

Interactions with dissolved metals

The tendency of glyphosate to form coordination compounds with metal ions is noticeable, and glyphosate is capable of forming stable complexes with cations in solution. It has been found that glyphosate forms a 1:1 chelate complex with copper involving the carboxylate, amino and phosphonate group (16). The stability of its complexes clearly shows the tridentate character of this ligand. Barja and dos Santos Afonso (7) using FTIR-ATR found that in aqueous solution Fe(III) forms a 1:1 Fe(III)/glyphosate complex. Barja et al. (17) also determined that the solid state complexes of glyphosate with cobalt, aluminium and iron(III) consist of a chelate in which the phosphonate and the carboxylic acid groups are coordinated to the metal.

The amino group remains protonated, and does not coordinate the metal while the phosphonate moiety always binds the metal ion. The carboxylic acid group of glyphosate only coordinates the Fe(III) in solid state, but not in aqueous solution. Thus, the coordination of Fe(III) with glyphosate in aqueous medium differs from the results obtained in solid state.

Stability constants of glyphosate-metal complexes have been reported in many publications (3,9,18,19), and the values of the stability constants follow the sequence:

$$Fe^{3+} > Al^{3+} > Cu^{2+} > Zn^{2+} > Fe^{2+} > Mn^{2+} > Ca^{2+} \approx Mg^{2+} \quad \text{(1:1 chelates)}$$

$$Fe^{3+} > Al^{3+} > Cu^{2+} > Zn^{2+} > Fe^{2+} > Mn^{2+} \quad \text{(2:1 chelates)}$$

where 1:1 means 1 glyphosate molecule per cation while 2:1 means two glyphosate molecules per cation.

More recently the stability constant values were reported in IUPAC Technical Report 2001 (20).

Interactions with pure minerals phases: adsorption on oxides and clay minerals

Iron and aluminium oxides

IR-spectroscopy and X-ray diffractograms have shown that glyphosate is adsorbed as an inner sphere complex through the phosphonic acid moiety on goethite and gibbsite (10,21-24). Other investigations with atomic force microscopy have shown that glyphosate may be adsorbed to goethite through coordination of both the carboxylic acid group and the phosphonic acid group (25). Other findings showed that the carboxylic acid group does not participate in the adsorption (10,21,22). A possible explanation for these contrasting observations may be that the carboxylic acid group participates in the adsorption at low pH values, and not at near neutral and higher pH values (25).

Adsorption of different mono- and polyphosphonates onto goethite showed a strikingly similar adsorption behavior despite the different number of phosphonate moieties present in the ligand and the dissimilarity in their protonation level and net charge, which confirms that the adsorption takes place through the phosphonic acid moiety (26).

The Langmuir and Freundlich models are the most frequently employed models to describe adsorption isotherms. The adsorption isotherms of equilibrated suspensions of glyphosate on goethite approximate a Langmuir shape. The Langmuir constant (K_L) and maximum coverage (Γ_{max}) of a glyphosate-goethite system is resumed in Table II (27). Γ_{max} values were normalized with the area of the goethite for a better comparison. The pattern for pH dependence of the adsorption is similar to those previously reported for anion adsorption on hydrous ferric oxides (27).

Table II. Maximum adsorption densities ($\Gamma_{max.}$) and Langmuir constants (K_L) for Glyphosate

pH	3.1	4.0	5.0	6.1	7.2	8.2	9.2
$\Gamma_{max.}$ (μmol/m^2)	2.4	2.2	2.1	1.9	1.7	1.0	0.6
K_L (L/μmol)	0.3	0.24	0.11	0.07	0.03	0.03-0.02	0.01

Table II shows that the change in affinity of glyphosate for the surface of goethite (given by the value of ΔK_L) for a constant $\Delta pH \sim 1$ is: $\Delta K_L/\Delta pH = -0.06$, -0.13, -0.04, -0.04, -0.01 and -0.01. For the ΔpH between 4.0 and 5.0, the affinities of glyphosate are approximately 3.25 and 2 times larger than for the ones between pH 5.0 - 7.2 and 3.1 - 4.0, respectively. K_L values for glyphosate decreases 37 % from pH = 3.1 to pH = 5. The $\Delta K_L/\Delta pH$ values decrease at pH close to 8, reflecting the fact that the adsorption of glyphosate onto goethite is not favored due to the negatively charged oxide surface.

With respect to the Γ_{max}, the highest values are obtained at pH < 5.0 and the value decreases smoothly up to pH = 7. Γ_{max} for pH = 8.2 and 9.2, are almost 60 % and 35 % of the value of pH = 7.2. This trend in the value of the Γ_{max} was also observed in the methylphosphonic acid/goethite system (28) since the values of their Γ_{max} are identical for similar pH. In view of these results, it is reasonable to think that the amino and carboxylate groups which are present in the molecule of glyphosate are not directly involved in the adsorption to the surface of the oxide.

The suspension pH affects the adsorption because the pH determines the charge of the glyphosate molecule and the surface charge of the adsorbent. When the pH changes, the glyphosate molecule changes its charge from +1 to -3 (see Scheme 1 and Table I). In the pH range commonly encountered in soils the charge of glyphosate is –1 to –2. At pH values around 2, the surfaces of the iron oxides are highly protonated whereas the glyphosate molecule is zwitterionic, and therefore the positively charged iron oxides attract the negative moiety of the glyphosate molecule. As the pH approaches the point of zero charge of the iron oxides, the positive surface charge decreases and so does the adsorption of glyphosate. When pH is above the point of zero charge, both glyphosate and the iron oxides are negatively charged, and therefore they repel each other.

Barja and dos Santos Afonso (22) have also reported ATR-FTIR spectra of glyphosate/goethite suspensions. Figure 1 shows interfacial ATR-FTIR spectra of aqueous suspensions of goethite with glyphosate together with the correspondent aqueous solution of the ligand at the same pH value for the phosphonate frequency range (1200-950 cm^{-1}). Differences in both the position and number of bands can be observed when comparing the two spectra, which indicate that glyphosate is coordinated to the surface of the goethite particles. Similar interfacial spectra of glyphosate have previously been reported (10).

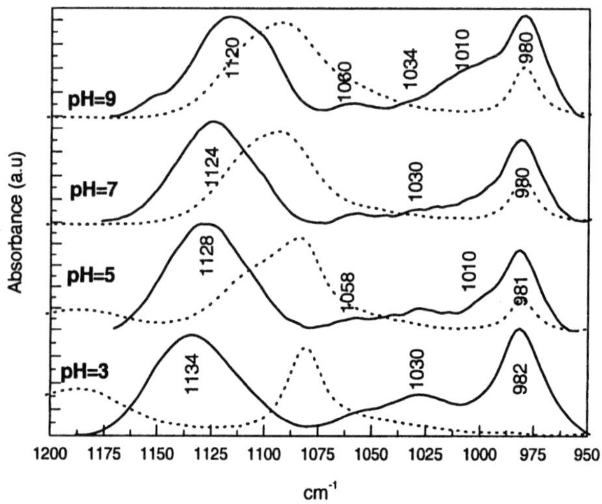

FIGURE 1. ATR-FTIR spectra of adsorbed glyphosate onto the surface of goethite at different pH values IR spectra were the result of 2000 co-added interferograms. Goethite suspensions and Glyphosate surface coverages were 60 g/L and 0.8 μmol/m² respectively, the ionic strength was I = 0.01 M in NaCl. The corresponding spectra in aqueous solution at the same pH values (dotted lines) are superimposed to the interfacial ones (solid lines).

The amino group of glyphosate is not coordinated to the surface of the goethite particles and the interaction of the carboxylate moiety of the herbicide with the surface of the oxide does not exist or it is extremely weak (10). These results suggest that the trend observed for the decreasing values of the ν_a stretching of the carboxylate group in aqueous solution is similar to what happens in presence of the oxide; the amino group remains protonated until the pH of the suspension gets close to its pKa value. These results also agree with what was observed for the 1:1 Fe(III)/glyphosate complex in aqueous solution (7) in which no important changes or shifts were observed for the $\nu_{a\ COO-}$, $\nu_{s\ COO}$

and δ_{NH2} modes when compared with the free ligand at similar pH values showing no evidence for amino or carboxylate coordination.

The phosphonic acid group in glyphosate may form both monodentate and bidentate surface complexes with goethite and gibbsite (10,21,22,24). According to Barja and dos Santos Afonso (21) the bidentate bridging complex ([(≡FeO)$_2$-P(O)-R] where R refers to the following group $^-OOCCH_2(NH_2)^+CH_2$) forms at all pH values and the monodentate complex ([≡FeO-P(O)(OH)-R]) forms at lower pH values (in their study it was found to form at pH 3 and 5). Sheals et al. (10) found that the binding mode was dependent of both pH and surface coverage. Thus, at pH 4.3 and 8.5 glyphosate was adsorbed as a monodentate complex irrespective of the surface coverage, whereas at pH 6 glyphosate could be adsorbed as a bidentate complex, if the surface coverage was low.

The adsorption process is characterized by the following reactions:

$$\equiv FeOH + H^+ \leftrightarrow \equiv FeOH_2^+ \qquad K_{s01}$$
$$\equiv FeOH \leftrightarrow \equiv FeO^- + H^+ \qquad K_{s02}$$
$$\equiv FeOH + R\text{-}P(O)(OH)\text{-}O^- + H^+ \leftrightarrow \equiv FeO\text{-}P(O)(OH)\text{-}R + H_2O \qquad K^s_1$$
$$\equiv FeOH + R\text{-}P(O)(OH)\text{-}O^- \leftrightarrow \equiv FeO\text{-}P(O)_2\text{-}R + H_2O \qquad K^s_2$$
$$2 \equiv FeOH + R\text{-}P(O)(OH)\text{-}O^- + H^+ \leftrightarrow (\equiv FeO)_2\text{-}P(O)\text{-}R + 2\, H_2O \qquad K^s_3$$

R is $^-OOCCH_2(NH_2)^+CH_2$. The acid-base surface constants and the stability constant of surfaces complexes of glyphosate are 7.47, -9.51, 10.5, 4.5 and 16.5 for log K_{s01}, log K_{s02}, log K^s_1, log K^s_2 and log K^s_3 respectively (29). K^s_1, K^s_2 and K^s_3 were calculated from the adsorption isotherms results using the Constant Capacitance Model and MINEQL 3.01 software. These values are comparable to those obtained by Nowack et al. (29) and Sheals et al. (10).

In conclusion, glyphosate forms surface complexes on goethite mainly through the phosphonic acid moiety, and the extent of the complexation is dependent on the ligand concentration in solution and the pH.

Clay minerals

Adsorption to clay minerals appear to be one of the most important factors affecting the behaviour and fate of pesticides and heavy metals in soils. Glyphosate can penetrate into the interlayer space of the clay silicates (30-33) and can also be adsorbed by clay silicates. It has been suggested that the primary adsorption mechanism is complexation of glyphosate to the exchangeable cations through water bridges (30-32) or by the mechanism of cation exchange (30,33). On the other hand, Morillo et al (35) suggested that the adsorption takes place only on the external surface of the mineral and not in its interlayer space. The adsorption of glyphosate seems to occur on the broken edges of the clay mineral (positions with variable charge) via the hydroxyl groups (35). It has been shown that glyphosate adsorption by smectite is due to hydrogen bonding of the acid functionalities to the exchangeable cations in the interlayer region (30,31).

Saturation of clay silicates with different cations affects the amount of glyphosate adsorbed. It is generally so that the highest stability constant for complex formation in solution between glyphosate and the metal ion correlates with the highest amount of glyphosate adsorbed (5,32,33). This may be seen as a confirmation of the hypothesis that sorption of glyphosate by clay is due to complex formation with the interlayer cations. Glass (34) found that the adsorption by cation-saturated montmorillonite increased following the sequence: $Na^{+}<Ca^{+2}<Mg^{+2}<Cu^{+2}<Fe^{+3}$. The ability of the cations to form coordination compounds or complexes is believed to be a factor responsible for the adsorption of glyphosate by these cation–saturated clays.

The minerals with pH-dependent charge probably adsorb glyphosate through an anion-exchange mechanism. The adsorption of glyphosate varied inversely with the pH of the clay suspensions. When glyphosate is adsorbed on clay silicates, low pH tends to increase the adsorption by suppressing the negative charge of glyphosate, and thereby increasing the attraction between the negatively charged clay silicate surfaces and the glyphosate molecule (33). When the pH is raised, glyphosate becomes more and more negatively charged and the attraction between clay silicate surfaces and glyphosate decreases, due to electrostatic repulsion. This repulsion decreases the adsorption.

The type of clay silicate may affect glyphosate adsorption. Thus, Dion et al. (35) found that glyphosate adsorption followed the sequence: illite > beidellite > kaolinite, and Glass (32) found that it followed the sequence: montmorillonite > illite > kaolinite. However, these differences may be due to the different clay silicate types, but it may also be due to different surface areas. Gimsing and Borggaard (36,37) reported that the adsorption by kaolinite was dependent on the surface area, whereas the clay type only had a limited effect on the adsorption, especially when compared with the high adsorption on oxides. In the studies by Dion et al. (35) and Glass (32) the surface areas were not reported. Barja et al. (23) showed that the adsorption sequences is iron oxides > kaolinite > illite when adsorption is normalized by surface areas.

Interactions with organic matter

Glyphosate not only interacts with the inorganic fraction of the soil. The organic matter in soil may also affect adsorption. Soil properties like organic carbon content, texture, pH and partly the phosphorus and oxide content ofdifferent soils from Southwestern Finland were tested against the adsorption coefficients (Koc) of ethofumesate, metamitron, phenmediphan, glyphosate and glufosinate-ammonium (39). The Koc values are also used as input values in many simulation models of pesticide fate. The most sensitive parameters of pesticide models are the sorption and degradation of the compound to be studied. The input form of the adsorption coefficient varies model by model and is often linked to the organic carbon or the organic matter content of the soil (39). The use of Koc values may not be feasible for herbicides with other sorption mechanisms than binding to the organic matter in soil. The greatest variation in the Koc values was observed with glyphosate, which was analyzed in 20 soil samples (39). The organic carbon content of the soils could not predict the mobility of glyphosate. However, for ionisable pesticides like glyphosate the normalization of the distribution coefficient to organic carbon may not be appropriate, as properties of the soil other than its organic carbon content, e.g. pH, mineral composition or phosphorus content, may be more important in regulating the sorption of such compounds (39,40). Furthermore, sorption coefficients determined under laboratory conditions may not be able to describe the leaching characteristics in the field (41).

Adsorption to organic matter occurs through hydrogen bonding (38) and is considered by several authors (44,45,46,47) to be comparable with or even stronger than adsorption to clay minerals. Although most authors indicate that glyphosate exhibits a low mobility in soils, some studies based on laboratory experiments show that this compound can be extensively mobile in certain soils (43,48,47). Maqueda et al. (47) found that glyphosate is highly adsorbed by natural fulvic acids, but the adsorption decrease in the presence of cations that form stable complexes with the herbicide, as is the case for Cu^{+2} and Al^{+3}, which forms a very stable complex with glyphosate (8,47), both cations are abundant in the soil studied. So soluble or adsorbed aluminium may decrease glyphosate adsorption and increase its mobility (42).

Piccolo et al. (38) found that the glyphosate adsorption capacity of four humic substances (humic substances extracted from peat, volcanic soil, oxidized coal and lignite) was extremely high compared to other adsorbents. The adsorption of glyphosate on the humic substances was suggested to be through hydrogen bonding between the various acidic and oxygen-containing groups (38). The adsorption could not be attributed to complexation with cations on exchange sites on the humic substances, because the humic substances had been purified and contained only 0,1 % polyvalent cations (38). On the other hand, the adsorption experiment was done in 0.01 M $CaCl_2$, so cations were present in the system.

Adsorption of glyphosate can also take place on metal-humic acid complexes, and Piccolo et al. (39) found that glyphosate adsorbed on an iron-humic acid complex was bound by ligand exchange, or that glyphosate in the solution could form hydrogen bonds with glyphosate already adsorbed. Glyphosate bound to an iron-humic acid complex by ligand exchange could not be desorbed suggesting strong bonding, whereas the hydrogen bonds are rather weak (39). On this basis it was concluded that humic acid complexes with polyvalent cations may be a major glyphosate adsorbent in soil.

Interactions with soils

Generally, the amounts of glyphosate adsorbed in soils are small compared with the amounts adsorbed by pure minerals. This fact is probably due to the sand present in soils, because quartz does not adsorb glyphosate (5,37). Table III show the adsorption on selected adsorbents.

In some studies it has been found that the content of iron and/or aluminium oxides are important for the adsorption of glyphosate, and that these soil constituents are the main adsorbents of glyphosate in soil (40-43). Other studies show that the clay content, clay type or CEC may be the most important soil factors for glyphosate adsorption (32,35).

The adsorbed percentage depends on the initial concentration of glyphosate. When the initial concentration is low, almost all of the applied glyphosate is adsorbed whereas the adsorbed percentage decreases as the initial concentration increases (40). This observation is in accordance with the reported isotherms for glyphosate, which are of the L-type (32,33,35), and have been fitted to Freundlich (32,33,35), the extended Freundlich (44) and the Langmuir isotherm (35,40).

There is some disagreement about the effect of pH on glyphosate adsorption in soil. Some studies state that adsorption of glyphosate is not strongly dependent on pH (5,32,45), whereas others find that there is a dependence on pH (33,34,42,44,46,47). The effect of pH may be due to the influence on the charge of the glyphosate molecule and the surface charge of the adsorbents.

The adsorption of glyphosate in soils was markedly affected by changes in the concentration of the background electrolyte (47), whereas glyphosate adsorption on goethite was only slightly influenced (48).

Adsorption kinetics

The adsorption/desorption of chemicals in natural soils is dependent not only on rapid and reversible equilibria but also on slow non-equilibrium time-dependent transport of the chemicals to soil surfaces and in the soil pore space (59). Although the batch equilibrium data cannot be used alone for predicting the leaching behavior of chemicals in soil, adsorption coefficients are needed as input data for simulations of the leaching potential with computer models (39).

The kinetics of glyphosate adsorption on pure minerals has been found to be fast. Adsorption of glyphosate on well-crystallized goethite, gibbsite and clay

Table III. Glyphosate adsorption on different adsorbents.

Adsorbent	Surface area [$m^2 g^{-1}$]	Glyphosate adsorption [mmol kg^{-1}]	Background electrolyte KCl; [M]	pH	Initial conc. [mM]	adsorbent gl^{-1}	Reference
Hematite	11.1	46.7		7.0	0.59		(33)
Goethite	50.7	127.8		7.0	0.59		(33)
Goethite	40	60	0.01	7.0	0.5	2.0	(48)
Goethite	43	91	0.01[a]	4.6	0.95	5.0	(49)
Gibbsite	43	85	0.1	7.0	0.5	2.0	(37)
Montmorillonite		23.7		4.0	1.2	20 - 40	(32)
Illite		4.7		4.0	1.2	20-40	(32)
Kaolinite		3.4			2.96	100	(35)
Illite		15.3			2.96	100	(35)
Montmorillonite	20	6.5	0.1	7.0	0.5	10	(36,37)
Illite	43	5.2	0.1	7.0	0.5	10	(36,37)
Kaolinite	12	3.9	0.1	6.0	0.24	20	(36,37)
Clay loam		4.3		7.5	0.6	80	(32)
Silt loam		3.3		5.8	0.6	80	(32)
Sandy loam		2.2		5.6	0.6	80	(32)
Illitic silty clay		12		6.78	2.96	100	(35)
Kaolinitic silty clay		7.5		7.77	2.96	100	(35)
Smectitic silty clay		13		7.5	2.96	100	(35)
Sandy loam		8.8	0.1	4.6	0.5	50	(42)
Sand		4.3	0.1	5.7	0.5	100	(42)
Sandy loam		3.1	0.1	5.7	0.5	100	(42)
Sand		1.8	0.1	5.9	0.5	100	(42)
Sandy loam		1.9	0.1	6.5	0.5	100	(42)

a) background electrolyte NaCl

minerals was virtually completed within a few hours, and there was no increase in adsorption after 2 days (37,48).

In contrast to this, the adsorption in soil has been found to be initially fast followed by a slower reaction, which may continue for days (5,42,43,50,51)

The increase in adsorption with time has been explained by clay-catalyzed chemical degradation, formation of kaolinite intercalates, and slow diffusion into the clay matrix (43). It is known that glyphosate forms an iron-glyphosate precipitate (6, 17), so a slow precipitation process may also explain the long-term reaction between glyphosate and soil (7).

The kinetics of the competitive adsorption and desorption of glyphosate and phosphate on goethite, gibbsite and in soils was recently modeled by Gimsing et al (62).

Competition with phosphate

The herbicide glyphosate and inorganic phosphate compete for adsorption sites in soil and on oxides. This competition may have consequences for the transport of both compounds in soil and hence for the contamination of groundwater (62).

In several studies it has been shown that glyphosate and phosphate compete for the same adsorption sites because glyphosate is mainly adsorbed through the phosphonic acid moiety in a way that resembles the adsorption of phosphate (5,35-37,42,44,51-53).

Phosphate has a very strong effect on the adsorption of glyphosate especially on pure oxides. It has been found that glyphosate adsorbed on an amorphous iron oxide was completely desorbed when an excess of phosphate was added (53). In another study with goethite and gibbsite it was shown that addition of phosphate before glyphosate resulted in a substantial reduction in the amount of glyphosate adsorbed, and phosphate was able to almost completely desorb glyphosate (36,37,54).

Phosphate also affects the adsorption of glyphosate on clay minerals. Dion et al. (35) found that as the initial level of phosphate increase, the adsorption of glyphosate on illite, kaolinite and in three clayey soils dominated by illite, kaolinite and smectite decreased. Other experiments with illite, montmorillonite and kaolinite showed that the competition for adsorption sites was not as pronounced on the silicates as on the oxides (37).

In soils the competition implies that the phosphate level appears to be the most important factor in determining the amount of glyphosate adsorbed in soils (5), and that glyphosate adsorption is correlated with unoccupied phosphate sorption capacity (52). It has also been shown that addition of 0.002 M orthophosphate strongly reduced the adsorption of glyphosate in a sandy loam soil, and that a long-term increase in the phosphorus status in agricultural soils lead to a reduction in glyphosate adsorption (44,47). Other studies with soils showed that the competition was much less pronounced in soils compared to oxides (42).

The competition between glyphosate and phosphate for adsorption sites supports the hypothesis that glyphosate is adsorbed through the phosphonic acid moiety.

Concluding remarks

Glyphosate interacts with dissolved metals, metal oxides, clays and soils, and it forms strong metal complexes with dissolved metals, some of which (like iron(III) complexes) have a very low solubility. Glyphosate adsorbs on solid phases such as metal oxides, clays and soils. The adsorption process is dependent on the mineral composition, pH, ligand concentration and temperature. The adsorption affinity follows the sequence: iron oxides >aluminium oxide > clays > soils > quartz. Experiments with illite, montmorillonite and kaolinite showed that the competition for adsorption sites was not as pronounced on the silicates as on the oxides. The adsorption in soils is lower than on pure minerals because soils generally contain many minerals, like quartz, that do not adsorb glyphosate. The adsorption of glyphosate is an inner sphere process with the formation of surface complexes with the phosphonate group bounded to the surface metal ion. Two types of surface complexes have been determined by ATR-FTIR analytical technique, and the adsorption constants have been calculated. Phosphate and dissolved metals may influence the environmental fate of glyphosate in two different ways. Phosphate competes with glyphosate for similar surfaces sites while dissolved metals have an effect on the sorption process by modifying the glyphosate concentration in the aqueous phase due to metal/glyphosate complex formation. Both of them, competition between glyphosate and phosphate and metal/glyphosate complex formation are important when modeling the transport of glyphosate in soil and water.

References

1. Franz, J. E.; Mao, M. K.; Sikorski, J. A. Glyphosate - A unique global herbicide. ACS Monograph 189. 1997. Washington DC, USA, American Chemical Society.

2. Kononova, S. V.; Nesmeyanova, M. A. *Biochem-Moscow* **2002,** *67*(2), 184-195.

3. Madsen, H. E. L.; Christensen, H. H.; Gottliebpetersen, C. *Acta Chem Scand Series A-Physicl Inorg Chem* **1978,** *32*(1), 79-83.

4. Wauchope, D. *J Agric Food Chem* **1976,** *24*(4), 717-721.

5. Sprankle, P.; Meggitt, W. F.; Penner, D. *Weed Sci* **1975,** *23*(3), 229-234.

6. Subramaniam, V.; Hoggard, P. E. *J Agric Food Chem* **1988,** *36*(6), 1326-1329.

7. Barja, B. C.; Afonso, M. D. *Environ Sci Techn* **1998,** *32*(21), 3331-3335.

8. Appleton, T. G.; Hall, J. R.; Mcmahon, I. J. *Inorg Chem* **1986,** *25*(6), 726-734.
9. Motekaitis, R. J.; Martell, A. E. *J Coord Chem* **1985,** *14*(2), 139-149.
10. Sheals, J.; Sjoberg, S.; Persson, P. *Environ Sci Techn* **2002,** *36*(14), 3090-3095.
11. Kobylecka, J.; Ptaszynski, B.; Zwolinska, A. *Monatshefte fur Chemie* **2000,** *131*(1), 1-11.
12. Daniele, P. G.; DeStefano, C.; Prenesti, E.; Sammartano, S. *Talanta* **1997,** *45*(2), 425-431.
13. Buglyo, P.; Kiss, T.; Dyba, M.; JezowskaBojczuk, M.; Kozlowski, H.; Bouhsina, S. *Polyhedron* **1997,** *16*(19), 3447-3454.
14. JezowskaBojczuk, M.; Kiss, T.; Kozlowski, H.; Decock, P.; Barycki, J. *J Chem Society-Dalton Transact* **1994,**(6), 811-817.
15. Castellino, S.; Leo, G. C.; Sammons, R. D.; Sikorski, J. A. *Biochemistry* **1989,** *28*(9), 3856-3868.
16. Sheals, J.; Persson, P.; Hedman, B. *Inorg Chem* **2001,** *40*(17), 4302-4309.
17. Barja, B. C.; Herszage, J.; Alfonso, M. D. *Polyhedron* **2001,** *20*(15-16), 1821-1830.
18. Smith, P. H.; Raymond, K. N. *Inorg Chem* **1988,** *27*(6), 1056-1061.
19. Dhansay, M. A.; Linder, P. W. *J Coord Chem* **1993,** *28*(2), 133-145.
20. Popov, K.; Ronkkomaki, H.; Lajunen, L. H. J. *Pure Appl Chem* **2001,** *73*(10), 1641-1677.
21. Barja, B. C.; Afonso, M. D. *Abstracts of Papers of the American Chemical Society* **2000,** *220* U363.
22. Barja, B. C.; dos Santos Afonso, M. *Environ Sci Techn* **2004,** submitted.
23. dos Santos Afonso, M.; Barja, B. C.; Pessagno, R. C.; Tevez, H. R. Biogeochemistry of Chelating Agents Symposium **2003**, 226 th ACS National Meeting. New York, USA, ACS.
24. Dubbin, W. E.; Sposito, G.; Zavarin, M. *Soil Sci* **2000,** *165*(9), 699-707.
25. Dideriksen, K.; Stipp, S. L. S. *Geochim Cosmochim Acta* **2003,** *67*(18), 3313-3327.
26. Nowack, B.; Stone, A. T. *J Colloid Interface Sci* **1999,** *214*(1), 20-30.
27. Stumm, W. Chemistry of the Solid-Water Interface. 1992. New York, USA, Wiley & Sons.
28. Barja, B. C.; Tejedor-Tejedor, M. I.; Anderson, M. A. *Langmuir* **1999,** *15*(7), 2316-2321.
29. Nowack, B.; Stone, A. T. *Envirol Sci Techn* **1999,** *33*(20), 3627-3633.
30. Mcconnell, J. S.; Hossner, L. R. *J Agric Food Chem* **1989,** *37*(2), 555-560.
31. Shoval, S.; Yariv, S. *Clays Clay Min* **1979,** *27*(1), 19-28.
32. Glass, R. L. *J Agric Food Chem* **1987,** *35*(4), 497-500.
33. Mcconnell, J. S.; Hossner, L. R. *J Agric Food Chem* **1985,** *33*(6), 1075-1078.

34. Morillo, E.; Undabeytia, T.; Maqueda, C. *Environ Sci Techn* **1997,** *31*(12), 3588-3592.
35. Dion, H. M.; Harsh, J. B.; Hill, H. H. *J Radioanal Nuclear Chem* **2001,** *249*(2), 385-390.
36. Gimsing, A. L.; Borggaard, O. K. *Intern J Environ Anal Chem* **2002,** *82*(8-9), 545-552.
37. Gimsing, A. L.; Borggaard, O. K. *Clay Min* **2002,** *37*(3), 509-515.
38. Piccolo, A.; Celano, G.; Conte, P. *J Agric Food Chem* **1996,** *44*(8), 2442-2446.
39. Piccolo, A.; Celano, G.; Pietramellara, G. *Sci Total Environ* **1992,** *123* 77-82.
40. Piccolo, A.; Celano, G.; Arienzo, M.; Mirabella, A. *J Environ Sci Health Part B-Pestic Food Contam Agric Wastes* **1994,** *29*(6), 1105-1115.
41. Morillo, E.; Undabeytia, T.; Maqueda, C.; Ramos, A. *Chemosphere* **2000,** *40*(1), 103-107.
42. Gimsing, A. L.; Borggaard, O. K.; Bang, M. *Europ J Soil Sci* **2004,** 55, 183-191.
43. Gerritse, R. G.; Beltran, J.; Hernandez, F. *Aust J Soil Res* **1996,** *34*(4), 599-607.
44. de Jonge, H.; de Jonge, L. W.; Jacobsen, O. H.; Yamaguchi, T.; Moldrup, P. *Soil Sci* **2001,** *166*(4), 230-238.
45. Cheah, U. B.; Kirkwood, R. C.; Lum, K. Y. *Pestic Sci* **1997,** *50*(1), 53-63.
46. Nicholls, P. H.; Evans, A. A. *Pestic Sci* **1991,** *33*(3), 331-345.
47. de Jonge, H.; de Jonge, L. W. *Chemosphere* **1999,** *39*(5), 753-763.
48. Gimsing, A. L.; Borggaard, O. K. *Clays Clay Min* **2001,** *49*(3), 270-275.
49. Maqueda, C.; Morillo, E.; Undabeytia, T. *Soil Sci* **2002,** *167*(10), 659-665.
50. Lafleur, K. S. *Soil Sci* **1979,** *127*(2), 94-101.
51. Sprankle, P.; Meggitt, W. F.; Penner, D. *Weed Sci* **1975,** *23*(3), 224-228.
52. Hance, R. J. *Pestic Sci* **1976,** *7*(4), 363-366.
53. Mcbride, M.; Kung, K. H. *Soil Sci Soc Am Jl* **1989,** *53*(6), 1668-1673.
54. Borggaard, O. K.; Szilas, C.; Gimsing, A. L.; Rasmussen, L. H. *Geoderma* **2004,** *118*(1-2), 55-61.

Chapter 17

Rates and Mechanisms of $Co(II)EDTA^{2-}$ Interactions with Sediments from the Hanford Site

Melanie A. Mayes, X. L. Yin, M. N. Pace, and Philip M. Jardine

Environmental Sciences Division, Oak Ridge National Laboratory, P.O. Box 2008, MS–6038, Oak Ridge, TN 37831

The migration of ^{60}Co from buried radioactive waste in the Hanford subsurface may be facilitated by chelation with organics such as EDTA. The goal of this research was to quantify the rates and mechanisms of $Co(II)EDTA^{2-}$ interactions with twenty Hanford subsurface sediments using kinetic batch experiments. Our results suggested that oxidation to $Co(III)EDTA^{-}$ occurred in all sediments, oxidizing up to 75% of the initial $Co(II)EDTA^{2-}$ concentrations. Oxidation was sustained over 30 days, though rates typically decreased as a function of time. Dissociation of the $Co(II)EDTA^{2-}$ complex and adsorption/precipitation of free Co^{2+} tended to increase with contact time. Correlations were derived between sediment Mn content and 1) the rate of $Co(II)EDTA^{2-}$ disappearance, and 2) the production of $Co(III)EDTA^{-}$. These correlations will provide an improved mechanistic understanding and predictive capability of the interactions of chelated metals in the Hanford subsurface.

Introduction

At the U.S. Department of Energy (DOE) Hanford Reservation in southeastern Washington, prolonged production of plutonium and other nuclear weapon components generated large quantities of radioactive waste. High-level wastes were co-disposed with complex, caustic mixtures of sodium hydroxide and nitrate in large underground storage tanks. Approximately 200 million liters of high-level waste in 177 storage tanks contained 400 million Ci of radioactivity *(1)*. Sixty-seven of the 149 single-shelled tanks have leaked to varying degrees into a deep vadose zone (~100 m) where recharge is minimal since the region experiences low annual precipitation (~170 mm). Past releases of radionuclides have altered the distribution of moisture in the vadose zone, causing groundwater levels to rise ~30 m below the tanks *(2)*. Further, the areal extent of contamination is great, consisting of 1900 waste sites and 500 contaminated facilities over the large area (1700 km^2) of the Reservation. Predictions of radionuclide migration have generally been unsuccessful in matching the observed subsurface distribution, thus transport of radionuclides has been described as "accelerated" *(3)*.

Both hydrological and geochemical factors contribute to such "enhanced" mobility of radionuclides at Hanford. The subsurface media consists of unconsolidated sediments of Miocene-Pleistocene age, which are diverse in terms of grain size and mineralogy. Hydrologic processes potentially contributing to accelerated transport include preferential finger and funnel flow *(4- 6)*, thin "film flow" along borehole walls *(7)*, and lateral transport within fine-grained sedimentary units *(3, 8-11)*. Geochemical factors contributing to enhanced migration include extreme temperatures within the tanks due to radioactive decay *(12)*, the concentrated, caustic nature of the tank waste ($pH \geq 10$, $I \cong 1\text{-}10$ M) which may dissolve and re-precipitate subsurface minerals *(13)*, and co-disposal with organic complexants *(14)*. During the REDOX method of fuel reprocessing, the procedure for separation of U and Pu involved the addition of chelating agents to ensure all constituents remained in solution *(2)*. Further, millions of kilograms of organic complexants including citrate, EDTA, HEDTA, and glycolate, were used to enhance the recovery of U and ^{90}Sr, and these solutions were subsequently discharged into the underground storage tanks. Analyses of tank wastes indicate that some carbon compounds have survived the high temperatures of the tanks, though high concentrations of carbonate suggest that considerable degradation has also occurred *(2)*. However, many tank leaks occurred early in the Hanford's history, so it is likely that chelators and metal-chelate complexes leaked into the vadose zone before and during high temperature events in the waste tanks. In the TX and TY Waste Management Areas (WMA), in particular, EDTA and ^{60}Co were both significant components of the waste stream, and some of these tanks have discharged waste into the subsurface. In addition, a large (~435,000 L) leak from the T-106 tank resulted in the discharge of ^{60}Co into the subsurface *(2)*. Observed ^{60}Co migration through the vadose zone is substantial, as observed from drywell gamma logs, which is probably linked to the formation of stable complexes with EDTA *(2)*. In the B-BX-BY WMA, the migration of ^{60}Co is occurring at a rate of ~0.5 m y^{-1} *(9)*, which is several orders of magnitude greater than the infiltration rate. These observations suggest that radionuclides are continuing to migrate through the vadose zone, and the relatively rapid rate of ^{60}Co migration suggests that the metal may be present in a chelated form.

The purpose of this study is to quantify the rates and mechanisms of $Co(II)EDTA^{2-}$ interactions using batch kinetic studies with twenty different Hanford subsurface sediments. Multiple sediment types were used because of the area, depth, and sedimentary diversity of the waste disposal area. Mass balance was used to quantify dissociation, adsorption, and oxidation reactions at natural sediment pH. Selective extractions were utilized to characterize soil chemical properties. The relationship between observed $Co(II)EDTA^{2-}$ reaction and soil chemical properties was quantified using statistical correlations. These results will allow the reactivity of $Co(II)EDTA^{2-}$ to be predicted using known sediment geochemistry in the Hanford subsurface.

Materials and Methods

The Hanford Reservation is located within a large bend in the Columbia River, consequently all subsurface deposits reflect past depositional environments of the river system. The vadose and saturated zones are comprised of three sedimentary formations with different depositional environments: fluvial (Middle Ringold Formation), alternating fluvial and lacustrine sedimentation (Upper Ringold Formation), periodic desiccation and soil development (Plio-Pleistocene Unit), and successive glacial outwash deposits (Hanford flood deposits) *(15)*. Subsequent weathering may have also altered the oxidation state and the distribution of some reactive minerals (e.g., Mn, Al, and Fe oxides), particularly in the older Ringold (Pliocene age) and Plio-Pleistocene units. The Hanford Hff and Hfc sediments were obtained from approximately 20m below ground surface in the Environmental Restoration Disposal Facility (ERDF) in the 200W area *(10, 15)*, while the five IDF sediments were composited from an exploratory borehole in the proposed Interim Disposal Facility (IDF) in the 200E area *(16)*. The sample number corresponds to the depth below ground surface (in feet) of the IDF composites (Table I). Samples of the Plio-Pleistocene Unit *(15, 17)* and the Upper Ringold Fm. *(11, 18)* were collected from various beds in the White Bluffs, which is located at the northern boundary of the Reservation. Samples of the Middle Ringold Fm. were

Table I. Hanford area sediments, extractable oxides, distribution coefficients of $Co^{(II/III)}$ and $Co(II)EDTA^{2-}$, and 1st order rate coefficients.

Sediment	*Formation*	*Mn* ($g\ kg^{-1}$)	*Al* ($g\ kg^{-1}$)	*Fe* ($g\ kg^{-1}$)	K_d ($cm^3\ g^{-1}$) $Co(II)EDTA^{2-}$	K_d ($cm^3\ g^{-1}$) $Co^{(II/III)}$	*Rate* (h^{-1}) *loss* $Co(II)EDTA^{2-}$	*Rate* r^2
Hfc	Hanford	0.13	0.95	6.62	2.41	0.73	2.01 e-4	0.602
Hff	Hanford	0.11	0.23	4.90	4.93	1.79	4.65 e-4	0.954
IDF45	Hanford	0.12	0.85	5.13	2.94	0	2.63 e-4	0.935
IDF110	Hanford	0.12	0.85	6.00	3.22	0	1.17 e-3	0.909
IDF150	Hanford	0.14	0.97	5.68	3.90	0	3.56 e-4	0.949
IDF200	Hanford	0.15	0.92	6.80	1.85	0	1.85 e-4	0.808
IDF215	Hanford	0.09	1.15	7.05	1.31	0	1.84 e-4	0.899
PP	Plio-Pleistocene	0.04	0.31	1.61	3.00	1.06	2.36 e-4	0.796
US	U. Ringold	0.18	0.36	7.49	18.9	1.49	1.17 e-3	0.992
LS	U. Ringold	0.15	0.25	6.62	8.61	0.39	8.67 e-4	0.967
SS4	U. Ringold	1.67	1.44	18.4	345	4.14	5.24 e-4	0.976
SS4a	U. Ringold	0.15	2.48	46.3	2.87	0	5.54 e-4	0.768
SS4b	U. Ringold	0.15	0.69	16.4	5.23	0	5.79 e-4	0.911
S5	U. Ringold	1.88	2.20	23.1	373	1.60	5.45 e-4	0.909
PS2	U. Ringold	0.40	1.14	5.25	110	0.87	4.40 e-3	0.975
PS1	U. Ringold	0.23	0.67	1.68	52.0	2.35	3.22 e-3	0.988
WB§	U. Ringold	4.00	n/a	25.3	147	2.24	3.64 e-3	0.930
TF1	M. Ringold	0.14	0.29	19.3	2.64	0	3.91 e-4	0.870
TF2	M. Ringold	0.04	0.01	6.54	0.89	0.63	1.17 e-4	0.817
TF3	M. Ringold	0.96	1.45	55.7	41.0	0	3.35 e-3	0.864
TF4	M. Ringold	0.91	0.01	19.4	41.0	0.92	2.68 e-3	0.967

§ Extraction data taken from Barnett et al., (2000).

collected from the Taylor Flats area, which is located to the east of the Reservation. The sediments, which are mostly sands with some silts and clays, were sieved to <2 mm, removing associated gravels. The total reactive oxides of Fe, Al, and Mn were quantified by the selective citrate-bicarbonate-dithionite (CBD) extraction method *(19)* and analyzed by ICP/MS (Perkin-Elmer Elan-6100) (Table I). Total organic and inorganic carbon were quantified on HCl-pretreated and untreated samples using standard combustion techniques (Table II). Surface area was measured using the Brunauer, Emmett and Teller (BET) method *(20, 21)* (Table II).

Table II. Total organic carbon (TOC), inorganic carbon (TIC), and surface area (SA) of select Hanford area sediments.

Sediment	*TOC (%)*	*TIC (%)*	*SA (m^2 g^{-1})*
Hff	0.02	0.22	6.13
Hfc	n/a	n/a	7.26
PP	0.32	2.68	11.21
US	0.02	0.26	9.11
LS	0.01	0.28	n/a

Influent $Co(II)EDTA^{2-}$ was made by adding 0.5mM $CoCl_2{\cdot}6H_2O$ to water, then slowly adding 0.5mM $Na_2H_2EDTA{\cdot}2H_2O$ in order to minimize the formation of $Co(III)EDTA^-$. Analyses confirmed 100% complexation of $Co(II)EDTA^{2-}$ and absence of $Co(III)EDTA^-$. After this was completed, 0.1 M NaCl was added and pH was adjusted using 1N NaOH to be equivalent to soil pH (~8). This high ionic strength was chosen for relevance to the concentrated waste stream at Hanford. Kinetic batch experiments were conducted in polypropylene centrifuge vials using 1g of soil with 20ml of reactant solution. Initial (C_o) concentrations were determined by the addition of solution to tubes without sediments. Samples and blanks (C_o) were both equilibrated on a reciprocal shaker for different time intervals up to 30 days. Samples were centrifuged at 2000 rpm and solutions were decanted for analyses. Subsamples were decanted and immediately analyzed for pH.

Direct analyses for $Co(II)EDTA^{2-}$ and $Co(III)EDTA^-$ were performed by using a liquid ion chromatograph (Dionex IC, model DX-600), and Dionex Peaknet data collecting software. A standard eluant mixture was passed through the IonPac AS4A anion exchange column. The Dionex AD25 Absorbance detector was used for the detection of UV/Vis wavelength emissions with an automated switch from 254 nm for the detection of $Co(III)EDTA^-$ to 190nm for $Co(II)EDTA^{2-}$. Detailed methodology is published in *(22)*. EDTA concentration was analyzed by Total Organic Carbon Analyzer (Shimadzu TOC-5000), and the total metal concentration was determined by ICP/MS. Analytical error was typically within (<5%) for the analytical instruments, which was determined by the use of standard curves. Improved methods exist for the detection of all chelated components on a single instrument *(23, 24)*.

Mass Balance and Statistical Correlations

Mass balance was used to quantify $Co(II)EDTA^{2-}$ reactivity by comparison of samples without soil (blanks) to samples with soil (Table III). The concentrations of the blanks were observed to be stable over the 30 d time period. The amount of Co and EDTA adsorbed to the soil was determined by observing a reduction in Co and/or EDTA concentrations as a function of time. Dissociation of the Co-EDTA complex was determined by two methods: a comparison of total ETDA (TOC) and chelated EDTA (IC) concentrations, and of total Co (ICP/MS) and chelated Co (IC) concentrations. Agreement between these two methods was within 5%. The EDTA data were generally used for the dissociation estimates shown in Table III. Percentages of $Co(III)EDTA^{-}$ produced, $Co(II)EDTA^{2-}$ lost, Co-EDTA dissociated, EDTA adsorbed, and Co adsorbed were determined at the 30-day endpoint of the experiments (Table III). The sum of these components at 30 days indicated ~100% recovery for all components between solid and solution phases (Table III).

The $Co(II)EDTA^{2-}$ 30-day concentration was used to generate a one-point distribution coefficient (K_d) representing the interaction of $Co(II)EDTA^{2-}$ with the sediments (Table I). The 30-day concentration of Co was used to generate a one-point K_d to explicitly represent the adsorption of Co (Table I). Significant differences in the two K_d arise because they are representative of two different processes. The $Co(II)EDTA^{2-}$ K_d represents the combined effects of adsorption, oxidation, and dissociation, and is thus indicative of contact and interaction of the anionic complex with sediment surfaces. The Co K_d, in contrast, specifically represents the adsorption of Co only. Both of these calculations assume adsorption as a function of concentration was linear, an assumption which has not been tested for these sediments. The rate of $Co(II)EDTA^{2-}$ loss from solution was fitted to first- and second-order equations, but in all cases the first order equation provided a superior fit (Table I). This contrasts with the second-order equation used on ORNL sediments *(25)*, a difference which may be a result of diverse geologic and climatic environments, and/or lower Mn concentrations in the Hanford sediments. The production rate of $Co(III)EDTA^{-}$ is expected to depart from the rate of loss of $Co(II)EDTA^{2-}$ because the two represent different processes. The production of $Co(III)EDTA^{-}$ represents the formation and desorption of the complex, whereas the loss of $Co(II)EDTA^{2-}$ represents

Table III. Mass balance (% mol/mol) of $Co(II)EDTA^{2-}$ reaction with Hanford sediments for 30 days.

Sediment	*$Co(II)EDTA^{2-}$*	*$Co(III)EDTA^{-}$*	*Dissociation*	*Co_{sorbed}*	*$EDTA_{sorbed}$*	*Total*
Hfc	85	4.8	9.4	3.5	5.0	104
Hff	70	12	18	8.2	0.0	100
IDF45	80	8.8	11	0	0	100
IDF110	77	10	12	0	0	99
IDF150	74	11	14	0	0	99
IDF200	84	7.9	7.6	0	0	100
IDF215	86	8.3	5.4	0	0	100
PP	84	3.8	5.3	5.0	0.0	¶98
US	29	43	19	6.9	7.0	98
LS	52	25	19	1.9	2.5	99
SS4	2.3	§66	21	17	17	99
SS4a	67	24	7.9	0.0	5.8	105
SS4b	64	19	8.0	0.0	7.6	99
S5	1.6	§76	24	7.4	11	105
PS2	4.1	73	19	4.2	8.2	104
PS1	9.6	66	11	11	13	100
WB	6.8	§51	33	10	16	99
TF1	74	16	9.2	0.0	4.8	104
TF2	90	5.7	2.6	3.0	0.0	98
TF3	7.9	76	17	0.0	3.5	104
TF4	12	63	19	4.4	11	105

§Percentages of $Co(III)EDTA^{-}$ taken at 7- or 15-day endpoint. Totals calculated with respect to 30-day endpoint.

¶Total was calculated with respect to Co, due to additional TOC leaching from the sediment (see Table II).

adsorption, oxidation, dissociation, and desorption. The production of Co(III)EDTA$^-$ was found to obey neither first- or second-order rate laws, and an appropriate rate law has not yet been identified.

The rate coefficient and endpoint percentages for all of the sediments were tested for correlation with the concentrations of Mn, Fe, and Al extracted by the CBD method. The Ringold WB sediment was not included in the statistical correlations because the sediment extractions were not performed using the same methodology *(26)*. The purpose of the statistical correlations was to provide a predictive tool for the reactions of Co(II)EDTA^{2-} in the Hanford subsurface.

Results and Discussion

Rates and Mechanisms

Influent Co(II)EDTA^{2-} concentrations decreased with increasing contact time for all 21 sediments, resulting in 1.6-90% of influent Co(II)EDTA^{2-} remaining at the 30 d endpoint (Table III). The dominant reaction observed in all sediments resulted in the oxidation of influent Co(II)EDTA^{2-} (log K = 16.8) to Co(III)EDTA$^-$ (log K = 39.8) *(27)*. Select sediments were chosen to provide graphical examples of Co(II)EDTA^{2-} reactivity (Figure 1a-d). The Hff sample (Figure 1a) represents the Hanford flood deposits, the PP sample represents the Plio-Pleistocene Unit (Figure 1b), the Upper Ringold Formation is represented by US (Figure 1c), and the Middle Ringold Formation by TF4 (Figure 1d). Distribution and reaction rate coefficients for all sediments are shown in Table I and overall mass balance is shown in Table III. Oxidation of influent Co(II)EDTA^{2-} to Co(III)EDTA$^-$ ranged from 3-76% (Table III). This is important from an environmental perspective since high stability and minimal adsorption of the anionic Co(III)EDTA$^-$ complex tend to promote sustained migration of ^{60}Co in the subsurface *(28-36)*. The oxidation reaction is most likely catalyzed by interaction with Mn(IV) and/or Fe(III) oxide surfaces, although Fe is a less effective oxidant than Mn *(31-32, 34)*. In addition, neutral pH tends to reduce the ability of Fe-oxides to promote the redox reaction *(32, 34-35)*, while Mn-oxides tend to remain active over a wider range of pH *(31-32)*. To determine which oxide was responsible for observed oxidation, the relationship between sediment Mn, Fe, and the rate of Co(II)EDTA^{2-} loss was tested. Only Mn exhibited a relationship with the rate of Co(II)EDTA^{2-} loss, which suggests that Mn oxides were at least partially responsible (Figure 2a). The production of Co(III)EDTA$^-$ was also correlated exclusively with Mn oxides (Figure 2b). No correlation with sediment Fe content was observed, which is

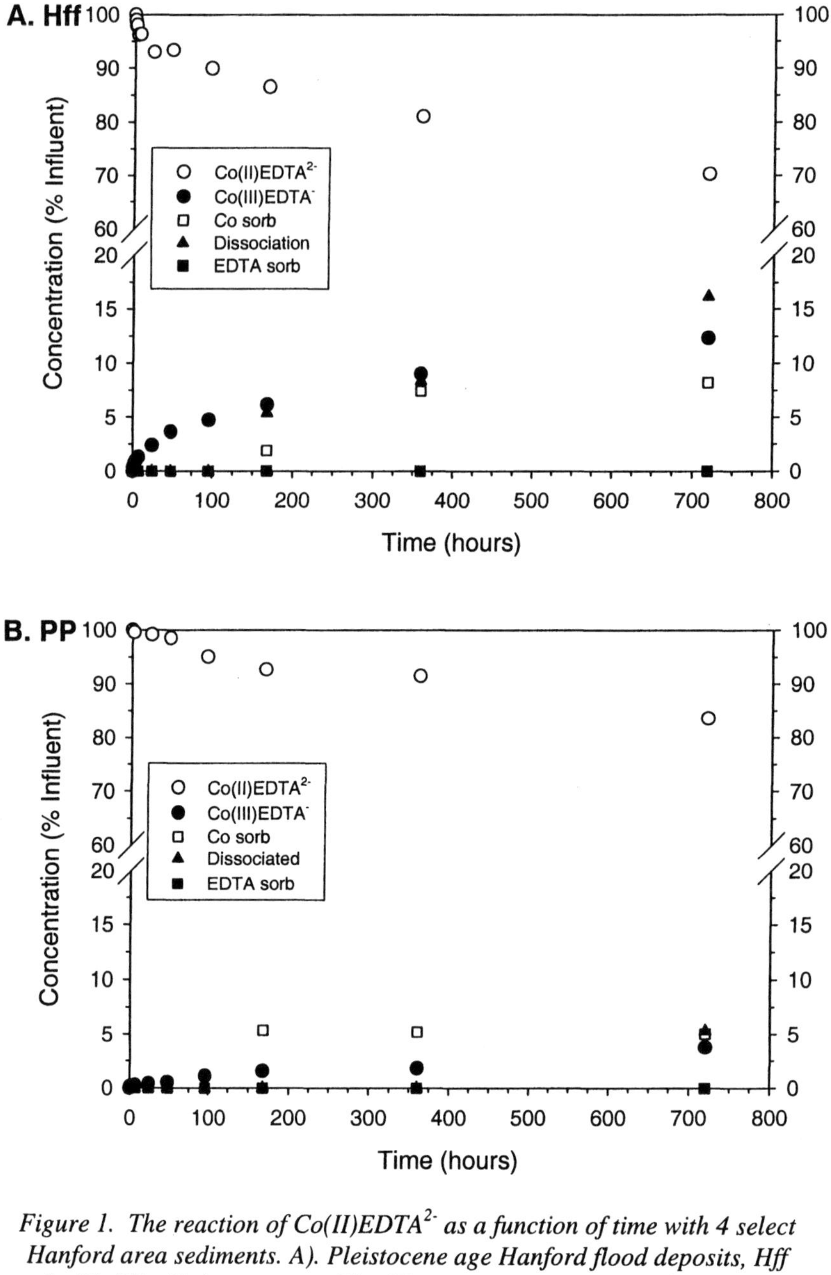

Figure 1. The reaction of Co(II)EDTA^{2-} as a function of time with 4 select Hanford area sediments. A). Pleistocene age Hanford flood deposits, Hff sample. B). Plio-Pleistocene age Plio-Pleistocene Unit, PP sample. C) Pliocene age Upper Ringold Formation, US sample. D). Miocene-Pliocene age Middle Ringold Formation, TF4 sample.

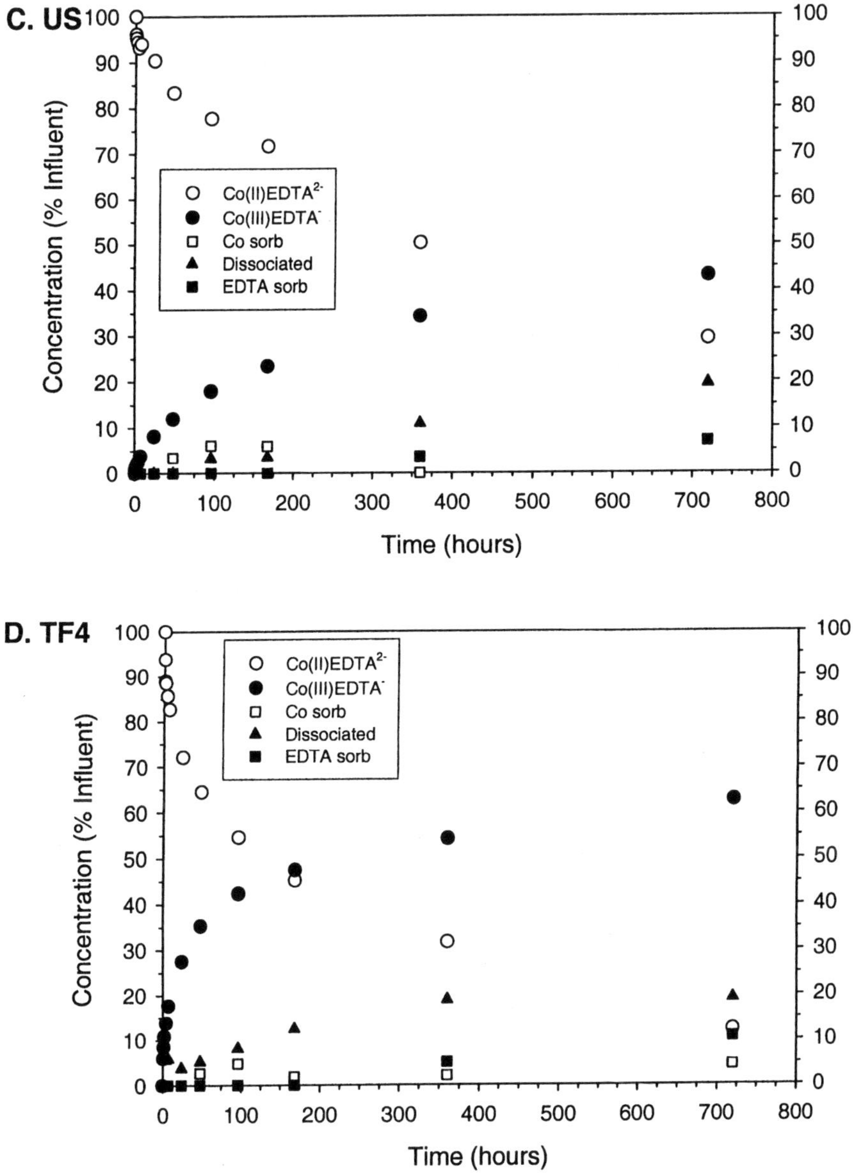

Figure 1. *Continued.*

most likely because all of the sediments contained some Mn (Table I), which is a more powerful oxidant than Fe(III). Such correlations may be utilized to provide predictions of $Co(II)EDTA^{2-}$ mobility in the Hanford subsurface. In sediments with mixed oxides, Mn oxides are dominant in controlling the oxidation reaction *(25, 30, 32, 36)*. Further, in quartz-rich (i.e. sandy) sediments such as these, aqueous Si is believed to dominate the exchange complex of Fe-oxides, resulting in decreased capacity for sorption of additional anions *(32)*. Aqueous Si concentrations up to 0.04 mM were observed, which supports the hypothesis that silicic acid could have competed with $Co(II/III)EDTA^{-}$ for sorption onto Fe-oxides.

Most sediments exhibited an initial rapid oxidative period, followed by a more gentle increase in $Co(III)EDTA^{-}$ production at longer times (Figure 1a-d). This is consistent with natural sediments in which oxidation was observed to occur within minutes *(32)*. However, oxidation was also sustainable over long periods of time, continuing throughout the duration of the experiments (30 d) for all twenty sediments (Figure 1a-d). Differential rates of Co(II) oxidation as a function of time could be a result of re-adsorption of reduced $Mn^{2+}{}_{(aq)}$ and the subsequent formation of intermediate solid phase Mn(III) oxides, such as α-Mn_2O_3 *(31, 37-38)*. The Mn(III) oxides are less effective oxidants, and slower oxidation rates have been observed when compared with Mn(IV) oxides. In synthetic pyrolusite-coated columns, initially rapid rates of oxidation dropped following the formation of this intermediate, which was confirmed using X-ray Absorption Near-Edge Spectroscopy *(37-38)*. Decreased rates as a function of time may also be an artifact of the batch system in which reaction products tend to accumulate in solution and/or on sediment surfaces where they have the potential to interfere with additional reactions *(39)*. Finally, variations in the surface area, crystallinity and reactivity of surficial oxides may result in differential rates of oxidation. Regardless, the observation of sustained oxidation may be significant for the Hanford subsurface in which residence times may be on the order of decades.

Dissociation of the chelate-metal complex ranged from 0-33% (Table III), and was observed to increase in importance as the duration of the experiments persisted (Figure 1a-d). This is especially true for the less redox reactive of the sediments, e.g., Hff (Figure 1a), in which dissociation and oxidation were equivalent at 30 d. Surface species of Al and Fe are known to aggressively promote the dissociation of divalent metal-EDTA complexes with the concomitant formation of $Fe(III)EDTA^{-}$ and/or $AlEDTA^{-}$ *(25, 30, 32-33, 35-36, 40-49)*. Aqueous Fe and Al, however, were not found in sufficient abundance to explain the observed dissociation, as the sum of aqueous Al and Fe were consistently an order of magnitude less than that of the dissociated EDTA. Further, no correlation was found between the amount of dissociation and sediment Al or Fe content. $AlEDTA^{-}$ is less likely to form due to decreased Al activity at neutral pH *(32, 41, 50)*. Ligand-promoted dissolution of Fe-oxides also occurs at slower rates at circum-neutral pH *(35, 40, 42-46)*. In addition, adsorption of $Fe(III)EDTA^-$ at pH 8 can be significant in equilibrium batch systems *(45-46)*, which is consistent with observed sorption of EDTA in many of the Ringold sediments (Table III). However, sorption of EDTA was minimal in the Hanford and Plio-Pliestocene sediments, while dissociation was significant. This suggests the formation of other stable, aqueous metal-EDTA complex(es) which have not been identified. Neutral pH favors the stability field of $CaEDTA^{2-}$ (log K=10.7) *(44)*, although its formation is thermodynamically unfavorable in the presence of $Co(II)EDTA^{2-}$ (log K = 16.8) *(27)*. However, thermodynamic predictions may depart from experimental observations due to kinetic limitations, and/or the presence of competing metals and ligands *(46-47)*. Finally, reorganization of the metal-EDTA complexes can continually occur in a closed system, as demonstrated by *(47)* in which $AlEDTA^-$ is believed to be an intermediate in the formation of $Fe(III)EDTA^-$. Finally, dissociated Co^{2+} could become oxidized to Co^{3+} by Mn-oxides, and subsequently scavenge EDTA from any of the less thermodynamically stable complexes.

Loss of Co and EDTA from solution was observed, ranging from 0-17% but was close to zero for many of the sediments (Table III). Intact adsorption of the Co-EDTA complex probably occurred, based upon similar percentages of adsorbed Co and EDTA (Table III). However, disparities in the percentages were also observed, indicating adsorption of Co^{2+}, Co^{3+}, and/or $EDTA^{4-}$ (Table III). A one-point K_d was obtained for Co that ranged from 0-4 $cm^3\ g^{-1}$ (Table I). This was much lower than the K_d of $Co(II)EDTA^{2-}$, which ranged from 1-370 $cm^3\ g^{-1}$ (Table I). The latter K_d suggests that the $Co(II)EDTA^{2-}$ complex interacts extensively with the sediment mineralogy, because it also considers surface-mediated oxidation and dissociation reactions. The former K_d (Co) suggests that Co^{2+} liberated from the dissociation of the Co-EDTA complex is not readily sorbed, which is possibly due to the formation of the anionic $HCoO_2^-$ complex which was predicted by geochemical speciation modeling. Our predicted Co^{2+} sorption is consistent with the K_d <3 expected for the Hanford subsurface *(2)*. Our Co^{2+} K_d is, however, 4-5 orders of magnitude lower than that predicted in Hanford sediments without the presence of organic ligands *(51)*.

Relation to Sediment Characteristics

Hanford flood deposits were less effective at oxidation than the Ringold sediments, exhibiting a maximum of 12% oxidation (Table III). Figure 1a shows the kinetic experiment of the Hanford fines (Hff), which is typical of most of the Hanford flood deposits (Hfc, IDFs) which were utilized in these experiments. Similar reactivity is consistent with similar Hanford mineralogy, since only minor variations in extractable Mn, Fe, and Al were observed (Table I). Grain size may be related to the capacity for oxidation of $Co(II)EDTA^{2-}$, as evidenced by the three more fine-grained sediments (Hff, IDF150, and IDF110) *(15-16)* which have somewhat greater $Co(III)EDTA^{-}$ production and dissociation of the Co-EDTA complex. This relationship could be a result of greater surface area and/or of different mineralogy of the fines (e.g., clays). The former hypothesis is not supported by surface area measurements which are similar for both Hfc and Hff (Table II). Dissociation in the Hanford deposits is of similar magnitude to the production of $Co(III)EDTA^{-}$ (Table III), which is encouraging from a contaminant transport perspective since the vadose zone at Hanford is mostly comprised of these flood deposits. Adsorption of Co and EDTA, however, was minor.

The Plio-Pleistocene Unit (PP) exhibited very low amounts of oxidation and dissociation (Figure 1b). This is consistent with its mineralogy which is approximately 25% secondary calcium carbonate precipitates formed during subaerial exposure and soil development *(3, 17)*. Lithic (rock) fragments are visible in thin section, but their proportion appears to be volumetrically small, and no clay minerals have been identified by XRD/XRF *(52)*. Extracted Fe and Mn were the lowest in the dataset, which most likely accounts for the minimal oxidation and dissociation observed (Table I). The lack of dissociation in the Ca-rich sample suggests that the thermodynamically unfavorable dissociation of $Co(II)EDTA^{2-}$ and subsequent formation of $CaEDTA^{2-}$ are not important at Hanford.

The Ringold sediments were the most effective oxidants, resulting in an average conversion of 46% influent $Co(II)EDTA^{2-}$ to $Co(III)EDTA^{-}$, compared to a 9% average in the Hanford sediments (Table III). Ringold sediments SS4, S5, WB, TF3, and TF4 (Figure 1d) produced large quantities of $Co(III)EDTA^{-}$ (51-76%) (Table III). Each of these has correspondingly high extractable Mn in comparison to the other sediments (Table I), which is consistent with the correlations derived in Figure 2. Each of these sediments also exhibited a high degree of dissociation, ranging from 17-33%. Iron content was also high for these sediments, which suggests that Fe-oxides may have contributed to the dissociation reaction and/or to the production of $Co(III)EDTA^{-}$. In addition, similar adsorption of Co and EDTA were observed, which most likely suggests intact adsorption of the Co-EDTA complex (Table III). Further, $Co(III)EDTA^{-}$ became adsorbed at longer times (> 7 d) for S5, WB, and SS4, which suggests the sediments have considerable affinity for sorption of anionic complexes. The high level of interaction of the sediments with $Co(II)EDTA^{2-}$ is also reflected in the calculated K_d for both Co and $Co(II)EDTA^{2-}$ (Table I).

Ringold sediments with intermediate Fe and Mn contents can also effectively oxidize $Co(II)EDTA^{2-}$ to $Co(III)EDTA^{-}$ (Figure 1c). The Mn content of the US sample, however, is similar to that of the Hanford sediments (Table I).

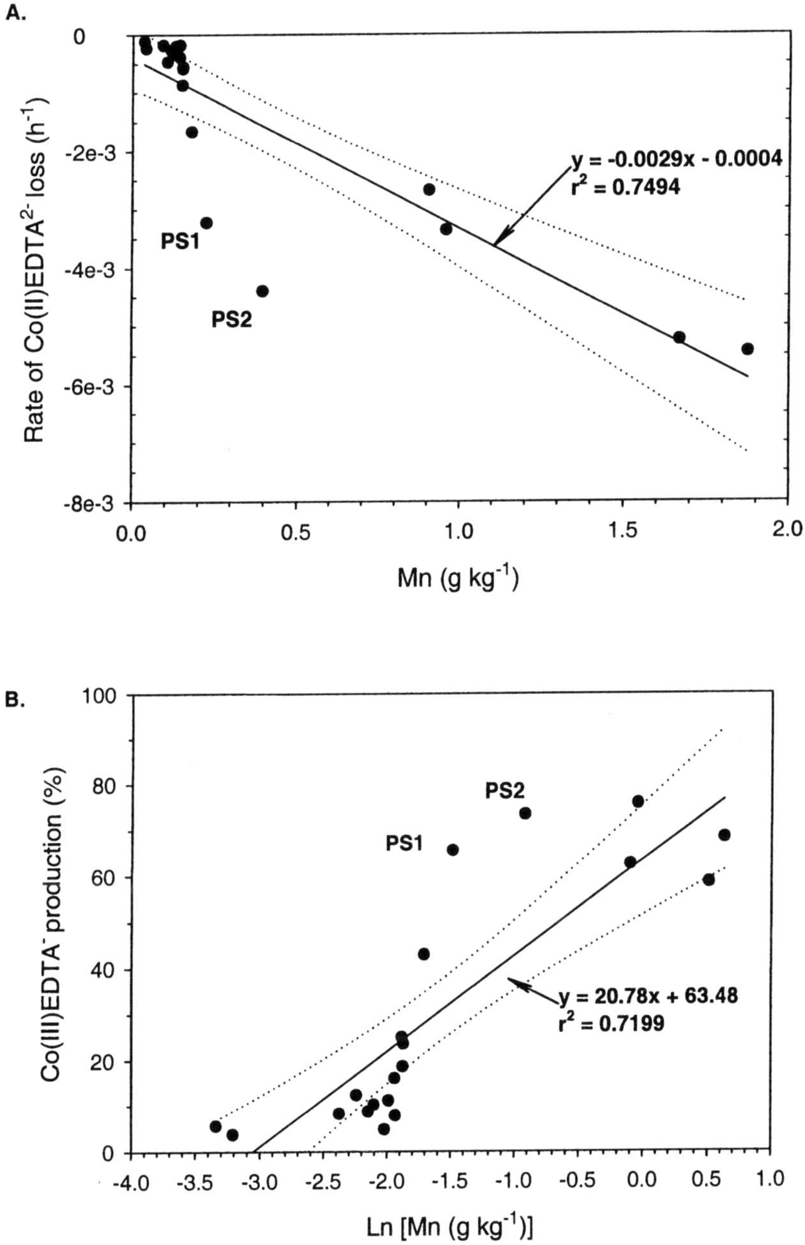

Figure 2. Results of statistical correlation of extracted Mn content with twenty Hanford sediments, with 95% confidence interval. A) Rate of Co(II)EDTA^{2-} loss versus extracted Mn. B). Production of Co(III)EDTA^{-} (= 100[Co(III)EDTA^{-}]$_{time}$/[Co(II)EDTA^{2-}]$_{influent}$) versus natural log of extracted Mn.*

The greater reactivity of the Ringold US in comparison to the Hanford may be related to grain size, surface area and/or sedimentary history. The US is more fine-grained, consisting of 58% silt/clays and 42% sand *(11)*, while the Hanford contains only minor (5%) silt/clays *(10)*. In addition, the surface area of US is greater than that of Hff (Table II). In the US, coatings of clay, Mn- and Fe-oxides are visible on sedimentary bedding planes *(11)*, which has not been observed in the Hanford. Such coatings are probably due to secondary precipitation of oxidized Fe- and Mn minerals by post-depositional soil- and/or groundwater alteration. The greater age of the Ringold (millions of years) versus the Hanford (tens to hundreds of thousands of years) may have resulted in the formation of reactive Fe- and Mn-minerals which are conducive to the oxidation of $Co(II)EDTA^{2-}$.

The PS2 and PS1 Ringold sediments are also intermediate in their Mn and Fe content (Table I). Despite this, both were robust in producing $Co(III)EDTA^{-}$ in comparison to other sediments such as US (Table III). This is also evidenced by their presence below the line of correlation in Figure 2a, which suggests that they were more reactive to $Co(II)EDTA^{2-}$ than would be expected based upon their Mn content. Removal of these samples results in a significant improvement in the correlation coefficient (r^2 = 0.72 to r^2 = 0.97). In addition, Figure 2b demonstrates that they are more effective oxidants than might be predicted. When these two samples are removed from the correlation, the r^2 increases to 0.87. Such results may be related to their depositional environment, which is significantly different in comparison to the other samples. It is likely that PS1 and PS2 represent a paleosol (fossil soil) horizon in which there was a significant break in deposition that allowed for weathering and alteration of ~1.5m thickness of sediments for a significant period of time. This is apparent by the uniform reddish-pink color of PS2, which suggests that the sediments became wholly oxidized during this period of exposure. PS1 is a gray, "gleyed" clay-rich horizon that probably represents an underlying, reduced hardpan layer. Both units also contain extensive fossilized root casts of calcium carbonate and other nodules, which strongly suggests that the units comprise a fossil soil. Such root casts and nodules (rhizoliths) were most likely formed by plant- and microbially-mediated processes. Such processes can result in high surface area, amorphous minerals which may be more reactive than those produced by inorganic processes *(53)*. This hypothesis might account for the observed greater reactivity of the PS1 and PS2 samples than would be expected based upon their Mn content (Figure 2). Multiple paleosols are present in the Hanford subsurface, and these results suggest that contaminant transport in such units may depart from predictions generated with ordinary sediments.

Some Ringold sediments such as SS4a, SS4b, TF1, and TF2 resulted in only minor oxidation of $Co(II)EDTA^{2-}$. TF1 and TF2 are representative of several clean, sandy lenses outcropping within the Middle Ringold which are mostly devoid of oxides and reactive minerals, which suggests that sedimentary heterogeneities may influence reactive transport of contaminants in the subsurface. SS4, LS, SS4a, and SS4b are all similar in terms of grain size and age, yet significant differences exist in their reactive oxide content (Table I) and in their observed reactivity towards $Co(II)EDTA^{2-}$ (Table III). This study suggests there is significant geochemical variation throughout the Ringold, which occupies the saturated zone at Hanford.

Implications Regarding the Mobility of Co-EDTA in the Hanford Subsurface

These experiments have quantified the adsorption, oxidation, and dissociation of $Co(II)EDTA^{2-}$ in twenty-one Hanford area sediments. It is apparent that the anionic $Co(II)EDTA^{2-}$ complex significantly interacted with the subsurface sediments at neutral pH. Oxidation of $Co(II)EDTA^{2-}$ to $Co(III)EDTA^{-}$ ranged from 3-76%, which is significant because the high stability constant of the latter (log K = 39.8) tends to promote its migration in the subsurface. Dissociation of the $Co(II)EDTA^{2-}$ complex was typically 10-30%, a mechanism which may result in decreased mobility of ^{60}Co. The dissociation reaction, however, proceeded at slower rates in comparison to the oxidation reaction. The soil moieties responsible for the dissociation reaction have not been completely identified. Loss of aqueous Co was minimal, which suggests that dissociation may not effectively prevent the migration of radioactive ^{60}Co. Adsorption of the Co-EDTA complexes ranged from 0-17%. Adsorption probably included both monovalent and divalent anionic complexes, Co^{2+}, Co^{3+}, and/or $EDTA^{4-}$. In the Hanford subsurface, however, contaminant transport may depart from that observed here due to differences in the rock:water ratio, open- versus closed-systems, competition by other metals or ligands, and kinetics imparted by flowing versus static conditions.

The oxidative capacity of the sediments was related to sedimentary Mn content, which provided a quantitative method to predict the oxidation of $Co(II)EDTA^{2-}$ in the Hanford subsurface using known geochemical parameters. Our results suggest that original sedimentary depositional characteristics (e.g., grain size, surface area, age) and secondary alteration of mineral compositions influenced the reactivity of the sediments. Vertical variations in the mineralogy of the Hanford subsurface sampled in these experiments suggest that contaminant transport will vary with depth. Consideration of such heterogeneities, especially between the saturated and unsaturated zones, should improve predictions of subsurface contaminant transport.

Acknowledgements

This research was sponsored by the U.S. Department of Energy Environmental Management Science Program. The authors would like to thank Yul Roh of ORNL for mineralogical characterization, Zuojiang Li of ORNL for surface area measurements, and two anonymous reviewers for their suggestions which improved this manuscript. Oak Ridge National Laboratory is managed by the University of Tennessee-Battelle, LLC, under contract DE-AC05-00OR22725 with the U.S. Department of Energy.

References

1. *Performance Management Plan for the Accelerated Cleanup of the Hanford Site;* U.S. Department of Energy: Richland, WA, 2002; Predecisional draft, Rev. 0.
2. *Supplement Analysis for the Tank Waste Remediation System, Appendix A;* U.S. Department of Energy: Richland, WA, 1998; DOE/EIS-0189-SA2.
3. *Field Investigation Report for Waste Management Area S-SX;* U.S. Department of Energy: Richland, WA, 2002; RPP-7884, Rev. 0.
4. Glass, R.J.; Steenhuis, T.S.; Parlange, J.-Y. *J. Contam. Hydrol.* **1988,** *3,* 207-226.
5. Glass, R.J.; Steenhuis, T.S.; Parlange, J.-Y. *Soil Sci.* **1989,** *148*, 60-70.
6. Kung, K.-J.S. *Geoderma* **1990**, *46,* 51-58.
7. Tokunaga, T.K.; Olson, K.R.; Wan, J. *Vadose Zone J.* **2003,** *2*, 322-329.
8. Gee, G.W.; A.L. Ward. *Vadose Zone Transport Field Study: Status Report.* U.S. Department of Energy: Richland, WA, 2001; PNNL-13679.
9. Knepp, A. *Field Investigation Report for Waste Management Area B-BX-BY;* CH2M HILL Hanford Group: Richland, WA, 2002; RPP-10098, Rev. 0.
10. Pace, M.N.; Mayes, M.A.; Jardine, P.M.; Mehlhorn, T.L.; Zachara, J.M.; Bjornstad, B.N. *Vadose Zone J.* **2003,** *2,* 664-676.
11. Mayes, M.A.; Jardine, P.M.; Mehlhorn, T.L.; Bjornstad, B.N.; Ladd, J.L.; Zachara, J.M. *J. Hydrol.* **2003,** *275*, 141-161.
12. Pruess, K.; Yabusaki, S.; Steefel, C.; Lichtner, P. *Vadose Zone J.* **2002,** *1*, 68-88.
13. Bickmore, B.R.; Nagy, K.L.; Young, J.S.; Drexler, J.W. *Environ. Sci. Technol.* **2001,** *35,* 4481-4486.
14. Jones, T.E.; Simpson, B.C.; Wood, M.I.; Corbin., R.A. *Preliminary Inventory Estimates for Single-Shell Tank Leaks in T, TX, and TY Tank Farms;* U.S. Department of Energy: Richland, WA, 2000; RPP-7218.

15. *Standardized Stratigraphic Nomenclature for Post-Ringold Formation Sediments within the Central Pasco Basin;* U.S. Department of Energy: Richland, WA, 2002; USDOE/RL-2002-39, Rev. 0.
16. Horton, D.G.; Schaef, H.T.; Serne, R.J.; Brown, C.F.; Valenta, M.M.; Vickerman, T.S.; Kutnyakov, I.V.; Baum, S.R.; Geiszler, K.N.; Parker, K.E. *Geochemistry of samples from borehole C3177 (299-E24-21);* U.S. Department of Energy: Richland, WA, 2003; PNNL-14289.
17. Slate, J.L. *Nature and Variability of the Plio-Pleistocene Unit in the 200 West Area of the Hanford Site;* U.S. Department of Energy: Richland, WA, 2000; BHI-01203, Rev. 0.
18. Lindsey, K. A. *Revised Stratigraphy for the Ringold Formation, Hanford Site, Couth-Central Washington;* U.S. Department of Energy: Richland, WA, 1991; WHC-SD-EN-EE-004.
19. Loeppert, R.H.; Inskeep, W.P. In *Methods of Soil Analysis – Part 3 Chemical Methods.* SSSA Book Series No. 5.; Soil Science Society of America: Madison, WI, 1996; pp 647-648.
20. Gregg, S.J.; Sing, K.S.W. *Adsorption, Surface Area and Porosity.* Academic Press: London, 1982.
21. Sing, K.S.W.; Everett, D.H.; Haul, R.A.W.; Moscou, L.; Pierotti, R.A.; Rouquerol, J.; Siemieniewska, T. *Pure Appl. Chem.* **1985,** *57,* 603-619.
22. Taylor, D.L.; P.M. Jardine. *J. Environ. Qual.* **1995,** *24,* 789-792.
23. Ammann, A. *J. Chromatogr. A.* **2002,** *947,* 205-216.
24. Ammann, A. *Anal. Bioanal. Chem.* **2002,** *372,* 448-452.
25. Jardine, P.M.; Taylor, D.L. *Geoderma* **1995,** *67,* 125-140.
26. Barnett, M.O.; Jardine, P.M.; Brooks, S.C.; Selim, H.M. *Soil Sci. Soc. Am. J.* **2000,** *64,* 908-917.
27. Parker, D.R.; Zelazny, L.W.; Kinraide, T.B. *Soil Sci. Soc. Am. J.* **1987,** *51,* 488-491.
28. Means, J.L.; Crerar, D.A.; Duguid, J.O. *Science (Washington, D.C.)* **1978,** *200,* 1477-1481.
29. Olsen, C.R.; Lowry, P.D.; Lee, S.Y.; Larsen, I.L.; Cutshall, N.H. *Geochim. Cosmochim. Acta* **1986,** *50,* 593-607.
30. Jardine, P.M.; Jacobs, G.K.; O'Dell, J.D. *Soil Sci. Soc. Am. J.* **1993,** *57,* 954-962.
31. Jardine, P.M.; Taylor, D.L. *Geochim. Cosmochim. Acta* **1995,** *59,* 4193-4203.
32. Zachara, J.M.; Gassman, P.L.; Smith, S.C.; Taylor, D.L. *Geochim. Cosmochim. Acta* **1995,** *59,* 4449-4463.
33. Nowack, B.; Sigg, L. *J. Colloid Interface Sci.* **1996,** *177,* 106-121.
34. Brooks, S.C.; Taylor, D.L.; Jardine, P.M. *Geochim. Cosmochim. Acta* **1996,** *60,* 1899-1908.

35. Szecsody, J.E.; Zachara, J.M.; Ashokkumar, C.; Jardine, P.M.; Ferrency, A.M. *J. Hydrol.* **1998,** *209,* 112-136.
36. Mayes, M.A.; Jardine, P.M.; Larsen, I.L.; Brooks, S.C.; Fendorf, S.E. *J. Contam. Hydrol.* **2000,** *45,* 243-265.
37. Fendorf, S.E.; Jardine, P.M.; Patterson, R.R.; Taylor, D.L.; Brooks, S.C. *Geochim. Cosmochim. Acta* **1999,** *63,* 3049-3057.
38. Fendorf, S.E.; Jardine, P.M.; Taylor, D.L.; Brooks, S.C.; Rochette, E. In *Mineral-Water Interfacial Reactions.* Sparks, D.L., Grundl, T.J., Eds.; ACS Symposium Series 715; American Chemical Society: Washington, D.C., 1999; pp 358-371.
39. Porro, I.; Newman, M.E.; Dunnivant, F.M. *Environ. Sci. Technol.* **2000,** *34,* 1679-1686.
40. Chang, H.C.; Matijevic, E. *J. Colloid Interface Sci.* **1983,** *92,* 479-488.
41. Girvin, D.C.; Gassman, P.L.; Bolton, H. *Soil Sci. Soc. Am. J.* **1993,** *57,* 47-57.
42. Bowers, A.R.; Huang, C.P. *Wat. Res.* **1987,** *21,* 757-764.
43. Zachara, J.M.; Smith, S.C.; Kuzel, L.S. *Geochim. Cosmochim. Acta* **1995,** *59,* 4825-4844.
44. Szecsody, J.E.; Zachara, J.M.; Bruckhart, P.L. *Environ. Sci. Technol.* **1994,** *28,* 1706-1716.
45. Nowack, B.; Sigg, L. *Geochim. Cosmochim. Acta* **1997,** *61,* 951-963.
46. Nowack, B. *Environ. Sci. Technol.* **2002,** *36,* 4009-4016.
47. Kent, D.B.; Davis, J.A.; Anderson, L.D.; Rea, B.A.; Coston, J.A. *Geochim. Cosmochim. Acta* **2002,** *66,* 3017-3036.
48. Davis, J.A.; Kent, D.B.; Rea, B.A.; Maest, A.S.; Garabedian, S.P. In *Metals in Groundwater;* Allen, H.E., Perdue, E.M., Brown, D.S., Eds.; Lewis Publishers: Chelsea, MI, 1993; pp 223-273.
49. Davis, J.A.; Kent, D.B.; Coston, J.A.; Hess, K.M.; Joye, J.L. *Water Resour. Res.* **2000,** *36,* 119-134.
50. Jardine, P.M.; Mehlhorn, T.L.; Larsen, I.L.; Bailey, W.B.; Brooks, S.C.; Roh, Y.; Gwo, J.P. *J. Contamin. Hydrol.* **2002,** *55,* 137-159.
51. Krupka, K.M.; Serne, R.J. *Geochemical Factors Affecting the Behavior of Antimony, Cobalt, Europium, Technetium, and Uranium in Vadose Zone Sediments;* U.S. Department of Energy: Richland, WA, 2002; PNNL-14126.
52. Mayes, M.A.; Pace, M.N.; Jardine, P.M.; Fendorf, S.E.; Farrow, N.D.; Yin, X..L.; Zachara, J.M. In *Subsurface Contamination Remediation: Accomplishments of the Environmental Management Science Program.* Zachry, T., Berkey, E., Eds.; ACS Symposium Series xxx; American Chemical Society: Washington, D.C., in press; pp xxx-xxx.
53. Nelson, Y.M.; Lion, L.W. In *Geochemical and Hydrological Reactivity of Heavy Metals in Soils;* Selim, H.M., Kingery, W.L., Eds.; Lewis Publishers: New York, 2003, pp 169-186.

Chapter 18

Coupled Hydrological and Geochemical Processes Influencing the Transport of Chelated Metals in the ORNL Vadose Zone and Groundwater

Melanie A. Mayes, Tonia L. Mehlhorn, and Philip M. Jardine

Environmental Sciences Division, Oak Ridge National Laboratory, P.O. Box 2008, MS–6038, Oak Ridge, TN 37831

Subsurface and surface water contamination at Oak Ridge National Laboratory (ORNL) is extensive due to legacy waste from Cold War era nuclear weapons production. Shallow land burial occurred within a weathered, oxidized and acidic saprolitic material, while the underlying groundwater is dominantly anaerobic and equilibrated to the neutral pH of the bedrock. Consequently, chelated metal transport ($Co(II)EDTA^{2-}$, $SrEDTA^{2-}$, and $CdEDTA^{2-}$) at ORNL is a function of the geochemistry and oxidation state of both the weathered upper unit and the underlying bedrock. In this paper, recent vadose zone and groundwater studies are reviewed and the implications regarding the offsite transport of chelated metals are discussed.

Introduction

Nuclear weapons production has resulted in a legacy of subsurface radionuclide disposal at U.S. Department of Energy (DOE) sites, and offsite migration represents a continuing risk at certain locations in the U.S. Historical waste disposal practices at the Oak Ridge Reservation (ORR), located in eastern Tennessee, involved the shallow land burial of intermediate- and low-level radioactive waste in unlined pits and trenches within the vadose (unsaturated) zone. This disposal practice was employed from the 1940s and continued through the middle 1980s. The waste stream was diverse and resulted in the co-disposal of a variety of sources and concentrations of radionuclides, toxic metals, and organics, including ^{3}H, ^{60}Co, ^{99}Tc, ^{90}Sr, ^{244}Cm, ^{233}U, Cd, trichloroethylene (TCE), and ethylenediamenetetraacetic acid (EDTA). In the late 1970's, the detection of radionuclides significantly downstream of ORNL raised alarms regarding the potential for offsite migration *(1)*. It was soon discovered that chelators were major contributors to the spread of radioactive ^{60}Co in surface waters *(1-2)*. Chelators, particularly EDTA, were extensively used during cleanup and decontamination procedures at ORNL and other DOE facilities *(3-4)*.

During the 1990's, considerable scientific attention was given to this issue, resulting in numerous publications relevant to field transport at ORR *(5-17)*, Hanford *(18-23)*, and the Cape Cod aquifer *(24-27)*. However, each of these case studies exists within a diverse geologic, hydrologic, and geochemical environment, resulting in both similar and dissimilar conclusions regarding the actual risks to the general population from widely-utilized radionuclide chelators. The ORNL case provides an excellent example of the complex transport pathways of chelated radionuclides from source to the public environment. The source of chelated metal contamination is within an extensively weathered, acidic vadose zone replete with oxides of Mn and Fe, where oxides may contain up to 35% Al substitution *(28)*. Subsequent discharge of contamination into surface waters involves anoxic groundwater flow through a fractured limestone and shale bedrock of neutral pH *(11-12)*. In this paper, the linkage of contaminant transport between the vadose zone and the groundwater will be explored, thus yielding new insights into the influence of subsurface geochemistry on the migration of the chelated metals $Co(II)EDTA^{2-}$, $SrEDTA^{2-}$, and $CdEDTA^{2-}$.

Hydrogeologic Setting

Eastern Tennessee resides in the Valley and Ridge Tectonic Province, in which a thick, diverse assemblage of Paleozoic marine and near-shore sedimentary rocks was deposited, then subsequently folded and faulted during the Alleghenian uplift of the Appalachian Mountains *(29)*. The bedrock of the ORNL Waste Area Grouping 5 (WAG 5) is the Dismal Gap member of the Upper Cambrian Conasauga Group, which is comprised of interbedded limestones, sandstones, siltstones, and shales (Figure 1a). The mineral composition of the WAG 5 bedrock is calcite, quartz, K-feldspar, biotite and small quantities of kaolinite *(11-12)*. The pH of WAG 5 groundwater is neutral (6.7-7.0), consistent with the influence of limestone bedrock (Table 1).

Table 1. Select, typical chemical properties of WAG 5 groundwater.

Component	*Concentration (μM)*	*Component*	*Concentration (μM)*
Al	2.2	NO_2^-	0-0.6
Ca	2.5-4.5 x 10^3	NO_3^-	0-6.5
Fe(II)	4-88	CH_4	63 -250
Fe(III)	0	DO	0.5-31
S^{2-}	0.3-16	HCO_3	11-15 x 10^3
SO_4^{2-}	10-52	pH	6.7-7.0

Currently, the Valley and Ridge Province is an intensely vegetated, high precipitation regime (150 cm y^{-1}). The upper exposed layers (0.5-3 m depth) have been extensively leached, though many sedimentary and geologic structures, including bedding, folds, and faults, have been preserved (Figure 1b). This intermediate status between soil and bedrock is referred to as "saprolite." Leaching has been extensive, while illuviation has redeposited reactive (Mn, Fe, and Al) oxides and clays within fractures and bedding planes *(28, 30)*. The pH of saprolite porewaters ranges from 3.5-7.0 *(5, 7-8, 17)*, depending upon the degree of calcite weathering. The pH of the saprolite experiments discussed here is 3.7-3.9 *(17)*. The clay mineralogy is dominated by illite, interstratified kaolinites and smectities, and vermiculite *(31)*.

In ORNL's humid climatic regime, precipitation often occurs during sustained intense rainstorms which generate rapid surface and subsurface flow. The topography is highly dissected by permanent and intermittent streams; consequently discharge of contaminated water to surface streams is the major pathway for contamination to enter the public environment *(32)*. Bedding plane partings and two sets of orthogonal fractures are typical pathways of preferential flow *(32-33)*. Fracture density in unweathered bedrock is 5 m^{-1} *(34)*, increasing to 200 m^{-1} in the weathered saprolite *(35)*. Fractures are not typically extensive in length, but there is generally a high degree of fracture interconnectedness in both the bedrock and the saprolite which allows for sustained preferential flow. Stormflow in the vadose zone moves preferentially along bedding plane fractures, whereas groundwater flow migrates preferentially along strike-oriented fractures *(32)*. Unweathered bedrock porosity is generally on the order of 10-25%, most of which is in the matrix rather than the fractures *(9)*. Similarly, fractures constitute a small proportion (1-5%) of total porosity (40-50%) in the saprolite *(36)*. Overall, ORNL is an extremely complex hydrogeologic setting that is dominated by coupled preferential fracture flow and slow but ubiquitous matrix diffusion *(10)*.

The Disposal Environment

At ORNL, subsurface trenches were typically emplaced in the residual saprolite to depths close to the bedrock interface (Figure 2). Thus, the weathered, fractured saprolite material represents the disposal environment. Both liquid and solid wastes of low to intermediate levels of radioactivity were disposed, and at times geochemical treatments were also employed to retard contaminant migration *(2)*. The records of disposal vary greatly in their extent and accuracy, and some records have been lost to fire and other events over the intervening five decades. During the early history, solids were not containerized *(12)*, whereas later some secondary containment was employed (Figure 2). Trenches were usually compacted and then backfilled, however subsidence is a recurring problem in the burial grounds, which suggests that significant voids still exist in the trench environment. The location of the unconfined water table at WAG 5 is subject to extreme seasonal and storm-related fluctuations, so that the elevation of the water table can be equivalent to the ground surface.

Approach

In an effort to understand the rates and mechanisms of contaminant migration in the saturated bedrock zone, a field facility was established along a discharge pathway in Waste Area Grouping 5 (WAG 5) on the ORR. The WAG 5 well-field is strategically located along a 35 m long, strike-parallel transect extending from two long trenches downslope to a prominent seep that discharges into a tributary stream *(9)*. The seep is responsible for a large proportion of ^{3}H discharges to surface streams on the ORR, where the ^{3}H concentrations are 20,000 times over the drinking water standard *(37-38)*. The well-field was designed to intersect the flowpath between the highest ^{3}H concentration, which is in the two long trenches, and that which discharges from the major groundwater seep *(9)*. Wells were installed to sample this suspected zone of preferential flow, which was found to be highly fractured, and to sample

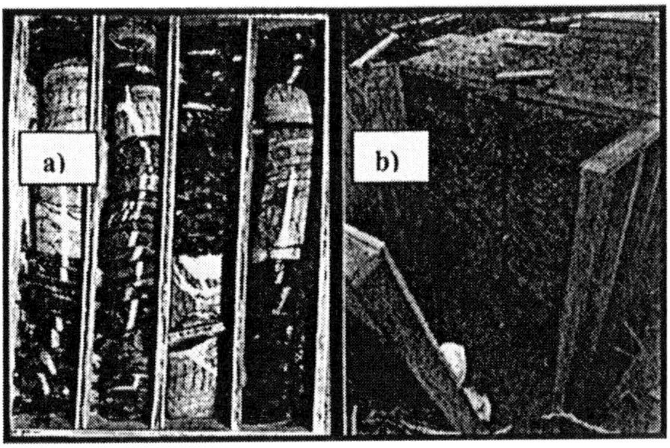

Figure 1. a) Example of bedrock in the saturated zone at WAG 5, showing interbeds of limestone (gray/white) and shale (black). (Reproduced with permission from reference 12. Copyright 2003 Lewis Publishers.) b) Example of fractured saprolite at Melton Branch Watershed, showing shallow soil zone and dipping beds of interbedded weathered limestones (dark) and shales (light).

Figure 2. Example of radioactive waste disposal in shallow trench at WAG 5 in the 1980s.

the surrounding matrix regime *(9, 11-12)*. Specific discharge measured in the 1-2 m thick fractured zone was as high as 300-500 m y^{-1}, which was an order of magnitude greater than that measured in the surrounding matrix regime. The aqueous geochemistry of the WAG 5 groundwater has been previously characterized as anaerobic, and is conducive to Fe and SO_4^{2-} reduction *(11-12, 39)* (Table 1).

Contaminant migration in the unsaturated vadose zone was studied using undisturbed columns isolated from the Melton Branch Watershed on the ORR, located parallel to strike from WAG 5. A fresh face of the B horizon was exposed, and stainless steel cylinders (15 cm length x 8.4 cm diameter) were hydraulically pressed into the saprolite using a modified drill rig *(17)*. A hand-sculpting technique was also utilized, in which a block of saprolite was exposed and coated with paraffin wax until the appropriate dimensions (40 cm length x 16 cm diameter) had been obtained *(7-8)*. PVC pipe was epoxied to the paraffin-coated core. Both types of columns were transported to the laboratory, and fitted with coarse (25-50μm pore diameter) fitted glass endplates in plexiglass endcaps. Saturated and unsaturated transport experiments were conducted following the methods of Jardine *et al.* *(7-8)* and Mayes *et al.* *(17)*.

After Tennessee state approval, injections of $^{57,58}Co(II)EDTA^{2-}$ and $^{109}CdEDTA^{2-}$ were performed over a period of six months in the WAG 5 bedrock aquifer *(11)*. Field monitoring continued for 1 y following a 6 mo injection. Injections of $^{57,58}Co(II)EDTA^{2-}$ $^{109}CdEDTA^{2-}$ and/or $SrEDTA^{2-}$ were performed over a similar time period in undisturbed cores of the ORNL weathered saprolite, which is representative of the trench environment *(17)*. In the following discussion, "saprolite" refers to the oxidized vadose zone or trench environment, and "bedrock aquifer" refers to the anoxic groundwater environment underlying the trenches.

Quantification of some metal-chelate complexes (Cd- and Co(II/III)EDTA) was accomplished using ion chromatography, using a modified method of Taylor and Jardine *(13)*. $Fe(III)EDTA^{-}$ was quantified by direct spectrophotometric analyses at 257 nm. At the time of these experiments, no methods existed for the quantification of $SrEDTA^{2-}$, $MnEDTA^{2-}$, or $AlEDTA^{-}$ as is presently available *(40-41)*. The existence of these latter species was predicted using mass balance and equilibrium geochemical speciation modeling *(11, 17)*. Total metal concentrations were quantified by Inductively Coupled Plasma spectroscopy (ICP), Atomic Absorption Spectroscopy (AAS), and/or by adding low concentrations of short-lived radiotracers to influent solutions, which were subsequently counted using gamma-ray spectroscopy with correction for radioactive decay *(11, 17, 42)*. Total organic carbon (TOC) was measured using standard combustion techniques.

Transport of Chelated Metals

The divalent metal-chelate complexes $CdEDTA^{2-}$ and $SrEDTA^{2-}$ were effectively dissociated in the saprolite. The transport of both was strongly influenced by the presence of Fe-oxides. For example, Cd^{2+} rather than $CdEDTA^{2-}$ was initially observed in the effluent of the saprolite columns, which suggested rapid dissociation of the $CdEDTA^{2-}$ complex (Figure 3). This was consistent with the transport of $SrEDTA^{2-}$ in the saprolite, in which it was clear that Sr^{2+} and $EDTA^{4-}$ were not co-transported, as evidenced by their temporal separation during elution *(8)*. In both experiments, evidence of the formation of $Fe(III)EDTA^{-}$ was provided by direct spectrophotometric analyses *(5, 8, 17)*. The dissociation reaction was rapid, and breakthrough of $Fe(III)EDTA^{-}$ preceded that of Cd^{2+} (Figure 3) and Sr^{2+} *(8)*. Ligand-promoted dissolution of the iron oxide coatings in the saprolite is most likely responsible for the dissociation of the Cd- and $SrEDTA^{2-}$ complexes *(8, 17, 24-27, 43-46)*. This reaction is promoted by the significantly greater stability constant of $Fe(III)EDTA^{-}$ (log K = 25.1) in comparison to the divalent metal-chelate complexes ($SrEDTA^{2-}$ log K = 11.0 and $CdEDTA^{2-}$ log K = 16.5) *(47)*.

Dissociation in the bedrock aquifer was very similar to that observed in the saprolite, as evidenced by the rapid formation of $Fe(III)EDTA^{-}$ (Figure 4). Again, the elution of $Fe(III)EDTA^{-}$ preceded that of $CdEDTA^{2-}$ and $Co(II)EDTA^{2-}$ (log K = 16.8) *(47)*. Further, the production of $Fe(III)EDTA^{-}$ continued with increasing distance from the source, indicating a constant source of Fe(III) (Figure 4). Transport of divalent Cd^{2+} and Co^{2+} was not observed in the bedrock aquifer, most likely because the groundwater is strongly buffered by the carbonate system (Table 1) *(11-12)*. Geochemical speciation modeling suggested that these divalent cations were removed from solution by precipitation of Co- and $CdCO_3$ *(11-12)*. In contrast, Cd^{2+} was completely recovered in the saprolite *(17)*, which is most likely related to the extensive leaching of carbonate from the weathered vadose zone saprolite *(28)*.

Though there are significant differences in the water and solid phase chemistry of the saprolite versus the bedrock aquifer, similar production of $Fe(III)EDTA^{-}$ was observed in both systems (Figures 3-4). Because the rate of formation of $Fe(III)EDTA^{-}$ decreases above pH 6 *(20-22)*, the neutral pH of the groundwater (6.7-7.0) is less conducive to the formation of $Fe(III)EDTA^{-}$ in comparison to the acidic pH of the saprolite (3.7-3.9). Oxygen levels in the groundwater are typically low (Table 1), and reduced Mn and Fe are observed in the groundwater *(11-12, 39)*. Further, long-term monitoring suggests that iron- and sulfate-reduction, methanogenesis, and anaerobic reductive dechlorination of trichloroethylene (TCE) are dominant biological processes in the groundwater *(39)*. In this environment, oxidized minerals (Fe-, Al- and Mn-oxides) are present in the overlying saprolite *(28)* will be thermodynamically

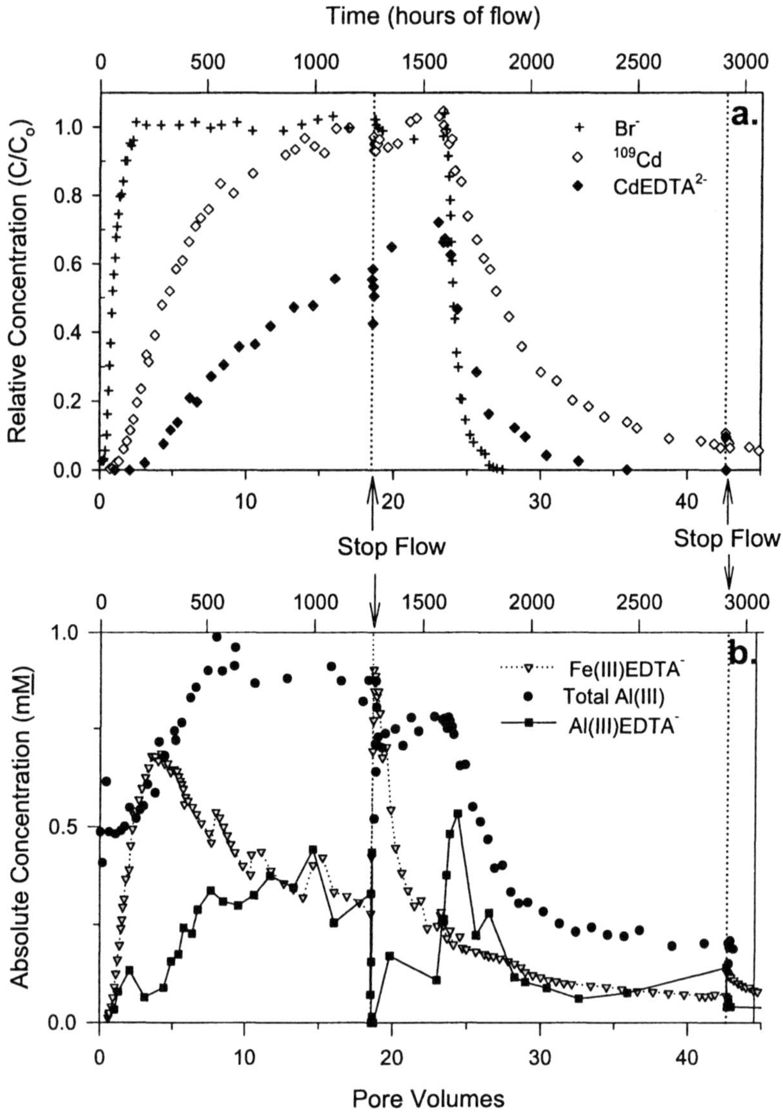

Figure 3. Observed solute effluent concentrations involving the displacement of Br^- and $CdEDTA^{2-}$ in undisturbed core of fractured saprolite. a) Relative effluent concentrations of Br^-, ^{109}Cd, and $CdEDTA^{2-}$. b) Absolute effluent concentrations of Al and $Fe(III)EDTA^-$, and predicted concentrations of $AlEDTA^-$. (Adapted with permission from reference 17. Copyright 2000 Elsevier Science B.V.)

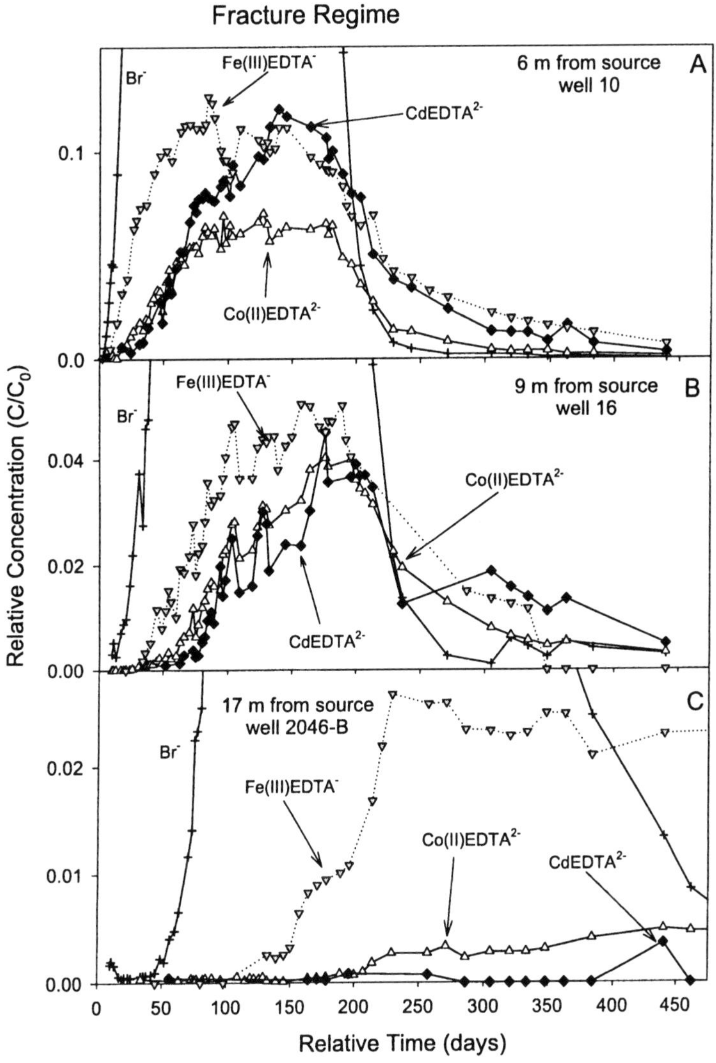

Figure 4. Breakthrough curves for the injection of Br^-, $Co(II)EDTA^{2-}$, $CdEDTA^{2-}$, and the formation of $Fe(III)EDTA^-$ in the fracture regime of the bedrock aquifer at WAG 5. For chelated metals, C_o is the concentration of influent EDTA species (e.g., $Co(II)EDTA^{2-}$ + $CdEDTA^{2-}$). (Reproduced with permission from reference 11. Copyright 2002 Elsevier Science B.V.)

unstable. Mineralogical analysis of the WAG 5 bedrock (x-ray diffraction coupled with scanning electron microscopy with energy-dispersive analysis) has identified biotite as the only iron-containing mineral *(11-12)*. Small amounts of Fe(III) can be present as structural impurities in the octahedral layers of biotite, and it has been theorized that adsorption of Cd- and Co(II)$EDTA^{2-}$ onto biotite may have promoted the release of unstable Fe(III) and the formation of Fe(III)$EDTA^{-}$ *(11-12)*.

The formation of Fe(III)$EDTA^{-}$ in the bedrock aquifer may also be a result of connectivity between the weathered trench environment and the underlying fractured zone in the bedrock aquifer. The wellfield is oriented within and parallel to a prominent fracture zone, and evidence suggests that the conductivity of the fracture zone exceeds that of the matrix by at least one order of magnitude *(9, 11-12)*. Oxygenated stormwater periodically enters the fractured regime of the bedrock aquifer *(9, 11-12, 32)*, resulting in dissolved oxygen (DO) concentrations as high as 100 μM *(39)*. Seasonal fluctuations in water levels are suspected to occasionally allow aerobic biodegradation of TCE at WAG 5, but such occurrences are only at very limited downgradient locations which are proximal to the ground surface *(39)*. Regardless, such temporal changes could result in the formation of poorly-ordered, nanocrystalline Fe(III) (hydr)oxides. Such oxides have been observed in the suboxic, mildly reducing zone of the Cape Cod aquifer *(48-49)* to contribute to the dissociation of divalent metal-EDTA complexes and the formation of Fe(III)$EDTA^{-}$ *(24-27)*. Further, nitrate reduction, which has been observed at WAG 5 *(11)*, is often associated with the formation of intermediate Fe(III) (hydr)oxides *(50-51)*. Sampling difficulties may prevent successful identification of highly disordered, nano-crystalline oxides in fracture zones; in addition, conventional techniques are often ineffective in identifying amorphous minerals *(52)*. The cause of the formation of Fe(III)$EDTA^{-}$ at WAG 5, however, remains unresolved without additional core analyses.

In addition to Fe-oxides, Al sources in the vadose zone effectively dissociated the Cd- and $SrEDTA^{2-}$ complexes. Four major sources of Al exist in the saprolite: in aqueous solution, on the exchange complex, in primary and secondary phyllosilicates, and co-precipitated (20-35%) with Fe-oxides *(28)*. Although the stability constant of $AlEDTA^{-}$ (log K = 16.3) is considerably less than that of Fe(III)$EDTA^{-}$ (log K = 25.1) *(47)*, our results suggest that it can compete effectively for divalent chelates in the natural subsurface environment *(8, 17)*. The formation of $AlEDTA^{-}$ was most important in the $SrEDTA^{2-}$ saprolite injection early in the experiment, which suggests formation from a readily available source of Al, such as from aqueous solution or from the exchange complex *(8)*. In the $CdEDTA^{2-}$ saprolite injections, however, $AlEDTA^{-}$ was only formed when the saprolite was primarily comprised of weathered limestones *(17)*, as shown in Figure 3. Because differences in their stability constants are significant, concomitant formation of $AlEDTA^{-}$ and

$Fe(III)EDTA^-$ suggests that other factors, such as kinetics and/or hydrology, might influence the dissociation reaction. The dynamics of competition between Al and Fe for the chelate was apparent following the flow interruption (Figure 3), which resulted in decreasing Al and increasing Fe concentrations. The formation of the more stable $Fe(III)EDTA^-$ has been observed to occur at a slower rate than the formation of $AlEDTA^-$ *(43-46)*, consequently the flow interruption may have been conducive to the production of $Fe(III)EDTA^-$. This suggests that the hydrologic flow regime, in which preferential fracture flow was significant *(7-8, 10, 14, 17)* may have inhibited the formation of $Fe(III)EDTA^-$. These observations could also imply that $AlEDTA^-$ became readsorbed to Fe-oxides during the interruption, where it was transformed to the more thermodynamically stable $Fe(III)EDTA^-$ *(25)*.

$AlEDTA^-$ was not formed in the bedrock injection *(11-12)*. Sources of Al identified in the bedrock are primary and secondary phyllosilicates such as biotite and kaolinite, which are stable and nonlabile *(12)*. The neutral groundwater pH is not as conducive to the formation of aqueous Al (Table 1) in contrast to the overlying saprolite *(18-21)*. The formation of $AlEDTA^-$ has been observed at neutral pH, but only in the absence of all other competing metal oxides *(19)*.

In the bedrock aquifer, mass balance for EDTA species was not always achieved between Co, Cd, and Fe(III) (Figure 5), the effects of which are more pronounced with distance from the source for both fracture and matrix flow regimes *(11)*. It is possible that $MnEDTA^{2-}$ (log K = 13.9) and/or $Fe(II)EDTA^{2-}$ (log K = 14.3) *(47)* were formed, but geochemical speciation calculations supported the formation of $MnEDTA^{2-}$ *(11)*. $MnEDTA^{2-}$ was tentatively identified using ion chromatography, but it was not possible with available methods to quantitatively confirm its existence at that time *(40-41)*. Neither complex is thermodynamically favored in the presence of $Co(II)EDTA^{2-}$ and $CdEDTA^{2-}$, which suggests that factors in addition to thermodynamic stability may influence metal speciation in natural waters *(25)*. No evidence of chelated Mn was observed in the saprolite experiments *(8, 17)*, most likely because of its relatively low stability constant in comparison to other readily available metals.

In the saprolite, transformation of $Co(II)EDTA^{2-}$ to $Co(III)EDTA^-$ was dominant, resulting in conversion of 88% of the influent $Co(II)EDTA^{2-}$ to the oxidized form (Figure 6). This is significant due to the high stability constant of $Co(III)EDTA^-$ (log K = 39.8) *(47)*, which greatly enhances its persistence and transport in the subsurface *(1-2)*. The role of Mn-oxides in mediating the redox reaction on the sediment surface has been confirmed in experiments with pyrolusite-coated sand by Jardine and Taylor *(5)* and Fendorf *et al.* *(53-54)*. The correlation with Mn-oxides in the intact cores was suggested by a large pulse of aqueous Mn, which accounts for 90% of the $Co(III)EDTA^-$ produced assuming a simple stoichiometry with MnO_2 as the mineral oxidant. Fe(III)oxides have also been identified as potential oxidants of $Co(II)EDTA^{2-}$

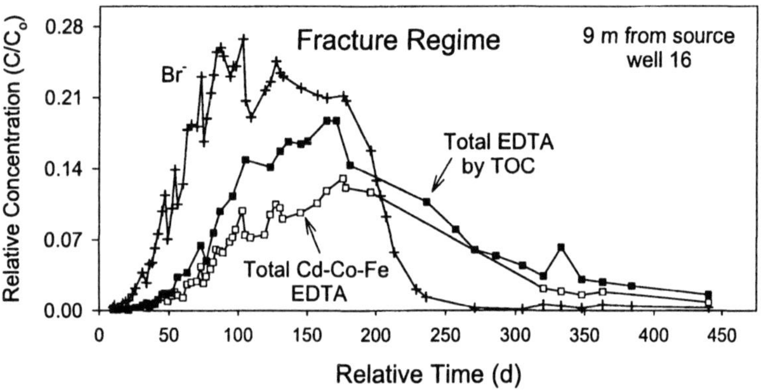

Figure 5. Mass balance of chelated species in the fracture zone 9 m from the source well. Note disparity between total EDTA and sum of Co-Cd-FeEDTA species. (Adapted with permission from reference 11. Copyright 2000 Elsevier Science B.V.)

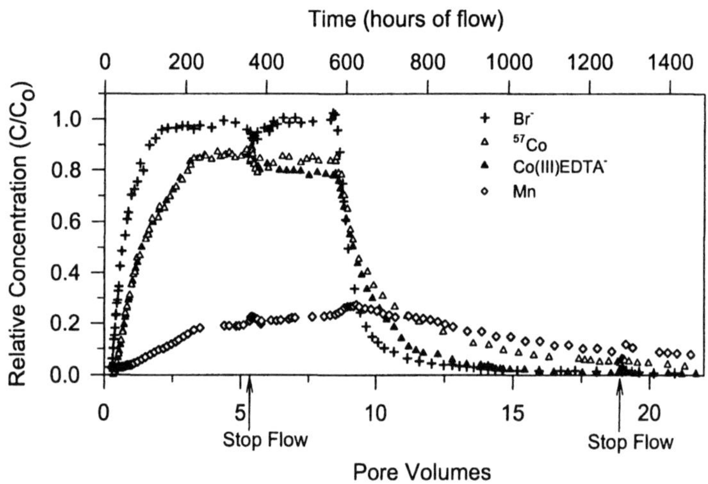

Figure 6. Observed solute effluent concentrations in undisturbed core of fractured saprolite. Effluent concentrations of Br^-, ^{57}Co, $Co(III)EDTA^-$, and Mn shown are relative to influent concentrations of Br^- and $Co(II)EDTA^{2-}$. (Adapted with permission from reference 17. Copyright 2000 Elsevier Science B.V.)

(15, 22-23), but this pathway has not been demonstrated in the saprolite experiments, most likely due to the presence of excess Mn-oxides which are stronger oxidants *(5, 8, 17, 19)*. Dissociation of the $Co(II)EDTA^{2-}$ complex was minimal (12%) in the saprolite, and was only observed following an interruption in flow as evidenced by separation between total ^{57}Co and $Co(III)EDTA^{-}$ (Figure 6). $Co(II)EDTA^{2-}$ was not present in the effluent, as confirmed by IC analyses *(17)*. This is similar to the $CdEDTA^{2-}$ injection (Figure 3), in which flow interruption resulted in an increase in the extent of ligand-promoted dissociation. Again, this suggests that rapid, preferential fracture flow tends to inhibit the rate-limited dissociation reaction.

In the bedrock aquifer, a pulse of $Co(III)EDTA^{-}$ was observed following a large precipitation event (Figure 7), which is most likely related to increasing DO in the fractured zone *(11-12, 39)*. Rapid, significant production of $Co(III)EDTA^{-}$ was observed in laboratory batch experiments with crushed bedrock from the site, but only in experiments in which oxygen was present (Figure 8). The presence of the solids clearly accelerates the oxidation reaction since $Co(II)EDTA^{2-}$ in an oxygenated aqueous environment devoid of mineral solids is stable for several months. Under anaerobic conditions, only small amounts of $Co(III)EDTA^{-}$ were formed, and at decreased rates in comparison to oxidizing conditions (Figure 8). The particular mineral phase catalyzing the reaction, however, is unknown because oxides of Fe and Mn *(5-8, 11-12, 15, 17, 19, 22-23)* have not been identified in the bedrock samples *(11-12)*. This suggests the formation of $Co(III)EDTA^{-}$ in the bedrock aquifer will only occur following significant and prolonged recharge events.

Implications for Offsite Migration of Chelated Radionuclides

The differences observed between the saprolitic trench environment and the underlying bedrock aquifer are related to the differing oxidation states of the water and the subsurface mineralogy. In the saprolite, oxidized Fe, Mn, and the presence of Al greatly influenced contaminant mobility, resulting in the dissociation of divalent chelated metals and rapid oxidation of redox-sensitive Co. The formation of $Co(III)EDTA^{-}$ has important environmental implications due to the high stability constant of the species. In the bedrock, dissociation of divalent metal chelates resulted in rapid and significant production of $Fe(III)EDTA^{-}$. The source of Fe(III) is probably the primary mineral biotite and/or poorly crystalline Fe(III) (hydr)oxides deposited in fractures. The oxidation of $Co(II)EDTA^{2-}$ to $Co(III)EDTA^{-}$ in the groundwater was only associated with storm events and increasing groundwater oxygen levels. This suggests that local hydrologic conditions can cause fluctuations in the oxidation

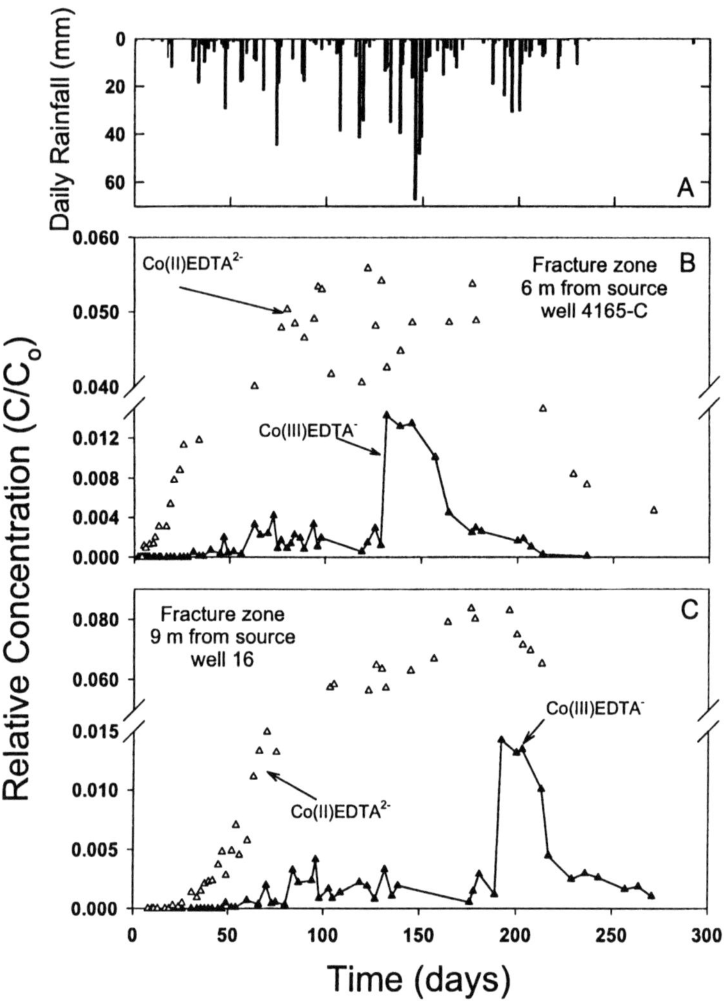

Figure 7. a) Daily rainfall during injection of Br^-, $Co(II)EDTA^{2-}$, and $CdEDTA^{2-}$ at WAG 5. b) Concentration of $Co(II)EDTA^{2-}$ and $Co(III)EDTA^-$ (relative to influent $Co(II)EDTA^{2-}$) in fracture zone 6 m from source, and c) 9 m from source. Note axis break. (Adapted with permission from reference 12. Copyright 2003 Lewis Publishers)

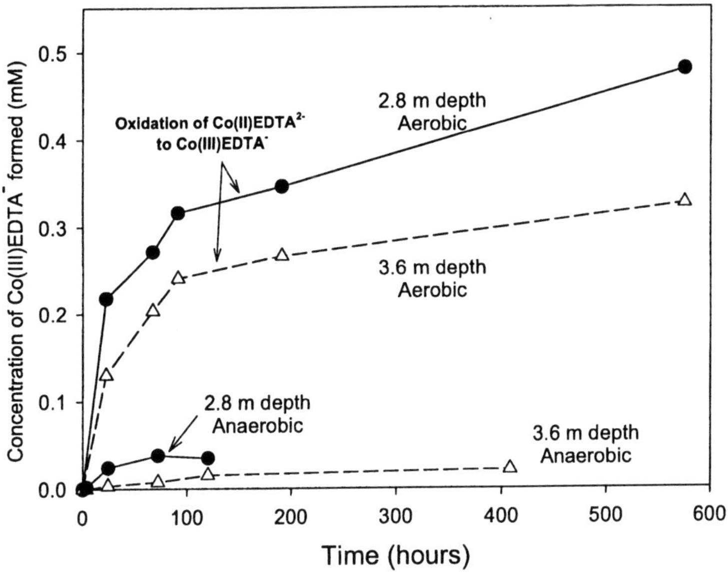

Figure 8. Batch kinetic studies investigating the oxidation of 1.0 mM Co(II)EDTA^{2-} to Co(III)EDTA$^-$ by bedrock from WAG 5, under aerobic and anaerobic conditions. (Reproduced with permission from reference 12. Copyright 2003 Lewis Publishers)

state of the dominantly anaerobic groundwater, which can subsequently impact the expected oxidation state of contaminants.

Extreme precipitation events, common in the humid southeastern U.S., and the hydrology of the fractured trench and groundwater environment, are expected to result in the dissemination of chelated metals to offsite locations. The bedrock injection seems to suggest that the risk for migration of metal-EDTA complexes offsite is minimal and is only associated with storm events. The oxidation and formation of highly stable, mobile $Co(III)EDTA^-$, however, was the dominant process in the saprolite which is representative of the disposal environment. Our results suggest that $Co(III)EDTA^-$ formed in the trenches would be conservatively transported through the bedrock. The hydraulic connection between an intensely weathered vadose zone and a structurally-controlled fractured bedrock, therefore, may contribute to the transport of contaminants from the saprolitic trench environment into the groundwater. This connection is supported by historical data from ORNL in which trench constituents including VOCs, ^3H, and $^{60}Co(III)EDTA^-$ discharged into surface waters *(37-39)* and were detected significantly downstream (1-2).

The WAG 5 site is currently undergoing installation of an engineered RCRA cap over the entire disposal area to prevent infiltration of storm water and reduce downgradient flux of contaminants. However, significant secondary contaminant sources have developed in the matrix regime over the last six decades, particularly in the deeper bedrock groundwater flow regime. The secondary sources initially formed in response to diffusional concentration gradients, in which contaminants were effectively sequestered into the slower-flowing matrix *(9, 11)*. Following installation of the cap, it is probable that these secondary sources will continue to contribute to groundwater flow and downstream contamination. In addition, reducing the influx of oxygenated waters may significantly alter the groundwater chemistry and predicted contaminant leaching.

Acknowledgements

This research was supported by the U.S. Department of Energy Office of Science and is a contribution from the Natural and Accelerated Bioremediation (NABIR) Program of the Office of Biological and Environmental Research. The authors would like to thank Mr. Paul Bayer and Dr. Anna Palmisano, program managers of DOE, for financially supporting this research. We would like to acknowledge D.B. Kent (USGS) and an anonymous reviewer for their suggestions which improved this manuscript. Oak Ridge National Laboratory is managed by the University of Tennessee - Battelle, LLC, under contract DE-AC05-00OR22725 with the U.S. Department of Energy.

References

1. Means, J.L.; Crerar, D.A.; Duguid, J.O. *Science (Washington, D.C.)* **1978,** *200,* 1477-1481.
2. Olsen, C.R.; Lowry, P.D.; Lee, S.Y.; Larsen, I.L.; Cutshall, N.H. *Geochim. Cosmochim. Acta* **1986,** *50,* 593-607.
3. Ayres, J.A. *Equipment Contamination with Special Attention to Solid Waste Treatment;* U.S. Department of Energy: Richland, WA, 1971; BNWL-B-90.
4. Riley, R.G.; Zachara, J.M. *Chemical Contaminants on DOE lands and Selection of Contaminant Mixtures for Subsurface Science Research*; U.S. Department of Energy. U.S. Government Printing Office: Washington, D.C., 1992; DOE/ER-0547T, pp 1-77.
5. Jardine, P.M.; Taylor, D.L. *Geoderma* **1995,** *67,* 125-140.
6. Jardine, P.M.; Taylor, D.L. *Geochem. Cosmochem. Acta* **1995,** *59,* 4193-4203.
7. Jardine, P.M.; Jacobs, G.K.; Wilson, G.V. *Soil Sci. Soc. Am. J.* **1993,** *57,* 945-953.
8. Jardine, P.M.; Jacobs, G.K.; O'Dell, J.D. *Soil Sci. Soc. Am. J.* **1993,** *57,* 954-962.
9. Jardine, P.M.; Sanford, W.E.; Gwo, J.P.; Reedy, O.C.; Hicks, D.S.; Riggs, R.J.; Bailey, W.B. *Water Resour. Res.* **1999,** *35,* 2015-2030.
10. Jardine, P.M.; Wilson, G.V.; Luxmoore, R.J.; and Gwo, J.P. *Conceptual Models of Flow and Transport in the Fractured Vadose Zone;* National Academy Press: Washington, D.C., 2001; pp 87-114.
11. Jardine, P.M.; Mehlhorn, T.L.; Larsen, I.L.; Bailey, W.B.; Brooks, S.C.; Roh, Y.; Gwo, J.P. *J. Contam. Hydrol.* **2002,** *55,* 137-159.
12. Jardine, P.M.; Mehlhorn, T.L.; Roh, Y.; Sanford, W.E. In *Geochemical and Hydrological Reactivity of Heavy Metals in Soils;* Selim, H.M.; Kingery, W.L., Eds.; Lewis Publishers: New York, 2003; pp 1-24.
13. Taylor, D.L.; P.M. Jardine. *J. Environ. Qual.* **1995,** *24,* 789-792.
14. Reedy, O.C.; Jardine, P.M.;Wilson, G.V.; Selim, H.M. *Soil Sci. Soc. Am. J.* **1996,** *60,* 1376-1384.
15. Brooks, S.C.; Taylor, D.L.; Jardine, P.M. *Geochim. Cosmochim. Acta* **1996,** *60,* 1899-1908.
16. Brooks, S.C.; Carroll, S.L.; Jardine, P.M. *Environ. Sci. Technol.* **1999,** *33,* 3002-3011.
17. Mayes, M.A.; Jardine, P.M.; Larsen, I.L.; Brooks, S.C.; Fendorf, S.E. *J. Contam. Hydrol.* **2000,** *45,* 243-265.
18. Girvin, D.C.; Gassman, P.L.; Bolton, H. *Soil Sci. Soc. Am. J.* **1993,** *57,* 47-57.

19. Zachara, J.M.; Gassman, P.L.; Smith, S.C.; Taylor, D.L. *Geochim. Cosmochim. Acta* **1995,** *59,* 4449-4463.
20. Zachara, J.M.; Smith, S.C.; Kuzel, L.S. *Geochim. Cosmochim. Acta* **1995,** *59,* 4825-4844.
21. Szecsody, J.E.; Zachara, J.M.; Bruckhart, P.L. *Environ. Sci. Tech.* **1994,** *28,* 1706-1716.
22. Szecsody, J.E.; Zachara, J.M.; Ashokkumar, C.; Jardine, P.M.; Ferrency, A.M. *J. Hydrol.* **1998,** *209,* 112-136.
23. Szecsody, J.E.; Ashokkumar, C.; Zachara, J.M.; Garvin, A.L. *Water Resour. Res.* **1998,** *34,* 2501-2514.
24. Kent, D.B.; Davis, J.A.; Anderson, L.D.; Rea, B.A. *Water-Resources Invest. Rep. 91-4034*; U.S. Geological Survey: Reston, VA, 1991; pp 78-83.
25. Kent, D.B.; Davis, J.A.; Anderson, L.D.; Rea, B.A.; Coston, J.A. *Geochim. Cosmochim. Acta* **2002,** *66,* 3017-3036.
26. Davis, J.A.; Kent, D.B.; Rea, B.A.; Maest, A.S.; Garabedian, S.P. In *Metals in Groundwater;* Allen, H.E.; Perdue, E.M.; Brown, D.S., Eds.; Lewis Publishers: Chelsea, MI, 1993; pp 223-273.
27. Davis, J.A.; Kent, D.B.; Coston, J.A.; Hess, K.M.; Joye, J.L. *Water Resour. Res.* **2000,** *36,* 119-134.
28. Arnseth, R.W.; Turner, R.S. *Soil Sci. Soc. Am. J.* **1988,** *52,* 1801-1807.
29. Hatcher, R.D.; Dreier, R.B.; Lemizski, P.J.; Lee, R.R.; Ketelle, R.H.; McMaster, W.M.; Lee, S.Y.; Lietzke, D.A.; Foreman, J.L. *Preliminary Summary of the Geology of the Oak Ridge Reservation;* U.S. Department of Energy: Oak Ridge, TN., 1992; ORNL/TM 12074.
30. Driese, S.G.; McKay, L.D.; Penfield, C.P. *J. Sediment Res.* **2001,** *71,* 843-857.
31. Kooner, Z.S.; Jardine, P.M.; Feldman, S. *J. Environ. Qual.* **1995,** *24,* 656-662.
32. Solomon, D.K.; Moore, G.K.; Toran, L.E.; Dreier, R.B.; McMaster, W.M. *A Hydrologic Framework for the Oak Ridge Reservation;* U.S. Department of Energy: Oak Ridge, TN, 1992; ORNL/TM-12026; pp 1-127.
33. Lemizski, P.J. *Mesoscopic Structural Analysis of Bedrock Exposures at the Oak Ridge K-25 Site, Oak Ridge, Tennessee;* U.S. Department of Energy: Oak Ridge, TN, 1995; ORNL/K-ER-259; pp 1-62.
34. Sledz, J.J.; Huff, D.D. *Computer Model for Determining Fracture Porosity and Permeability in the Conasauga Group;* U.S. Department of Energy: Oak Ridge, TN, 1981; ORNL/TM-7695; pp 1-150.
35. Dreier, R.B.; Solomon, D.K.; Beaudoin, C.M. In *Flow and Transport Through Unsaturated Rock;* Evans, D.D.; Nicholson, T.J., Eds.; Geophysical Monograph 42; American Geophysical Union: Washington, D.C., 1987; pp 51-59.

36. Rothschild, E.R.; Huff, D.D.; Spalding, B.P.; Lee, S.Y.; Clapp, R.B.; Lietzke, D.A.; Stansfield, R.G.; Farrow, N.D.; Farmer, C.D.; Munro, I.L. *Characterization of Soils at Proposed Solid Waste Storage Area (SWSA) 7;* U.S. Department of Energy: Oak Ridge, TN, 1984; ORNL/TM-9326; pp 1-151.
37. Hicks, D.S.; Solomon, D.K.; Farrow, N.D. *Investigation of Groundwater Flow Zones and Contaminant Transport in Solid Waste Storage Area 5 at Oak Ridge National Laboratory, Oak Ridge, Tennessee;* U.S. Department of Energy: Oak Ridge, TN, 1992; ORNL/ER-Rep. 154; pp 1-25.
38. Wickliff, D.S.; Solomon, D.K.; and Farrow, N.D. *Preliminary Investigation of Processes that Affect Source Term Identification;* U.S. Department of Energy: Oak Ridge, TN, 1991; ORNL/ER-Rep. 59; pp 1-31.
39. Lenczewski, M.; Jardine, P.; McKay, L.; Layton, A. *J. Contam. Hydrol.* **2003,** *64*, 151-168.
40. Ammann, A. *J. Chrom. A.* **2002,** *947*, 205-216.
41. Ammann, A. *Anal. Bioanal. Chem.* **2002,** *372,* 448-452.
42. Larsen, I.L. *Radioactivity and Radiochem.* **1998,** *9*, 4-12.
43. Chang, H.C.; Matijevic, E. *J. Colloid Interface Sci.* **1983,** *92*, 479-488.
44. Nowack, B.; Sigg, L. *Geochim. Cosmochim. Acta* **1997,** *61,* 951-963.
45. Nowack, B. *Env. Sci. Tech.* **2002,** *36,* 4009-4016.
46. Ludwig, C.; Casey, W.H.; Rock, P.H. *Nature (London)*, **1995,** *375*, 44-47.
47. Parker, D.R.; Zelazny, L.W.; Kinraide, T.B. *Soil Sci. Soc. Am. J.* **1987,** *51,* 488-491.
48. Coston, J.A.; Fuller, C.C.; Davis, J.A. *Geochim. Cosmochim. Acta* **1995,** *59,* 3535-3547.
49. Banfield, J.F.; Hamers, R.J. In *Geomicrobiology: Interactions Between Microbes and Minerals;* Banfield, J.F.; Nealson, K.H., Eds.; Reviews in Mineralogy; Mineraological Society of America: Washington, D.C., 1997; Vol. 35, pp 81-122.
50. Senn, D.B.; Hemond, H.F. *Science (Washington, D.C.)* **2002,** *296,* 2373-2376.
51. Postma, D.; Boesen, C.; Kristiansen, H.; Larsen, F. *Water Resour. Res.* **1991,** *27,* 2027-2045.
52. Roh, Y., Oak Ridge National Laboratory, personal communication, 2004.
53. Fendorf, S.E.; Jardine, P.M.; Patterson, R.R.; Taylor, D.L.; Brooks, S.C. *Geochim. Cosmochim. Acta* **1999,** *63*, 3049-3057.
54. Fendorf, S.E.; Jardine, P.M.; Taylor, D.L.; Brooks, S.C.; Rochette, E. In *Mineral-Water Interfacial Reactions*; Sparks, D.L.; Grundel, T., Eds.; ACS Symposium Series 715; American Chemical Society: Washington, D.C., 1999; pp 358-371.

Chapter 19

Transport and Reactions of ESTA in Soils: Experiments and Modeling

Iris Vogeler[1] and Thabo Thayalakumaran[2]

[1]Enviroment and Risk Management Group, HortResearch, Palmerston North, New Zealand
[2]Davies Laboratory, CSIRO Land and Water, Private Mail Bag, P.O. Aitkenvale, QLD, Australia

Abstract

The transport of EDTA and EDTA-metal complexes in soils is influenced by various rate-limited geochemical and hydrological processes. The following chapter discusses some of these interactive processes, including the effects of the EDTA concentration, the metal-EDTA stability in soils, and the pH and organic matter content of the soils on the transport of EDTA, copper and iron through two contrasting soils. Several column leaching experiments were perfomed using different amounts and concentrations of EDTA, using Cu-contaminated and uncontaminated soil, and varying the lengths of time EDTA was left in the soil. The results were successfully simulated with a simple model, based on the CDE and including a source/sink term for the time-dependent reactions of Cu and Fe with EDTA, and the reversion of $CuEDTA^{2-}$ to adsorbed Cu^{2+} and $Fe(III)EDTA^{-}$ in solution. However, model parameter values needed to be adjusted, as the various rate constants are soil speficic, the dispersivity depends on the soil structure, and the retardation of EDTA on the pH.

Introduction

Heavy metal (HM) contamination of soil is a serious issue in most industrialised and many emerging countries. The sources of heavy metal contamination include activities from mining, smelting, agriculture/industrial applications, and incineration processes.

Ethylendiaminetetraacetic acid (EDTA) is widely used in the photographic industry, in textile and paper manufacturing, and for industrial cleaning (*1*). It is also used in agriculture to increase the solubility and plant availability of micronutrients (*2*). Hence, EDTA has been measured in industrial and domestic wastewater and other aquatic environments (*3,4*). Furthermore it has been suggested that EDTA be added to HM contaminated soils to facilitate remediation both by means of *ex situ* chemical washings (*5*) and *in situ* phytoremediation (*6,7,8*). *In situ* application of EDTA for phytoremediation of HM-contaminated soil however involves the risk of heavy metals polluting the ground water because of the formation of water-soluble metal-chelate complexes (*9*).

The transport of EDTA and EDTA-metal complexes is influenced by various rate-limited or kinetic geochemical and hydrological processes operating in the soil. Important geochemical processes, which are discussed in other chapters of this book, include sorption-desorption of EDTA and/or Me-EDTA, chelate-enhanced dissolution of Fe and Al-oxides, oxidation of Mn and Fe-oxides, and the stability of the EDTA-Me complex that are affected by pH, EDTA concentration and soil constituents. Hydrological processes include convection, diffusion, and dispersion. Furthermore, during transport of solutes through the soil non-equilibrium, characterized by concentration differences between different regions of the soil, has often been observed. Non-equilibrium can either be due to physical or chemical processes. Physical non-equilibrium is the result of heterogeneous water flow regime and preferential flow (*10*), while chemical non-equilibrium is caused by kinetic adsorption (*11*).

The aim of this chapter is to discuss the transport of EDTA and heavy metals in soils, as influenced by these multiple, coupled processes. Examples from specific cases are used to illustrate the impact of individual and combinations of processes on transport. Various models developed to simulate the transport of EDTA and HM in soils are discussed. This chapter is based in part on material presented in recent articles (*12,13*).

Methods and Materials

Various column leaching experiments were carried out on two contrasting soils, the Manawatu fine sandy loam, a Dystric Fluventric Eutrochrept and the Opotiki sandy loam, a Typic Udivitrand Ashy Thermic.

The Manawatu soil was collected from the A horizon of Massey University's Fruit Crops Unit, and had an acid-extractable copper content of 60 mg kg^{-1}. Some soil was artificially contaminated to Cu concentrations of about 400 mg kg^{-1} (high Cu soil) by spraying the soil with $CuSO_4.5H_2O$ solution. The soil was stored for 10 days prior to further experimentation, and regularily remixed during this period. The soil contains 63% sand, 21% silt, and 16% clay, the organic matter content ranges from 5 to 9%, and the CEC is 15.5 Meq%. The mineralogical clay and sand fraction are M_{65} C_{12} V_{15} and Q_{40} F_{20} M_4 C_4 A_{30}, where Q = quartz, M = mica, C = chlorite, V = vermiculite, F = feldspar, A = aggregates.

The Opotki soil was collected from the topsoil of a a site near Opotiki, N.Z. most of which had been used to grow passionfruit for the preceding 10 years. The top 20 mm of the soil had become contaminated with copper due to the heavy use of fungicide sprays, with acid-extractable copper contents between 265 and 290 mg kg^{-1}, and a pH of 5.6 Soil was also collected from an adjacent area that had not been subject to copper sprays, and had a Cu content of 13 mg kg^{-1}, and a lower pH of 4.5. The soil has a relatively high cation exchange capacity of 22 cmolc kg^{-1}, and an organic matter content of 12-16%. The clay mineralogy of A horizon consists of 15% allophane plus imogolite, 25% vermiculite, 18% kandite, and 40% VG Am. SiO_2.

For the leaching experiments the soil was packed into acrylic columns which had nylon mesh at the base, and were 45 mm in internal diameter. Soil was packed to a depth of 100 mm at a bulk density of about 0.9 and 0.65 Mg m^{-3} for the Manawatu and Opotiki soil, respectively. A peristaltic pump was used to apply the leaching solutions to the columns at about 44.5 ml h^{-1}, corresponding to a Darcy flux density of 28 mm h^{-1}. The leachate dripped from the bottom of the columns at atmospheric pressure and was collected in aliquots. The volumetric water contents in the various experiments for the Manawatu soil ranged from 0.48 to 0.56 and for the Opotiki soil from 0.58 to 0.61, which corresoneds to liquid-filled pore volumes ranging from 48-56 mm for the Manawatu soil, and 58 to 61 mm for the Opotiki soil.

The influent EDTA solutions used in the experiments were prepared by mixing equal molarities of EDTA as Na_2H_2EDTA and either Cl as $CaCl_2$ or Br as $CaBr_2$. The Cl or Br were used as inert tracers. In all experiments, the inert tracer concentration was the same as the EDTA concentration. The solutions were then brought to pH 6.4 by the addition of KOH. The columns were pre-leached with a solution of $CaCl_2$ to induce steady-state flow conditions, to leach freely available and exchangeable metals including copper, and also to leach any DOC present in soil. The leachates were analysed for Br or Cl using a Dionex HPLC, for Cu and Fe using an atomic absorption spectrophometer, and dissolved organic carbon (DOC) was analysed using a Shimadzu TC-5000 analyser. It was assumed that after substracting background DOC (DOC cconcentration in the last leachate collected during preleaching), the remaining DOC was EDTA.

At the conclusion of each experiment the soil was extruded from each column, cut into sections and weighed. A sub-sample was used for gravimetric water content determination and another sub-sample was analysed for copper.

Results

The effect of pH on Adsorption of EDTA Complexes

Me-EDTA anionic complexes are known to adsorb weakly (ligand-like) on Fe and Al oxidyhroxide minerals (*14*), with adsorption decreasing with increasing pH. To investigate the effect of the pH on EDTA adsorption to soil under flowing conditions we compare results from our leaching experiment performed on repacked columns of Opotiki sandy loam. In the first experiment the columns were entirely packed with Cu contaminated soil, whereas in the second experiment the top 30 mm of the columns were packed with Cu contaminated soil, and the remaining 70 mm of the column was packed with uncontaminated soil. Note that the pH of the contaminated soil is 5.6, whereas the uncontaminated soil had a pH of 4.5. In both experiments, the columns were preleached with a 0.0025 M $CaSO_4$ solution, then a pulse of 0.01 M EDTA, plus Br was applied to the soil, which was leached again with a 0.0025 M $CaSO_4$ solution. The movement of EDTA and Br was monitored by collecting the effluent. In the first experiment EDTA eluted simultaneously with Br, and thus EDTA was not adsorbed by this soil. However in the second experiment the EDTA breakthrough showed retardation compared to the Br. This difference is probably due to the low pH of 4.5 of the uncontaminated soil relative to that of the contaminated soil with a pH of 5.6. It is noteworthy the difference in the soil pH is not due to the copper contamination but due to the lime application to the contaminated soil.

Similar pH effects on EDTA and Me-EDTA complexes have been found in other studies. For example Brooks *et al.* (*15*) reported greater retardation of $Co(II)EDTA^{2-}$ and $Co(III)EDTA^{-}$ transport through ferrihydrite when they decreased the pH. Kedziorek *et al.* (*16*) found no evidence of EDTA or Cd-EDTA and Pb-EDTA adsorption in their leaching experiments. They suggest that because the effluent pH of 7.8 was well above the point of zero charge (PZC) there was no reason to expect adsorption.

Me-EDTA Stability and its Exchange Reactions

The effectiveness of EDTA mobilisation of HM depends on the ability to keep the HM in a soluble, mobile form (Me-EDTA). We performed experiments on the Opotiki sandy loam and the Manawatu fine sandy loam to look at the stability of the Cu-EDTA complex. Pulses of 50μmol EDTA were applied to repacked soil columns and left in the soil for varying periods of time (0-30 days) before leaching with a 0.0025 *M* $CaSO_4$ solution. Table 1 shows the amounts of Cu and Fe leached with the different EDTA residence times in the Opotiki soil. The decreasing amount of copper leached with increasing residence time, coupled with a corresponding increase in the extracted iron, shows transformation of $CuEDTA^{2-}$ to $Fe(III)EDTA^{-}$ in the soil. We found similar results for the Manawatu soil (*13*). Several other authors have observed similar transformations when they conducted experiments involving flow interruption of metal-EDTA for different periods of time (*17*), slower flow rates (*15*), or injected metal-EDTA pulses into aquifers (*18*).

Table 1. Fe and Cu leached with varying EDTA times in Opotiki soil

Residence Time (d)	Cu leached (μmol)	Fe leached (μmol)
0	35	5
1	28	10
7	23	18
30	12	24

Spource: adapted with permission from reference (*12*). Copyright CSIRO Publishing

Concentration of EDTA

In a leaching experiment using steady flow through repacked soil column of Manawatu fine sandy loam, we investigated the effect of the EDTA concentration on the leaching of copper. A soil column packed with Cu contaminated soil (425 mg kg^{-1}) was preleached with a 0.0025 *M* solution of $CaSO_4$, followed by subsequent pulses of 10^{-3}, 10^{-2} and 10^{-1} *M* EDTA leached with $CaSO_4$. The EDTA applied to mobilise Cu also mobilised significant amounts of Fe, and the extraction of both Cu and Fe increased with increasing EDTA concentration (Fig. 1). The 10-fold concentration increase in the 2nd EDTA pulse resulted in a 7-fold increase in Cu leached. However, the 3rd pulse of EDTA resulted in an un-proportional low increase in Cu, with only a 10-fold increase compared to the 1st pulse. This suggests limited availability of Cu for complexation. In the first pulse the combined molar amounts of copper and iron almost added up to the amount of EDTA applied. During the second and third pulses, however, the amount of copper plus iron was lower than the amount of EDTA, suggesting limited availability of both Cu and Fe for complexation.

Effect of Organic Matter

To investigate the effect of the organic matter content of the soil on Cu-EDTA complexation and transport we compare results from leaching experiments performed on Cu-contaminated Manawatu fine sandy loam and

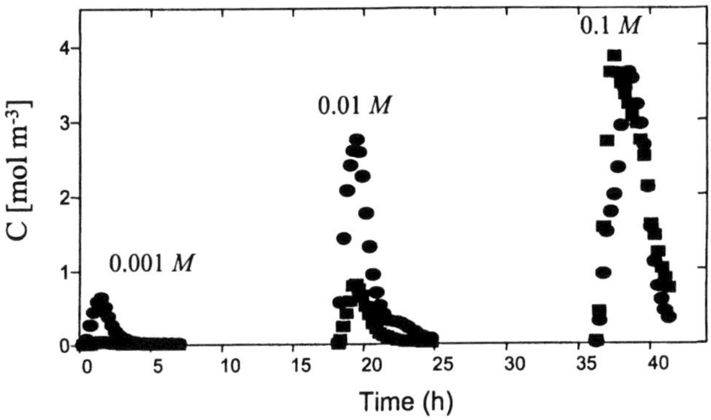

Figure 1. Measured effluent concentrations of Cu (circles) and Fe (squares) eluted with pulses of different EDTA concentrations

Opotiki sandy loam. In the Manawatu soil the organic matter content was 5-9%, while for the Opotiki soil 12-16%. EDTA solution was applied in excess of the amount needed to complex all the copper present in the 2 different soils, with 60 and 420 mg kg^{-1} Cu in the Manawatu soil (both in low and high Cu soil), and 265-290 mg kg^{-1} Cu in the Opotiki soil. Whereas in the Manawatu soil only 40 mg kg^{-1} Cu remained resident in the soil, about 90 mg kg^{-1} Cu remained in the Opotiki soil. This demonstrates the strong relationship between the organic matter content and the binding strength of copper (*19*). The clay mineralogical differences between soils would have also contributed to this. Thus HM complexation with organic matter and their sorption on Fe/Al oxides and clay minerals have a direct influence on the Me-EDTA complexation and hence the mobility of Me-EDTA complex.

Kinetics of Geochemical Processes affecting Heavy Metal-EDTA Transport

Past research indicates that heavy metal-EDTA transport is influenced by the kinetics of interacting geochemical processes, such as time-dependent sorption/desorption and dissolution (16, 20). To specifically investigate the time dependent reaction of $CuEDTA^{2-}$ with the iron in soil, we applied a $CuEDTA^{2-}$ with $CaBr_2$ solution to a column packed with uncontaminated Opotiki soil. Each day for 83 days, 28 mm of 0.001 *M* $CuEDTA^{2-}$ solution (44.5 μmol) was added. The application of the $CuEDTA^{2-}$ for only about an hour per day, with a 23 h rest period in between applications, and the collection of each day's leachate in three aliquots, allowed the rate dependence of the chemical reactions to be demonstrated.

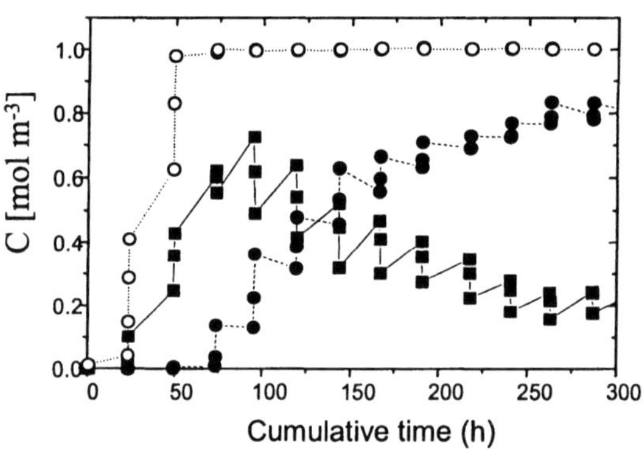

Figure 2. Measured effluent concentrations of bromide (O), copper (●), and iron (■) *following daily pulses of CuEDTA.* Source: reproduced with permission from reference (*13*). Copyright CSIRO Publishing

Figure 2 shows the bromide, copper and iron leachate concentrations in each aliquot during the first twelve days (300 h) of the experiment. As expected the copper concentration in the three aliquots collected on each day increased with time. However the concentration in the first aliquot on the following day is lower than in the last aliquot on the previous day. After the first 50h the iron concentration shows complementary behaviour, decreasing during each flow event, and then increasing immediately after the non-flow periods. This behaviour is explainable as the gradual replacement of $CuEDTA^{2-}$ in the soil solution with $Fe(III)EDTA^{-}$ during the 23 h non-flow period each day. The behaviour became less pronounced with increasing time, as the system approached equilibrium with the invading $CuEDTA^{2-}$ solution. On the other hand, bromide concentration in the three aliquots collected on each day increased with time, and also increased after the no flow period.

Non-Equilibrium during Transport

Flow interruption has been one approach to discriminate between the physical (multiple pore region flow) and chemical processes (rate limited chemical reactions) responsible for non-equilibrium (*21,22*). Thus we applied a pulse of 61 mm (0.7 PV) of 0.001 M EDTA mixed with $CaBr_2$ to an intact column of Opotiki sandy loam and interrupted the flow for a month before leaching with a solution of 0.0025 M $CaSO_4$.

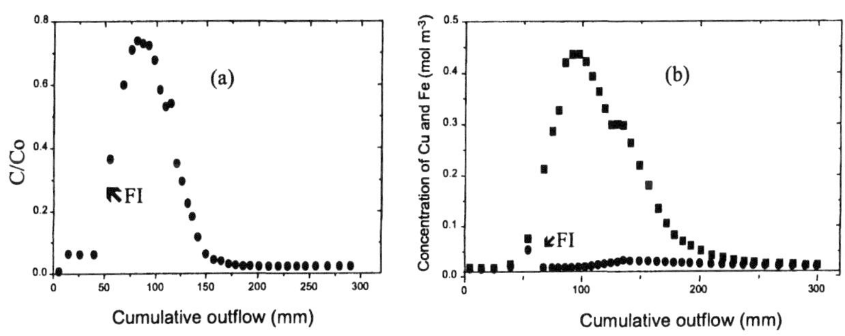

Figure 3. Measured effluent concentrations of (a) Br and (b) Cu (circles) and Fe (squares) from an undisturbed column of Opotiki soil with flow interruption. FI indicates flow interruption. Spource: adapted with permission from reference (*13*). Copyright CSIRO Publishing

While the one-month pause in leaching did not cause any noticeable discontinuity in the bromide concentration (Fig. 3a), the pause induced a pronounced drop in the copper concentration when leaching resumed, and a large increase in the iron concentration (Fig. 3b). These changes reflect the time-dependent chemical reaction of $CuEDTA^{2-}$ in the soil. The absence of a concentration drop in the non-reactive bromide tracer concentration shows that physical non-equilibrium was not a major factor in this experiment, verifying that the discontinuities in the copper and iron concentrations were due to the slow chemical transformation of $CuEDTA^{2-}$ to $Fe(III)EDTA^{-}$ (chemical non-equilibrium).

Mayes et al. (*17*) conducted laboratory leaching experiments using $CdEDTA^{2-}$, a $CdEDTA^{2-}$ and $Co(II)EDTA^{2-}$ mixture, and a $CdEDTA^{2-}$, $Co(III)EDTA^{-}$ and $HCrO_4^{-}$ mixture to investigate the geochemical and hydrological processes affecting metal-EDTA transport. Flow interruption confirmed that the time-dependent surface-mediated dissociation of $CdEDTA^{2-}$ resulted in chemical non-equilibrium.

A long term field experiment by Jardine et al. (*23*) where non-reactive bromide and reactive Co(II)EDTA, CdEDTA and $H^{51}CrO_4$ was injected into a contaminated fractured shale bedrock showed that physical non-equilibrium occurred within the soil matrix, and geochemical non-nonequilibrium within the fracture regime.

Modelling

The use of mechanistic models for simulating the fate and transport of heavy metals as Me-EDTA complexes is a key tool for an environmental impact assessment, and for planning the rates and timing of EDTA applications for cleanup and remediation purposes. The model has to be able to describe quantitatively the main physical and chemical processes involved in Me-EDTA transport from contaminated soils.

A number of models have been developed to describe the interaction and transport of metals and EDTA in soil. Some models, such as PHREEQC (*24*) and HYDROGEOCHEM (*25*) have been developed to simulate the transport of multiple metal species, and are quite complex. Jardine *et al.* (*20*) used the convection-dispersion equation (CDE) and the mobile immobile approach (MIM) with a retardation factor to account for oxidation of Co(II) to describe the transport of Co(II)-EDTA through fractured saporolitic shale. They found that Co(II)-EDTA transport was controlled by time-dependent sorption processes, and thus found the non-equilibrium CDE model more appropriate.

Dissolution of Fe-oxides was found to be minor in their subsurface media and thus not included. Szecsody *et al.* (*26*) developed a 2 D model for $Co^{II/III}$-EDTA transport through sediments including not only oxidation of Co^{II}, but also Fe dissolution. Samani *et al.* (27) developed a model for Pb-EDTA transport, based on equilibrium and kinetic dissolution of Pb. They found that the dissolution has a fast and a slow release phase.

Several other studies have shown that of transformation of HM-EDTA complexes to $Fe(III)EDTA^-$ occur in soils (*18,28,29*). And our results on Cu-EDTA transport through two different topsoils, discussed in the previous section, also show the importance of competition of soil iron for EDTA. Thus, we developed a simple conceptual model for copper transport through soils, based on the widely used convection-dispersion equation, and considering competition of soil iron for EDTA. The model is in part based on the model of Kedziorek *et al.* (*16*), and was tested using results from our leaching experiments on repacked and undisturbed soil columns of Opotiki sandy laom and Manawatu fine sandy loam under various flow conditions.

Convection-Dispersion Equation for Modelling Cu/Fe-EDTA Movement

The most widely used approach for modelling solute transport in the soil is the convection-dispersion equation, which is here briefly outlined for coupled Cu/Fe-EDTA transport. We assume that adsorption is instantaneous, with equilibrium between the solute in the soil solution and the adsorbate. Under steady-state one-dimensional water flow the CDE for reactive solute transport is then given by:

$$\theta R \frac{\partial C}{\partial t} = \theta D_h \frac{\partial^2 C}{\partial z^2} - q \frac{\partial C}{\partial z} + S \qquad [1]$$

Where *C* is the concentration of the halide, or of the free EDTA-, Cu-EDTA or Fe-EDTA in the soil solution (mol m^{-3} of solution), θ is the volumetric water content (m^3 m^{-3}), *t* is the time (s), D_h is the dispersion coefficient (m^2 s^{-1}) accounting for hydrodynamic dispersion and molecular diffusion, *q* is the Darcy flux density (m s^{-1}), *S* is a source/sink term accounting for any chemical reactions bringing that chemical species into or out of solution (mol m^{-3} of soil s^{-1}), and *R* is a dimensionless retardation constant accounting for any reversible instantaneous adsorption as described by a linear adsorption isotherm. *R* equals one if there is no adsorption to the matrix, and is greater than one if there is adsorption.

For diffusion-dispersion we assume

$$D_h = \tau D_o \lambda q / \theta \qquad [2]$$

where τ is a dimensionless tortuosity factor, D_o is the molecular diffusion coefficient in solution ($m^2 s^{-1}$), and λ is the dispersivity (m).

For the inert halide tracers $S = 0$. For EDTA, three source/sink reactions were considered. The first two involve the added EDTA, which denoted as EDTA, reacting with copper and iron in the soil to form complexes as $CuEDTA^{2-}$ and $Fe(III)EDTA^-$. The third reaction was $CuEDTA^{2-}$ reacting with iron in the soil to form $Fe(III)EDTA^-$ and copper ions. These copper ions would then be adsorbed by the soil. In all reactions it was assumed that one mole of metal ion (Cu^{2+} or Fe^{3+}) reacts with one mole of EDTA (*16,30*). Therefore,

$$S_{Cu} = k_1(C_o \rho_b M_{Cu})^n - k_3(C_{Cu} \rho_b M_{Fe})^n$$

$$S_{Fe} = k_2(C_o \rho_b M_{Fe})^n + k_3(C_{Cu} \rho_b M_{Fe})^n \qquad [3]$$

$$S_o = - S_{Cu} - S_{Fe}$$

In the above equations S_o, S_{Cu} and S_{Fe} are the source/sink terms for EDTA, $CuEDTA^{2-}$ and $Fe(III)EDTA^-$ respectively, and n is a dimensionless constant indicating the order of the reaction. The rate constant k_1 is for the reaction between EDTA and the extractable copper, k_2 is for the reaction between EDTA and the extractable iron, and k_3 is for the reaction between $CuEDTA^{2-}$ and the extractable iron. All rate constants have units of s^{-1}. The soil solution concentrations of C_o, C_{Cu} and C_{Fe} are for EDTA, $CuEDTA^{2-}$ and $Fe(III)EDTA^-$ respectively. M_{Cu} is the EDTA-extractable copper concentration in the soil (mol kg^{-1}), and M_{Fe} is the EDTA-extractable iron concentration (mol kg^{-1}). Note that in our experiments chemical equilibrium was not obtained, so stability equilibrium constants are not considered in the model. However, Eq.3 implies that at equilibrium all the EDTA in the soil would be complexed with iron, provided enough iron is available. This is consistent with the fact that the stability constant for the Fe^{3+} complex with EDTA is much greater than that for Cu^{2+}. The log $K_{M\text{-}Y}$ values given in GEOCHEM PC version 2.0 (*31*) are 20.5 for $CuEDTA^{2-}$ and 27.7 for $Fe(III)EDTA^-$.

Equations were solved numerically, using the appropriate boundary and initial conditions. This involved simultaneous solution for the halide tracer, for free EDTA and for EDTA complexed with both copper and iron. An explicit finite-difference scheme, written in Visual Basic™ within Excel™ was used to do this, and employed a forward-difference approach, with numerical dispersion taken into account. A few examples for modelling Cu-EDTA transport are given below.

Model Parameter Evaluation

In all cases model parameter values were either directly measured (q, θ), obtained from the literature ($\tau = 0.66$, Penman, 1940; $D_o = 1 \times 10^{-9}$ m^2 s^{-1} for Cl and Br (*32*), or determined indirectly from the experimental results (λ, R). A molecular diffusion coefficient value for EDTA was not found in the literature, so this value was also used for EDTA (*23*). For both, the repacked Opotiki soil and the repacked Manawatu soil an average of 3 mm was found for λ (for details see (*12*)). The initial value for M_{Cu} was calculated from the acid-extractable value minus the amount of Cu that could not be removed with leaching of excess EDTA. For M_{Fe} the amount leached with excess EDTA was used. Values found for the retardation (R) of chloride and bromide in our experiments were always close to unity, indicating that neither adsorption nor exclusion occurred. Depending on the pH EDTA was in some cases not adsorbed (R = 1), and in other cases adsorbed (R >1), see below.

Modelling the Movement of EDTA

To see if our simple model, based on the CDE and including a source/sink term for the various forms of EDTA can be used to simulate the movement of EDTA under the various experimental conditions, we use the results from our experiments on repacked and undisturbed column experiments of Opotiki and Manawatu soil. Some of the experiments have already been described in the previous section. Overall the Cu-Fe EDTA reactions over a 10-fold difference in EDTA concentration (Eqs. 3) could be best described with an n value of 0.5. Thus this value was used for all the modelling.

The model with the above parameter values successfully simulated the movement of EDTA, when pulses of different concentrations, 0.001 *M* and 0.01 *M* were applied to repacked columns of Manawatu soil, Fig. 4. No adsorption of EDTA was found at a pH of 6.1, thus R = 1. The model also successfully simulated the movement of 0.01 *M* EDTA applied continuously to repacked columns of Cu-contaminated Opotiki soil, see Fig. 5, closed symbols and full line. Again no retardation of EDTA was found at the pH of 5.6, thus R = 1. However in the leaching experiment, where uncontaminated Opotiki soil was used, EDTA was adsorbed, which is probably due to the lower pH of 4.5. This experiments is discussed in more detail in the results section on "The effect of pH on Adsorption of EDTA complexes". With an *R* value of 2 the model closely simulated the measured BTC of EDTA, see open symbols and broken line in Fig. 6. However when EDTA was applied at a lower concentration to the soil with the low pH, the model underestimated the amount of EDTA adsorbed (not shown here). This indicates more adsorption of EDTA than implied by the assumed *R* value of 2. Our simple model implies a linear adsorption isotherm, whereas isotherms are usually non-linear.

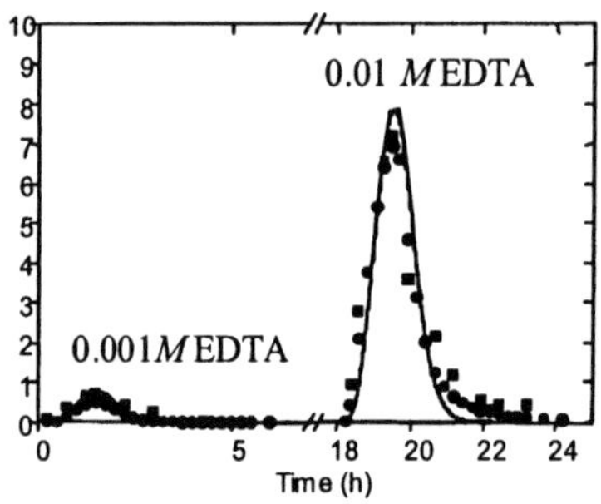

Figure 4. Measured and simulated concentrations of EDTA applied as a pulse at concentrations of 0.001 M and 0.01 M to repacked columns of Manawatu fine sandy loam.

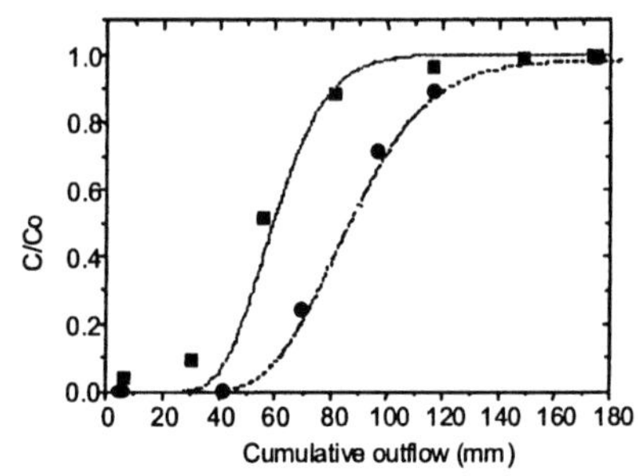

Figure 5. Measured and simulated concentrations of EDTA for repacked columns of Opotiki sandy loam at a pH of 5.6 with R = 1 (squares) and a pH of 4.5 with R = 2 (circles).

The model was also used to simulate the movement of EDTA, in the case where pulses of 31 mm of 0.001 *M* EDTA (0.5 PV) were applied to repacked columns of the Opotiki soil and then leached with a solution of 0.0025 M $CaCl_2$, either immediately, or with a delay of 1 month, see Fig. 6. This experiment is discussed in more detail in the results section on "Me-EDTA Stability and its exchange reactions". With increasing residence time of EDTA in the soil the amplitude of the EDTA pulse is reduced and the spreading is increased. This is due to molecular diffusion.

Modelling Cu/Fe-EDTA Transport

To model the transport of Cu/Fe-EDTA transport with the above equations, values for the rate constants in equation set 3 had to be determined. To obtain values for the rate-coefficients k_1 and k_2 we used results from a leaching experiment on repacked soil column of Opotiki soil, where the Cu contaminated soil was leached continuously with a solution of 0.01 M $EDTA_o$. The values of k_1 and k_2 were found by visually fitting the model predictions to the BTC's of Cu and Fe. The third rate coefficient k_3 was assumed to be zero, as little transformation of $CuEDTA^{2-}$ to $Fe(III)EDTA^-$ was expected during the relatively

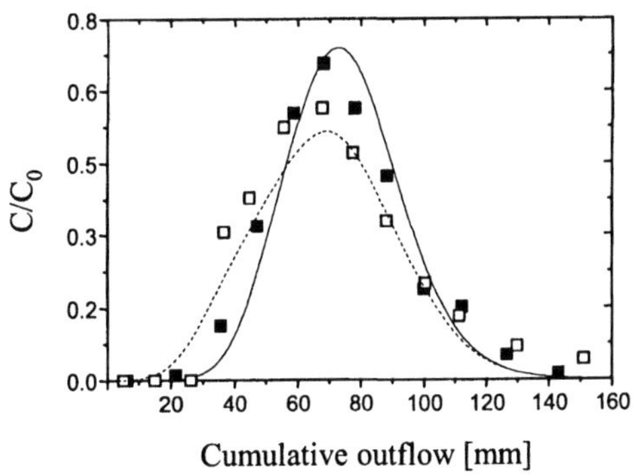

Figure 6. Measured and simulated concentrations of EDTA applied as a pulse and then either leached immediately (closed squares) or with a delay of 1 month (open squares). Source: adapted with permission from reference (12). Copyright CSIRO Publishing

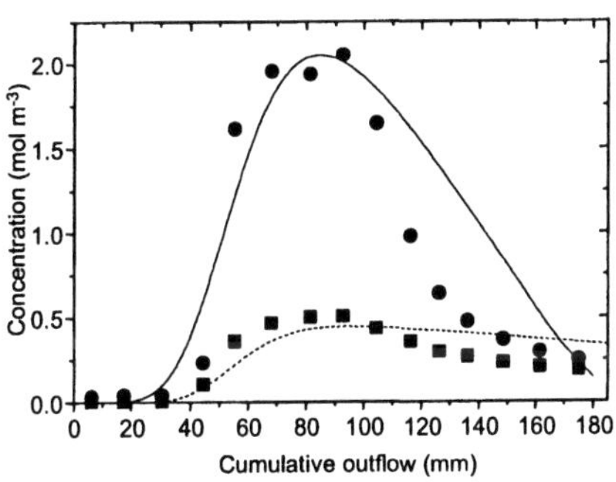

Figure 7. Measured and simulated concentrations of Cu (circles) and Fe (squares) eluted with continuous leaching 0.01 M EDTA through repacked columns of Opotiki sandy loam.

short resident times. The values obtained were 4 x10^{-5} s^{-1} for k_1, and 9 x 10^{-6} s^{-1} for k_2, see Fig. 7.

To obtain the value for k_3 the experiment with pulse application of EDTA to Opotiki soil, and leached immediately or with a 1 month delay, described before was used. A value of 2.2 x10^{-7} s^{-1} was found for k_3, see Fig. 8. Note that this value is two orders of magnitude lower than k_1, justifying the assumption that relatively little transformation from Cu-EDTA to Fe(III)-EDTA occurs within a period of 1 day. The relative good simulations of the Cu and Fe concentrations in the leachates for both immediate leaching of the EDTA pulse and leaching with a month delay demonstrate that the model, with the above model parameters is capable of modelling the transformation of CuEDTA to FeEDTA over relative long time periods.

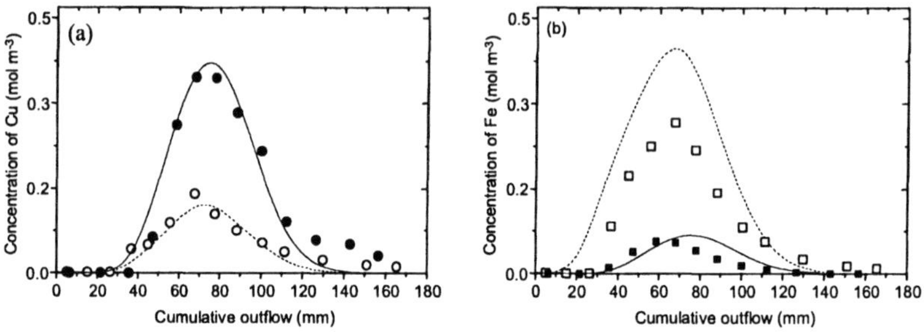

Figure 8. Measured and predicted concentrations of (a) Cu and (b) Fe following a pulse application of EDTA to repacked columns of Opotki sandy loam and either immediate leaching (closed symbols and full lines) or leaching with a 1 month delay (open symbols and broken lines). Source: reproduced with permission from reference (12). Copyright CSIRO Publishing

To see if the values for the rate constants are soil specific, we look at results from leaching experiments with continuous application of 0.01 M EDTA on repacked columns of the Manawatu fine sandy loam, which has a lower OM content, and a different mineralogy. Simulation of the BTC of Cu and Fe using the values for the rate constants, k_2 and k_3 used for the Opotiki soil grossly underestimates the Fe concentration in the leachate, and also predicts slower leaching of copper than was observed (Fig 9). This suggests that the copper and iron is adsorbed with different strengths in the Opotiki and Manawatu soils. We found that assuming all three rate constants were 2.5 times greater in the Manawatu soil than in the Opotiki soil greatly improved the simulations. Thus the values used for k_1, k_2 and k_3 were 1 x10^{-4} s^{-1}, 2.3 x 10^{-5} s^{-1} and 5.5 x 10^{-7} s^{-1} respectively. The faster rate constants in the Manawatu soil are probably due to its lower organic matter content.

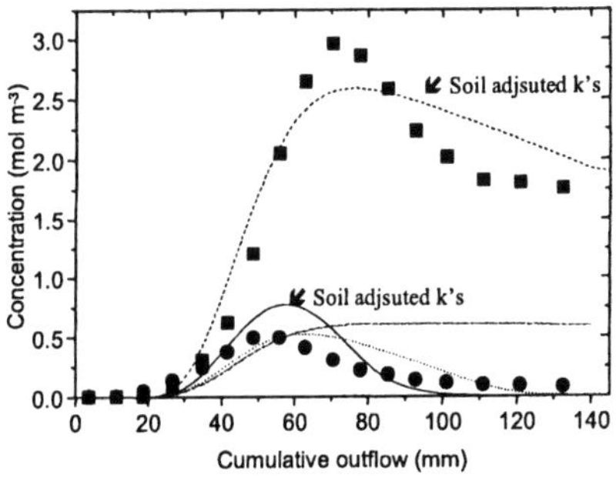

Figure 9. Measured and predicted concentrations of Cu (circles) and Fe (squares) following continuous application of 0.01 M EDTA to repacked columns of Manawatu fine sandy loam using either the rate constants obtained from the Opotiki soil, or soil adjusted valus. Source: reproduced with permission from reference (13). Copyright CSIRO Publishing

To further see if the model could simulate the reaction kinetics of Cu-EDTA with soil iron we examined the experiment where daily pulses of 0.001M $CuEDTA^{2-}$ were applied to an uncontaminated, repacked column of Opotiki soil for about 3 months, see Fig. 2. The only significant reaction that would have taken place during the experiment was the transformation of $CuEDTA^{2-}$ to $Fe(III)EDTA^{-}$. When the model was run with the values for M_{Fe}, and k_3 found for Cu-contaminated Opotiki sandy loam, it grossly underestimated the amount of iron leached and so the reaction rate of the $CuEDTA^{2-}$, resulting in an overestimation in the amount of copper leached. This and the shape of the iron breakthrough curve with an initial sharp peak of 0.6 mol m^{-3}, followed by a long plateau of around 0.2 mol m^{-3}, suggests that two EDTA-extractable iron fractions are present in the soil; a smaller and rapidly reacting fraction and a larger more-slowly reacting one. We further developed our model to take these two Fe-fractions into account, which are however not discussed here. For details see Thayalakumaran et al. (*13*).

Model Conclusions

In the studies described above, EDTA-enhanced transport of copper and iron species through contaminated soils was modelled using the CDE coupled with a source/sink term for the time-dependent reactions of Cu and Fe with EDTA, and the reversion of $CuEDTA^{2-}$ to adsorbed Cu^{2+} and $Fe(III)EDTA^{-}$ in solution. The simple model was quite robust, as it was capable of predicting the copper and iron leaching successfully in both the Opotiki and the Manawatu soil, despite the different amounts and concentrations of EDTA applied, and the varying lengths of time it was left in the soil. However, model parameter values need to be adjusted, as the rate constants are soil speficic, the dispersivity depends on the soil structure, and the retardation of EDTA on the pH.

Another drawback of our simple model is that the simple concept of reversible equilibrium sorption of HM assumed is not always appropriate (*33*). Instead, heavy metal retention and release reactions in the soils have been observed to be strongly time-dependent, as real soils contain many different sorptive components (soil minerals, organic matter, iron and aluminium oxides) that may react with HM via different mechanisms and at a wide range of rates. As a result, kinetic adsorption and two-adsorption site models have been developed (*34,35,36,37*).

Our model has been specifically developed for Cu transport in soils, where Cu is the only heavy metal present in significant amounts. For other metals different metal-EDTA and soil reactions have to be considered, such as Fe- and Mn dissolution, and oxidation-reduction reactions. Thus, for modelling HM-EDTA transport through soils, different models need to be developed or used, as the reactions are soil-and metal specific.

Conclusions

The effectiveness of EDTA in mobilising HM depends on the ability to keep the HM in a soluble, mobile form. This depends on many factors, including the organic matter content, the stability and adsorbability of the HM-EDTA complex, the EDTA concentration, the accessibility of the EDTA to the metal, the contact time of the percolating EDTA with the soil, and thus the flow rate, and the pH. Our and several other studies have shown that HM-EDTA complexes are, during their transport through soils, often unstable with time. The stability depends on the pH, the metal, and the binding strength with the soil constituents. Fe and Al oxides, which are abundant in many soils, decouple HM-EDTA complexes, which results in the more stable Fe(III)EDTA (or AL(III)EDTA), plus free metal. Re-adsorption of free metal cations by the soil is then likely to occur. Thus if EDTA is used for off site remediation the applied EDTA should be leached immediately or within a few days. Furthermore, as EDTA is not selective in extracting metals, it also increases the mobility of alkine earth metals such as Ca and Mg. This also decreases the remediation efficiency of EDTA for HM.

Our simple model, based on the CDE with a source/sink term for the rate-limited geochemical reactions, described Cu-EDTA transport well for two different soils, provided that the model parameters are adjusted for different soils and the pH.

The studies described here showed the ability of EDTA to enhance the HM concentration in the soil solution of contaminated soils, and so indicate that the HM are potentially available for uptake by plants used for phytoremediation. Thus, chelate-enhanced phytoremediation , as discussed in the Chapters by Song, and by Hong of this book, seems to have a great potential for cleaning-up HM contaminated sites. However environmentally safe methods by using biodegradable chelates, with focus on minimising the leaching and optimising the uptake of HM by plants must be developed. Models of coupled Me-EDTA transport and uptake in the soil-plant-atmosphere continuum, such as the model by Thayalakumaran et al. (38) for EDTA-enhanced Cu uptake, need to be developed and tested for various metals and soils, in order to develop effective chelate induced phytoremediation practices that are environmentally sound.

References

1. Ghestem, J.P., and Bermond, A., *Environ. Technol.,* **1998**, 19, 409-416.
2. Li, Z., and Shuman, L.M. *Soil Sci.,* 1996, 161, 226-232.
3. Kari, F.G., and Giger, W.*Water Res.,* **1996**, 30:122-134.
4. Nowack, B., Xue, H., and Sigg, L. *Environ. Sci. Technol.,* **1997**, 31, 866-872.
5. Yu, J., and Klarup, D. *Water Air Soil Pollut.,* **1994**, 75, 205-225.
6. Cunningham, S.D., Anderson, T.A., Schwab, A.P., and Hsu, F.C. *Adv. Agr.,* **1996**, 56:55-114.
7. Blaylock, M.J., Salt, D.E., Dushenkov, S., Gussman, C., Kapulnik, Y., Ensely, B.D., and Raskin, I. *Environ. Sci. Technol.* **1997**, 312(3), 860-865.
8. Kirkham, M.B. Intern J Phytorem, **2000**, 2, 159-172.
9. Lombi, E., Zhao, F.J., Dunham, S.J., and McGrath, S.P. *J. Environ. Qual.*, **2001**, 30, 1919-1926.
10. Nielsen, D.R., van Genuchten, M.T., and Biggar, J.W. *Water Resour. Res.*, **1986**, 22, 89S-108S.
11. Selim, H.M., Davidson, J.M., and Mansell, R.S. *Proceedings of Summer Computer Simulation Conference,* **1976**, Washington, D.C.
12. Thayalakumaran, T., Vogeler, I., Scotter, D.R., Percival, H.J., Robinson, B.H., and Clothier, B.E. *Aust.r J. Soil Res.*, **2003**, 41,323-333.
13. Thayalakumaran, T., Vogeler, I., Scotter, D.R., Percival, H.J., Robinson, B.H., and Clothier, B.E. *Austr. J. Soil Res.*, **2003**, 41, 335-350.
14. Nowack, B., Lutzenkirchen, J., Behra, P., and Sigg, L. *Environ. Sci. Technol.,* **1996**, 30: 2397-2405.

15. Brooks, S.C., Taylor D.L., Jardine, P.M. *Geochim. Cosmochim. Acta,* **1996**, 60:1899-1908.
16. Kedziorek, M.A.M., Dupuy, A., Bourg, A.C.M., and Compere, F. *Environ. Sci. Technol.,* **1998**, 32:1609-1614.
17. Mayes, M.A., Jardine, P.M., Larsen, I.L., Brooks, S.C., and Fendorf, S.E., *J. Cont. Hydr.,***2000**, 45:243-265.
18. Davis, J.A., Kent, D.B., and Coston, J.A. *Water Resour. Res.,* **2000**, 36, 119-134.
19. McGrath, S.P., Sanders, J.R., Shalaby, M.H. *Geoderma,* **1998**, 42, 177-188.
20. Jardine, P.M., Jacobs, G.K., and O'Dell, J.D. *Soil Sci. Soc. Am. J.,* **1993**,57,954-962.
21. Brusseau, M.L., Rao, P.S.C., Jessup, R.E., Davidson, J.M. *J. Contam. Hydr.,* **1889**, 4:223-240.
22. Jardine, P.M., O'Brien, R., Wilson, G.V., and Gwo, J.P. In H.M. Selim, and L. Ma (ed.) *Physical Nonequilibrium in soils Modelling and Application,* **1998**, Chapter 9, pp 243-271. Ann Arbor Press, Chelsea, Michigan.
23. Jardine, P.M., Mehlhorn, T.L., Larsen, I.L., Bailey, W.B., Brooks, S.C., Roh, Y., Gwo, J.P. *J. Contam. Hydr.,* **2002**, 55, 137-159.
24. Parkhurst, D.L., and Appelo, C.A.J. *US. Geological Survey Water Resources Investigation Report,* **1995**, 95-4227, 143.
25. Yeh, G.T., and Salvage, K. HYDROGEOCHEM 2.1: A coupled model of hydrologic transport and geochemistry with both equilibrium and kinetic reactions. Technical report, **1995**, Department of Civil and Environmental Engineering, Pennsylvania State University, State College.
26. Szecsody, J.E., Zachara, J.M., Chilakapati, A., Jardine, P.M., and Ferrency, A.S., *J. Hydrol.* **1998**, 209, 112-136.
27. Samani, Z., Hu, S., Hanson, A.T., and Heil, D.M. *Water Air Soil Pollut.,* **1999**, 102:221-238.
28. Nowack, B., and Sigg, L. *Geochim. Cosmochim. Acta,* **1997**, 61: 951-963.
29. Jardine, P.M., and Taylor, D.L. *Geoderma,* **1995**, 67, 125-140.
30. Sparks, D.L. Environmental Soil Chemistry, **1995**, Academic Press: London.
31. Parker, D.R., Norvell, W.A., and Chaney, R.L. In R.H. Loeppert, P.A. Schwab, and S Goldberg (ed.) *Chemical equilibrium and reaction models,* **1995**, SSSA Special Publication no. 42:253-269. SSSA, Madison, WI.
32. Robinson, R.A., and Stokes, R.H., **1959**. Electrolyte Solutions, 2nd Ed., p.513, Buttersworth, London.
33. Wagenet, R.J., and Chen, W. In H.M. Selim and L. Ma (ed) Physical nonequilibrium in soils, **1998**, Chapter 1, pp1-36, Ann Arbor Press, Chelsea, Michigan.
34. van Genuchten, M.Th., Davidson, J.M., and Wierenga, P.J. *Soil Sci. Soc. Am. Proc.,* **1974**, 38, 29-35.

35. Amacher, M.C.; Selim, H.M.; and Iskandar, I.K. *Soil Sci. Soc. Am. J.*,**1988**, 52:398-408.
36. Selim, H.M., Amacher, M.C., Iskandar, I.K. *Soil Sci. So.c Am. J.*, **1989**, 53, 996-1004.
37. Selim, H.M., and Ma, L. In H.M. Selim and D.L. Sparks (ed.) *Physical and Chemical processes of water and solute transport/ retention in soil*, **2001**, SSSA Special Publication No. 56, Inc. Madison, Wisconsin.
38. Thayalakumaran, T., Robinson, B.H., Scotter, D. R., Vogeler, I., Clothier, B.E. and Percival, H.J. *Plant and Soil*, **2003**, 254, 415-423

Chapter 20

Overview of the European Risk Assessment on EDTA

Otto J. Grundler, Arnold T. M. van der Steen, and Joel Wilmot

EAC, European Aminocarboxylates Committee, 4 Avenue van Nieuwenhuyse, B–1160 Brussels, Belgium (mailto:jwi@cefic.be http://eac.ceflc.org)

EDTA has been elected by the European authorities as part of the priority substances for extensive evaluation. Consequently the German Rapporteur drafted a comprehensive and detailed Risk Assessment Report that was completed in 2003. Their conclusions show that EDTA has a low toxicity profile for both human health and environment, that there is no risk for human health and that risks for the environment are limited to some local extreme cases of high emissions to small surface waters.

EDTA is a well-known chelating agent used to control the effect of metals in chemical processes for more than 50 years. Due to its excellent chelating performance it is used in many applications. It improves the efficiency of several industrial processes from pulp bleaching to the cleaning of dairies. It also improves the quality of food and many other product formulations and is able to increase the quality and yield of crops.

In the late 1980's the environmental impact of EDTA was scrutinized in Europe because of its widespread presence in the aqueous environment and its alleged role in the solubilization of heavy metals. Presumably due to these concerns, EDTA and Na_4-EDTA were chosen as priority substances for extensive evaluation and control of possible risks under European Regulation 793/93. The German authorities, BAuA (Federal Institute for Occupational Safety and Health), prepared a comprehensive risk assessment report on these substances (1, 2). More than 350 studies available from the open literature, scientific institutions, authorities and industry were evaluated by the authorities and the final Risk Assessment Report includes a critical review and discussion of about 250 validated references. The technical experts of the EU Member States endorsed the conclusions in 2003. The European risk assessment process follows the instructions described in detail in the extensive Technical Guidance Document (TGD) of the European Union. Both TGD and Risk Assessment Reports are available to the public on the site of the European Chemicals Bureau (3).

Conclusions of the European Risk Assessment

Human health

The risk for human health is determined by comparing the expected exposures to man (exposure assessment) with available data on toxicity and critical concentrations (effect assessment). EDTA has a relatively low toxicity and the estimated (reasonable worst-case) exposure levels are below the acceptable limits.

EDTA and its tetrasodium salt are not considered to be mutagenic for humans and there is no carcinogenic potential. Furthermore there are no product classification and labeling requirements for carcinogenic, mutagenic, reproductive (CMR) or sensitizing effects as approved by the EU Classification and Labelling Committee.

The member state experts of the European Union came to the following conclusions in relation to human health:

- No concern for consumers in any application
- No concern to workers
- No concern to the public who may be exposed via the environment

These conclusions are in line with the conclusions of the World Health Organization (WHO) on the quality standard of 600 $\mu g/l$ for EDTA in drinking water (4) and with several approvals for the use of EDTA as food additive in the USA (5) and Europe (6). EDTA has been successfully used in food for many years.

Environment

Due to its relatively low toxicity and physical properties there is no concern of possible EDTA emissions into the atmosphere. Also a risk to terrestrial organisms is not being expected. The potential risk for the aqueous environment is evaluated by comparing the predicted no-effect concentration (PNEC) with predicted environmental concentrations (PEC's) as a result of the emissions from the various application areas. Potential risk for the aqueous environment is indicated when the PEC/PNEC ratio for a certain emission scenario is > 1.

The PNEC is derived from ecotoxicity test results on various water species in combination with a safety factor that compensates for possible uncertainties in the effect assessment. The PNEC for EDTA is derived from 3 long-term toxicity tests on fish, daphnia and algae in combination with a safety factor of 10.

The low aquatic toxicity is expressed by the relatively high PNEC of 2.2 mg/l for EDTA in the aqueous environment. In addition after the evaluation of a large number of studies it is concluded that there is no risk to the aqueous environment due to the influence of EDTA on the mobility of heavy metals, eutrophication and nutrient deficiency. EDTA has no potential for bioaccumulation and therefore biomagnification via the food chain is not expected either.

It is well known that EDTA is not readily biodegradable. However several recent studies demonstrated that EDTA is effectively biodegraded under slightly alkaline conditions and that EDTA can be readily eliminated in wastewater treatment facilities when operated at slightly alkaline conditions. Removal rates of 90% or more have been established. Biodegradation of EDTA in the aqueous environment depends on concentration, pH and the species of complexed ions. EDTA is also eliminated from the aqueous environment by photochemical degradation of some of its metal complexes. Primarily the ferric complex breaks down upon the exposure to sunlight in natural aquatic environments. Biodegradable and non-toxic metabolites have been detected in the surface waters as a result of the degradation of EDTA. EDTA and its metabolites are not expected to be persistent in the aqueous environment. For the purpose of risk assessment a worst-case assumption of no biodegradation was used. Though a simple classification of 'readily biodegradable' and 'persistent' chelating agents is not possible due to the influence of metal ions, pH and concentration on the biodegradation of complexing agents in general. Therefore we recommend that EDTA is to be considered as 'inherently biodegradable'. Also the European Scientific Committee on Toxicity, Ecotoxicity and the Environment (CSTEE) indicated in their official opinion on the risk assessment report on EDTA (7) that the assumption of no biodegradation (persistent) should be taken as conservative worst case.

Environmental risks

The detailed sections of the Risk Assessment Report clearly indicate that the risk for the local aqueous environment is limited to a restricted number of 'extreme cases' that may appear in relatively small surface waters near effluent release points of large sites with no effective EDTA removal. The extensive monitoring data available for EDTA indicate lower concentrations than expected from the applied 'worst-case' exposure scenarios in the Risk Assessment Report. On the basis of the 'worst-case exposure scenarios' the risk assessment report indicates 4 areas where EDTA may pose a risk for aquatic organisms in the local aqueous environment;

- Industrial cleaning in large dairy and beverage sites
- Pulp and paper bleaching
- Circuit board production
- Recovery of EDTA containing (photographic) waste

The Risk Assessment Report concludes that appropriate risk reduction measures should be considered to reduce any potential risk (concentrations >2.2 mg/l) for the local aqueous environment. We recommend to use the experience of countries like Sweden where local measures have been taken for more than 20 years. Permitting has lead to substantial emission reductions, proving it to be an efficient tool in managing any possible local risks of EDTA.

Details on the Human health section

Justification for cross-reading from different EDTA compounds

In general, edetic acid and tetrasodium EDTA show similar properties and exposure pattern. However, with respect to acute toxic and local effects both substances behave differently. Thus, the hazard effects of the two substances have been evaluated separately only for the endpoints acute toxicity and irritation. For systemic effects any conclusions on H_4EDTA or Na_4EDTA have been derived from consideration of the overall available database due to the dissociation of these compounds under physiological conditions (1, 2).

Risk characterization

Specific chapters on workers, consumers and men indirectly exposed via the environment are discussed. No risks have been identified in any of these categories.

In detail the following effect assessment has been discussed. As mentioned before, H_4EDTA and Na_4EDTA show different patterns of irritancy and acute toxicity.

Irritation

The substance has shown weak irritant properties on rabbit skin but irritations to the rabbit eye. H_4EDTA is thus considered to be irritating to the eyes. Na_4EDTA may result in serious damage to the eye. However workers are protected by using adequate equipment and consumer products contains only very low levels of EDTA. Therefore no risk for irritating effects are to be considered for both consumers and workers.

Acute toxicity

As no valid data on inhalation toxicity are available, assessments are based on oral data. In summary for inhalation, dermal and oral exposure, tests have shown no concern for either workers or consumers.

Sensitization

Based on the intensive use in industry and consumer products for many decades where no significant problems on skin sensitization were reported, the substance is claimed to be non-sensitizing to humans. This was supported by mandatory testing in guinea pigs. In the absence of valid indications EDTA is not considered to be a respiratory sensitizer.

Repeated dose toxicity

In different studies after oral and dermal applications, EDTA was evaluated to be of no concern.

Exposure of children via use of cleansers for tooth brackets has been concluded to be of no concern.

Mutagenicity

Bacterial mutation tests are negative, but mutations and DNA damage were found in mouse lymphoma cells after exposure to very high concentrations. For somatic cells in mice (bone marrow cells) negative results with respect to the endpoints micronuclei, aneuploidy and sister chromatid exchanges were described. In germ line cells negative results were obtained for induction of structural chromosomal aberrations in spermatogonia, for induction of aneuploidy in primary and secondary spermatocytes, and also for induction of dominant lethals. A positive result was obtained in a micronucleus test with spermatids, indicating that aneugenic effects may be induced in specific phases of spermatogenesis (late spermacytoge-nesis). The effect was bound to the use of an extremely high dose in the LD_{50} range. Altogether, EDTA and its sodium salts have a low mutagenic potential at extremely high doses. On the basis of the various negative findings and the assumption of a threshold mode-of action for aneugens, it can be concluded that EDTA and its sodium salts are not mutagenic for man (1, 2).

Carcinogenicity

A bioassay of Na_3EDTA for possible carcinogenicity was conducted by administrating of test material in the diet to Fischer 344 rats and B6C3F1 mice. The studies did not report specific data on kidney toxicity in either species.

Although a variety of tumors occurred among test and control animals of both species, no tumors were related to treatment. Thus, there is no concern on a carcinogenic potential of EDTA (1, 2).

Fertility

Based on a multigeneration study on rats, the substance was concluded of no concern to affect the reproductive performance.

Developmental toxicity

Fetotoxic and teratogenic effects occurred in rats at exposure levels of around 1000 mg/kg bw/d, EDTA is considered of no concern with regard to fetotoxic and teratogenic effects for humans.

Details on the Environment section

Environmental exposure

EDTA is mainly used as a complexing agent for industrial purposes. We describe some details from the extensive European Risk Assessment Report and give the PECs only for the uses identified to be of concern.

According to the TGD (3) the predicted environmental concentration (PEC) near emission sources is estimated from the daily consumption of EDTA in its various applications. The concentration in the effluent is calculated from the estimated emission into a waste water treatment plant with a hypothetical effluent flow of 2000 m^3/hr. Subsequently the PEC is estimated from the effluent concentration assuming a hypothetical dilution factor of 1 to 10 upon release into the local surface waters and taken into account the estimated regional background concentration of EDTA.

The Risk Assessment Reports show a disagreement between the calculated PECs and measured EDTA concentrations in the environment. Extensive monitoring suggests actual EDTA concentrations in the aqueous environment to be lower than the PEC's estimated by the very conservative worst-case assumptions of the TGD (3). For instance extensive monitoring demonstrated that the concentrations of EDTA in large rivers were far below the presented PECregional of 95 µg/l. In addition monitored effluent concentrations presented for many application areas are always below the effluent concentrations used for the PEC determinations.

Industrial detergents

EDTA is used to prevent precipitation of calcium, magnesium and heavy metals. It has various uses within the industrial and institutional detergents market. The main application area is the dairy and beverage industry. Only one of the possible scenarios with a high consumption of 10 t/a in combination with no effective EDTA removal in the wastewater treatment plants leads to a PEClocal of 2.6 mg/l. Other scenarios with less consumption and/or effective elimination in wastewater treatment plant lead to no concern.

Photochemicals

In photo industry EDTA is mainly applied in the bleachfix process which is a combination of bleaching and fixing. The exposure scenario represents a large photofinisher, for which a PEClocal of 0.57 mg/l is calculated.
Wastes from photo industry are collected by special disposal companies. Bath residues are either incinerated or evaporated and deposited. Some monitoring

data demonstrated the presence of EDTA in the waste water at their disposal sites. TGD (3) default values result in an estimated worst-case PEClocal of 2.4 mg/l.

Pulp and paper

EDTA is used to chelate metal ions during the chlorine free bleaching process applied by paper mills. EDTA is not fixed onto the paper and is thus emitted into the sewage. Two scenarios have been used.

A removal factor of 90% has been chosen for sites reflecting the best available techniques. The monitoring data for effluents of such mills give a PEClocal of 0.5 mg/l.

Monitoring data from Swedish and Finnish sites (representing the majority of the European market) reveal that a PEClocal of 4.1 mg/l resulted from the worst case scenario is not reached for any site. Therefore, this value is not used in the risk characterization. Instead, the highest estimated local concentration at a mill without an effective EDTA removal (leading to a PEClocal of 2.6 mg/l) is considered as "worst case (1, 2).

Textile industry

EDTA is used in textile finishing to produce easy care fabrics, to support oxidative bleaching and to prevent catalytical damages of the fibers.

Metal plating

EDTA is used for the production of printed circuit boards. EDTA is mainly used in electroless copper plating, when copper is deposited on the board by catalytic reduction of complexed copper compounds. The exposure scenario, based on the average EDTA consumption of one site and TGD default values for the dilution model, results in a PEClocal of 12 mg/l (1, 2).

Water treatment

Only a limited volume is yearly used in Europe for this purpose and it is considered to be widespread with no high local exposure.

Polymer and rubber production

EDTA is used in the production of Styrene Butadiene Elastomers (SBR) which are mainly manufactured by emulsion polymerization.

Oil production

EDTA is used for well cleaning processes at oil platforms. A very high dilution factor leads to no concern.

Fuel gas cleaning

In Europe there are no emissions into the waste water from this use.

Disposal (Waste recovery)

Some monitoring data indicate that EDTA is released into the waste water at waste recovery sites. Results from photochemicals recovery were used as default value with a PEClocal of 2.4 mg/l.

Household sewage

Household detergents, cosmetics, pharmaceuticals and food are using EDTA for complexing of trace metals. The diffuse emissions of these EDTA applications in products are not a risk for the aqueous environment.

Effects assessment

According to the TGD the predicted no-effect concentration (PNEC) should be determined from the no-effect level of a long-term test on the most sensitive species and an extra assessment factor of 10 (3).

For EDTA the PNEC is determined from the no-effect concentrations of 3 long-term eco-toxicity tests on Fish, Daphnia and Algae. This resulted in a PNEC of 2.2 mg/l for EDTA in the aqueous environment derived from the no-effect level of 22 mg/l from a 21-d study with Daphnia magna (1, 2).

Based on environmental behaviour properties, EDTA was estimated not to be of concern for sediment, atmosphere and terrestrial compartment and the food chain. No formal PNEC was established for these compartments because of lack of appropriate studies (8).

The influence of EDTA on heavy metals in sediments and surface waters has been investigated in a large number of studies. EDTA has the tendency to prevent the adsorption of recently emitted heavy metals to sediments and suspended solids and to remobilize heavy metals from highly loaded sediments. Because of the complexity of the EDTA-heavy metals interactions it is not possible to present a no-effect level for metal remobilization by EDTA. The concentration of solubilized heavy metals in the aqueous phase is affected much more by natural causes -such as pH, precipitating anions and high molecular humic and fulvic acids- than by trace amounts of EDTA. In wastewater from a German municipal wastewater treatment plant and in surface water from a Spanish river it was demonstrated that the complexation of Zn, Cd and Cu was mainly determined by organic substances of high molecular weight like humic and fulvic acids and that the contribution of EDTA was limited to a few percent only.
The Risk Assessment Report concludes that significant remobilization can only occur in extreme cases of high local concentrations. (1,2)

The decrease of the toxicity of heavy metals by the influence of EDTA has been demonstrated in various studies. Complexation with EDTA decreased the toxicity of heavy metals to Daphnia by a factor of 17 to 1700 (1,2)

Risk characterization

Aquatic compartment

The risk assessment for aquatic organisms resulted in a PNECaqua of 2.2 mg/l.

A need for limiting risks has been identified in some extreme cases where the PEClocal is > 2.2 mg/l:

- the use of EDTA in industrial detergents by large sites within dairy and beverage industry, where no effective waste water treatment is applied,
- paper mills where no effective waste water treatment is applied,
- metal plating (circuit board production),
- releases at waste recovery sites.

Risk reduction measures already in place are to be taken into account.

Influence on the distribution of heavy metals:
Significant remobilization by EDTA can only occur in extreme cases where local high concentrations of EDTA can be found, for example when large point sources are emitted into small rivers. At high concentrations, EDTA could prevent the adsorption of heavy metals to sediments and could remobilize metals from highly loaded sediments. Both effects lead to increased heavy metal concentrations in the water phase. Simultaneously the sediment is deloaded. At

the same time EDTA complexes of heavy metals are less toxic than uncomplexed metals. Considering all these facts the overall effect for the aquatic environment will be limited and a risk for the aquatic environment due to the influence of EDTA on the mobility of heavy metals is not expected.

Atmosphere

EDTA is emitted into the atmosphere as dust during production. However because of the relative low toxicity of EDTA, a risk to the environment is not expected.

Terrestrial compartment

No effect tests on terrestrial organisms are available. Therefore the risk characterization has been based on calculated pore water concentrations. No risk to terrestrial organisms was expected.

Non compartment specific effects relevant to the food chain

As there is no bioaccumulation, a biomagnification via the food chain is not expected.

References

1. German Federal Institute for Occupational Safety and Health Notification Unit, Risk Assessment of Edetic acid (EDTA) – CAS # 60-00-4 (Draft of 15 December 2003)
2. German Federal Institute for Occupational Safety and Health Notification Unit, Risk Assessment of Tetrasodium ethylenediaminetetraacetate (Na_4EDTA) – CAS # 64-02-8 (Draft of 15 December 2003)
3. The European Chemicals Bureau (ECB), Technical Guidance Document and Risk Assessment Reports, http://ecb.jrc.it/existing-chemicals/
4. World Health Organization, Guidelines for Drinking-water Quality, Second edition, Addendum to Volume 2 : Health Criteria and other Supporting Information, 1998, pages 97-107, http://www.who.int/water_sanitation_health/dwq/en/2edaddvol2a.pdf
5. Title 21 - Food and drugs, Chapter I – Food and Drugs Administration, Department of Health and Human Services (Continued), part 172 – Food Additives permitted for Direct Addition to Food for Human Consumption

6. European Parliament and Councel Directive 95/2/EC of 20 Februari 1995 on food additives other than colours and sweeteners, Official Journal L061, 18/03/1995, pages 1-40
7. German Federal Institute for Occupational Safety and Health Notification Unit, Summary Risk Assessment of Edetic acid (EDTA) – CAS # 60-00-4 (Draft of 15 December 2003)
8. Scientific Committee on Toxicity, Ecotoxicity and the Environment (CSTEE), Opinion on the results of the Risk Assessment of Ethylenediamine Tetraacetate (EDTA), Environmental part, Adopted by the CSTEE during the 39th plenary meeting of 10 September 2003, http://europa.eu.int/comm/health/ph_risk/committees/sct/sct_opinions_en.htm
9. German Federal Institute for Occupational Safety and Health Notification Unit, SIDS Initial Assessment Profile for EDTA, December 2003

Chapter 21

Theoretical Modeling and Reactivity of the Iron Chelates in Agronomic Conditions

Juan J. Lucena, Sonia Garcia-Marco, Felipe Yunta, and Lourdes Hernández-Apaolaza

Department of Agricultural Chemistry, Universidad Autónoma de Madrid, Campus Cantoblanco, 28049 Madrid, Spain

O,oEDDHA/Fe^{3+} and its analogues (o,oEDDHMA/Fe^{3+}, EDDHSA/Fe^{3+}) are chosen as the main iron chelates used to correct iron chlorosis in crops grown in calcareous and alkaline soils. In order to test new chelating agents, prior to their application, theoretical modeling using the speciation program MINTEQA2 is proposed. Also interaction tests between the chelates and the main soil components (organic matter, metal oxide/hydroxide, clay, carbonate ...) are a useful tool to study the sorption processes on the soil surfaces. In addition, percolation column tests provide information on the mobility of these chelates depending on the soil type.

Iron chelates are used in agriculture to solve iron chlorosis, which is a nutritional disorder occurring in susceptible plants growing on calcareous soils. It presents as an intervenial yellowing of young leaves caused by the decrease of iron participating in the chlorophyll synthesis and a decrease in yield and fruit quality (*1*, *2*, *3*, *4*). In calcareous soils, the high pH and the bicarbonate buffer immobilize the iron (*5*), despite the fact that it is normally the most abundant plant nutrient in the soils (*6*). The best solution to iron chlorosis is the application of synthetic iron chelates to the soil, since they can increase the soluble iron concentration (*4*).

The most efficient chelating agents are polyamino carboxylic acids derived from EDDHA (see fig 1). (*7*).

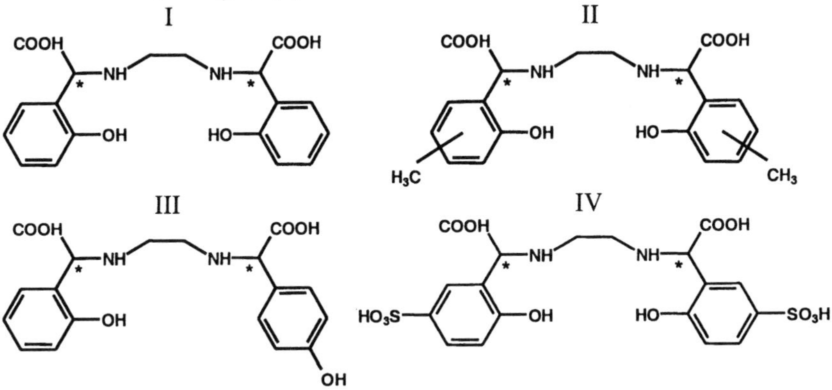

Figure 1. Chelating agents present in commercial Fe fertilizers. I: EDDHA, ethylendiamine-di(ortho-hydroxy phenyl)acetic acid, II:EDDHMA ethylendiamine-di(ortho-hydroxy methyl phenyl)acetic acid, III: o,pEDDHA ethylendiamine-N(ortho-hydroxy phenyl acetic)-N'(para-hydroxy phenyl acetic) acid and IV: EDDHSA ethylendiamine-di(2-hydroxy 5 sulfonate phenyl) acetic acid. Commercial products with EDDHSA also include condensation products.
** Chiral carbons*

The efficacy of iron sources used to correct chlorosis depends on the soil type and the chemical properties and/or purity of these products. Principally, biological tests are used in order to evaluate the behavior of these iron chelates in agronomic conditions (*8*) but they depend on many factors (e.g., soil properties, plant development, weather conditions) that can greatly affect the effectiveness of the iron chelates under study. Then laboratory tests are normally used to evaluate their efficacy (*9*, *10*, *11*). These tests are based on the determination of the soluble iron under different conditions: an adequate iron chelate should maintain Fe in the soil solution in a concentration sufficient to supply iron plant requirements, while minimizing leaching losses. Four processes may be involved in the chelate behavior in soils that reduce its presence in the soil solution:

- Chelate or chelating agent degradation

- Replacement of the iron by other competing metals
$$FeY^{-} + M^{2+} + 3H_2O \leftrightarrow Fe(OH)_3 + MY^{2-} + 3H^{+}$$
- Sorption of the chelate onto the soil surfaces
- Leaching

While biodegradation has not been detected in these EDDHA/Fe^{3+} analogous and the chemical degradation of the chelate fertilizers is slow in the dark (*12, 13*) the other three processes are important in soils.

Metal competition can be studied theoretically by means of speciation programs, providing thermodynamic data are available for all the possible reactions (*14*). The early works of Lindsay's group (*6*) are a good example.

Since adsorption-desorption reactions tend to be faster than precipitation-dissolution processes, adsorption at solid interfaces can be a dominant factor in regulating micronutrient concentration in solution (*15*). The adsorption of metal chelates by soils remains poorly understood. Full understanding of these reactions will always be impeded by the complexity of soil surfaces, continuing changes in chelate speciation and concurrent degradation of the chelating ligand. In a previous work (*16*) we studied the sorption isotherms of the meso and racemic isomers of o,oEDDHA/Fe^{3+} and o,oEDDHMA/Fe^{3+} chelates on peat, Ca montmorillonite, and ferrihydrite. Results suggested that the retention was quite different between isomers of the same chelate and among soil materials. However no previous work deals with differences in sorption of isomers of the same chelates in soils.

The fate of the fertilizer that it is not used by the plant is an environmental concern. In the case of anionic species leaching is the main lose from the upper soil layer that can cause pollution. Leaching columns tests are normally used to study the migration of such pollutants (*17*) and have been also used to study chelates movement (*18*).

The aim of our ongoing research is to measure the reactivity and mobility of the iron chelates (o,oEDDHA/Fe^{3+}, o,oEDDH4MA/Fe^{3+}, EDDHSA/Fe^{3+}, o,pEDDHA/Fe^{3+} and EDTA/Fe^{3+}), through interaction and leaching columns tests respectively, and compare their behavior with theoretical modeling.

Theoretical Modeling

The species distribution of the iron chelates o,oEDDHA/Fe^{3+}, o,oEDDH4MA/Fe^{3+}(methyl substitution at the benzene ring in position 4), o,oEDDH5MA/Fe^{3+}(methyl substitution at the benzene ring in position 5) EDDHSA/Fe^{3+}, o,pEDDHA/Fe^{3+} and EDTA/Fe^{3+}, at pH range 4-13 was determined using three theoretical models considering their agronomic use: in solution, in hydroponics and in soil conditions respectively.

The first model is used to determine the behavior of these iron chelates in solution conditions. Species distribution was obtained through the equilibrium speciation model MINTEQA2 program (*19*) using Hoagland nutrient solution composition (*20*), and the stability constants determined for the ligands (*12*). No solids were allowed to control metal solubilities in the model.

The second model is valid to determine the behavior in hydroponics. In this case Hoagland nutrient solution was also considered but $Fe(OH)_3$(amorphous) equilibrium was introduced in the system as a solubility controller for Fe.

The third model is used to study the behavior in soil conditions. In this model all soil components that could have some effect on iron chelates stability were considered. Due to the competition observed between Fe^{3+} and Cu^{2+} for *o,p*-EDDHA, two soil types with unlimited and limited Cu^{2+} availability respectively were tested in order to predict the stability of *o,p*-EDDHA/Fe^{3+} with high (i.e. fungicide application) and low (normal) Cu^{2+} levels in soil. In table 1, the solids controlling the solubility of metals are presented. In all cases a CO_2 partial pressure of 0.0003 atm were considered.

Table 1. Solution equilibria of solids controlling metal solubilities considered in the theoretical study in soil conditions.

Soil component	Equilibria	Log K°
*Soil-Ca	Soil-Ca $\leftrightarrow Ca^{2+}$	-2.50
*Soil-Mg	Soil-Mg $\leftrightarrow Mg^{2+}$	-3.00
Soil-Cu	Soil-Cu + $2H^+ \leftrightarrow Cu^{2+} + 2H_2O$	2.80
Soil-Fe	Soil-Fe + $3H^+ \leftrightarrow Fe^{3+} + 3H_2O$	2.70
Soil-Zn	Soil-Zn + $2H^+ \leftrightarrow Zn^{2+} + 2H_2O$	5.80
* these equilibria were replaced at pH over 7.5 by the following		
Calcite	$CaCO_3 \leftrightarrow Ca^{2+} + CO_3^{2-}$	-8.41
Dolomite	$CaMg(CO_3)_2 \leftrightarrow Ca^{2+} + Mg^{2+} + 2CO_3^{2-}$	-3.00

Figure 2 shows the chelated Fe versus pH for the different iron chelates and for the three models tested. In solution and hydroponics EDDHA/Fe^{3+}, EDDH4MA/Fe^{3+}, EDDH5MA/Fe^{3+}, EDDHSA/Fe^{3+} and *o,p*-EDDHA/Fe^{3+} maintain the total Fe as chelated iron at agronomic pHs (5-9). However, EDTA/Fe^{3+} is decomposed at pH above 8.5 and 7 in solution and hydroponics respectively. In soil conditions, although for EDDHA/Fe^{3+}, EDDH4MA/Fe^{3+}, EDDH5MA/Fe^{3+} and EDDHSA/Fe^{3+} the stability is enough to maintain all the iron in solution at pH below 8.5, which is a normal limit for agronomic purposes, the stability sequence is the following EDDHSA > EDDH4MA > EDDH5MA > EDDHA >>> EDTA. EDTA/Fe^{3+} can not be used in calcareous soil because the iron chelate is broken at pH above 6, due to the Zn, Mn or Ca substitution and causing iron precipitation. Two soil models are tested for *o,p*-EDDHA/Fe^{3+} depending on the Cu^{2+} level soil. With limited Cu^{2+} availability [(2)] *o,p*-EDDHA/Fe^{3+} is stable, but when the soil presents large Cu^{2+} availability[(1)], it can displace the Fe^{3+} from the chelate, reducing its availability to plants.

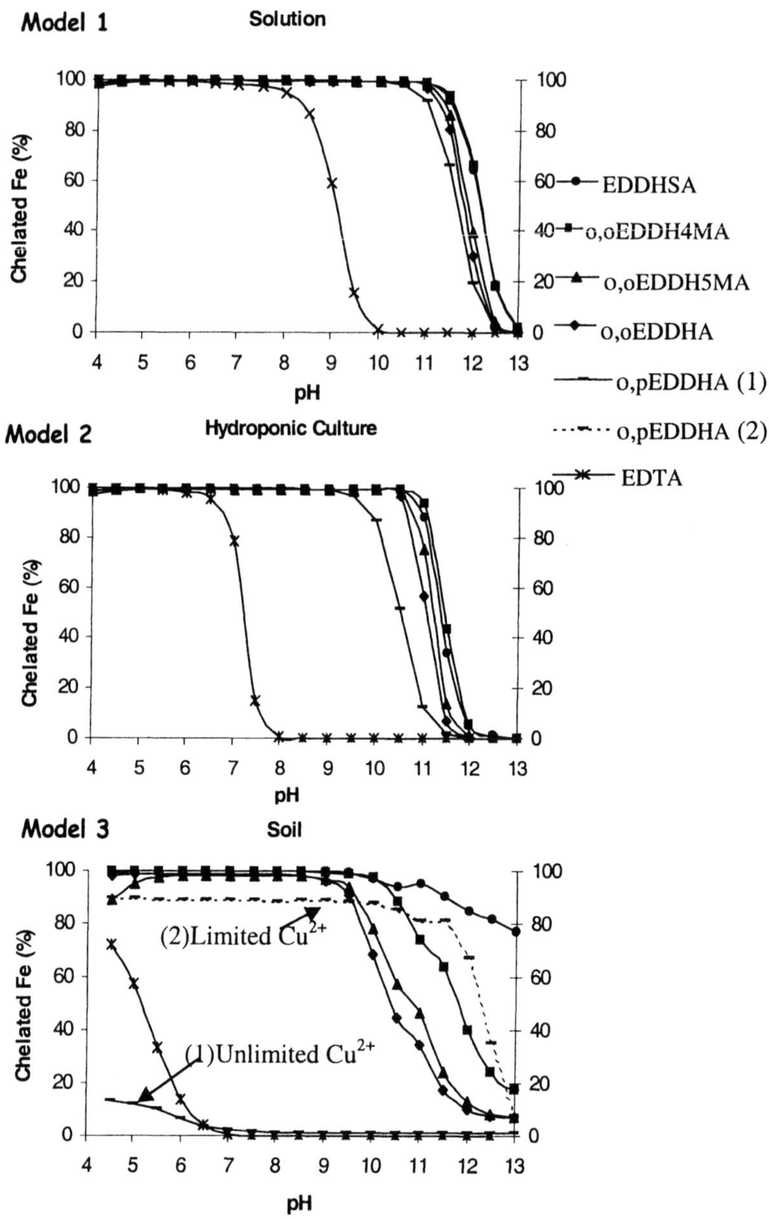

Figure 2. Percentage of chelate that remains in solution respect pH

We can conclude that theoretical modeling is a useful tool to predict the behavior of iron chelates in agronomic conditions prior to their application. However, when the main competitor is a micronutrient, with a low availability in the soil, small variations in that micronutrient availability may produce different Fe-chelate speciation.

Sorption and leaching column studies

In order to confirm the results hypothesized by the theoretical modeling, laboratory experiments were performed. The first set of experiments (sorption studies) consisted of the interaction between the iron chelates and several soils and soil materials, to determine the chelate adsorption on them, and consequently, the amount of chelated iron that remains in solution after the application of the chelate to the soil. The second set of experiments tried to match a more realistic situation (in terms of soil/iron chelate ratio), and for this purpose soil columns were settled and the iron chelate was added to the soil. Fe-chelate in column leachates was analyzed.

Soil materials

The following soil materials were used for the interaction studies:

Ferrihydrite was chosen among the soil iron sesquioxides because its high reactivity *(21)*. Ferrihydrite ($5Fe_2O_3{\cdot}9H2O$) was prepared in the laboratory following the procedure of Sims and Bingham (*22*). The resulting Fe(III) oxide corresponded with a ferrihydrite phase, presenting 6 lines at the X-Ray Difraction analysis, similar to that described by Drits et al. (23) and had a surface area value of 220 $m^2{\cdot}g^{-1}$.

The acidic mountain Sphagnum peat was provided by Tolsa S.A. (Buyos, Lugo, Spain). Its chemical characteristics were: pH (saturated paste): 4.0, dichromate oxidizable O.M. (%): 85.4, total O.M. (%) (determined by loose of weight by ashing): 99.5, C in humic acids (%): 30.2, C in fulvic acids (%): 18.3, N by the Kjeldahl method (%): 1.4%, C/N: 35.4, C.E.C.($cmol_c{\cdot}kg^{-1}$): 150, and DTPA extractable Fe and Mn (*24*) 295 $mg{\cdot}kg^{-1}$ and 8.2 $mg{\cdot}kg^{-1}$, respectively.

Calcium montmorillonite (STX-1, González County, Texas) was obtained from Clay Minerals Society source (Clay Minerals Repository, Department of Geology, University of Missouri, Columbia, MO). This reference material has been well characterized elsewhere, (*25*)

Illite Silverhill Montana (Imt-2) was also obtained from Clay Minerals Society source.

Calcium carbonate was obtained from Panreac (analytical reagent grade).

Standarized calcareous soil (SCS) composition was 50% acid washed quartz sand (diameter 1-3 mm), 15% Ca-montmorillonite, 5% illite, 2% organic matter (peat), 19% calcium carbonate, 1% dolomite and 8% oxides (7% ferrihydrite, 0.5% $Al(OH)_3$, 0.1% MnO_2, 0.004% $Cu(OH)_2$, 0.006% $Zn(OH)_2$ and 0.39% $CaHPO_4$). The Al, Cu and Zn oxides were prepared from $AlK(SO_4)_2$ (Probus), $Cu(SO_4)$ and $Zn(SO_4)$, respectively, using the same method as for the Fe(III) oxide. The Mn (IV) oxide was prepared by a warm reaction of $KMnO_4$ with ethanol.

Soils were sampled in the Valencia region (soil 1) and at two different locations at northeastern of Spain. The main physicochemical characteristics of the three calcareous soils tested are shown in Table 2. In the field, soils 1 and 3 were observed to grow chlorotic peach trees and soil 2 was observed to grow chlorotic pear trees.

Soil 2 was also used for the leaching columns experiment.

Table 2. Main physicochemical characteristics of the soils tested.

	Texture	pH Saturated paste		E.C. 1:5	N_{kj}
		H_2O	KCl	$\mu S \cdot cm^{-1}$	$g \cdot kg^{-1}$
Soil 1	Sandy loam	7.69	7.15	235	0.79
Soil 2	Clay Loam	7.75	7.20	286	1.2
Soil 3	Sandy Clay Loam	7.82	7.23	188	1.4

	O.M.ox $g \cdot kg^{-1}$	$CaCO_3$ ($g \cdot kg^{-1}$) Total	$CaCO_3$ ($g \cdot kg^{-1}$) Active	Fe	Mn	Cu	Zn
				$mg \cdot kg^{-1}$ (DTPA Extractable) (*26*)			
Soil 1	8	150	40	14.3	13.0	1.46	1.08
Soil 2	15	430	140	12.1	4.11	6.86	4.45
Soil 3	24	179	52	26.7	5.38	47.1	27.4

Iron chelate solutions

For preparing standard solution of Fe(III)-chelates, H_4o,oEDDHA (Sigma Lot#117f50221), H_4o,p-EDDHA (Syngenta Crop Protection), EDDHSA (NAC Química S.A.), Na_2H_2EDTA (Merck) and EDDH4MA (Sierra et al., 2000) (*27*) were dissolved in sufficient NaOH (normally 1:3 molar ratio). Then an amount of $Fe(NO_3)_3$ that was calculated to be 5% in excess of molar amount of ligand was added, the pH was adjusted with NaOH 1M to 7.0, and the solution was left to stand overnight to allow excess Fe to precipitate as oxides. The solution was filtered through Whatman No.2 filters and made up to volume with water.

Also a commercial chelate solution containing o,oEDDHA/Fe^{3+} and o,pEDDHA was prepared by simple dissolution and filtration of the solid.

Experimental

For the sorption studies 10 mL chelate solutions 10^{-4} M (pH 8.0 HEPES, 0.1 M $CaCl_2$) were added to soil materials: 0.25 g of ferrihydrite, 0.50 g of acid peat, 0.50 g of Ca-montmorillonite, 0.50 g of Illite, 2.00 g of calcium carbonate, 5.00 g of the calcareous standard soil, 5.00 g of Soils 1, 2 and 3. Also controls (chelate without soil material or soil material without chelates) were prepared. All interactions were prepared by triplicate. After 1 hour of agitation in a horizontal shaker (56 s^{-1}, 25°C) samples were let to stand for three days at 25°C (*28*). Then, the supernatant was filtered through 0.45 μm Millipore membranes and pH, total Fe by flame-Atomic Absorption Spectrophotometry (Perkin-Elmer Analyst 800, hollow cathode lamp, 248.3 nm, slit 0.2, spoiler and air-acetylene flame) and iron-chelate concentration by ion-pair HPLC (29) were determined. For the HPLC determination 0,03 M tetrabutyl ammonium as ion-pair reagent in 30% acetonitrile, pH=6, was used as mobile phase. Also a column Symmetry RP=18, dp=5μm, d.i. 3.9x150mm, a separation module Waters 2695 Alliance, a PDA detector Waters 996 Photodiode Array, and Empower software were used.

The leaching columns system was designed similarly to Elgala and Maier (30). In this experiment only soil 2 was used since it was the one with a more active $CaCO_3$ content (see table 2). A 50:50 soil:quartz sand (acid washed) mixture (w/w) was placed in plastic columns of 50x6 cm, up tol 20 cm height. At the bottom, 10 cm of sand was placed in order to help the leaching process, as well as 2 cm at the top to prevent evaporation. The columns were situated on the top of Büchner funnels. Between each funnel and column a cellulose filter (Whatman, No.42) and a plastic grid were placed in order to avoid soil losses. Five replicated columns per treatment were prepared. Each column was covered with aluminum foil to avoid the photodecomposition of the iron chelate. The soil was brought to its water-holding capacity by adding deionized water. After 24 hours, a single 2 mL dose of 500 mg Fe·L^{-1} solutions of o,o-EDDHA/Fe^{3+}, EDDH4MA/Fe^{3+} and EDDHSA/Fe^{3+} was added to each column and then irrigated with 50 ml of deionized water every 48 hours during the experiment. The experiment lasted for 32 days, and percolates were collected and analyzed for pH, total Fe and Cu by flame-Atomic Absorption Spectrophotometry (as before for Fe and 324.8 nm, slit 0.7 and impact bead for Cu) and iron-chelate concentration by HPLC as before.

Results of the Interaction Experiment

For this set of experiments iron chelates of EDTA, o,oEDDHA, o,pEDDHA, and a commercial product that contains both o,o- and o,pEDDHA were compared. The percentage of the chelated Fe remaining in (determined by HPLC) respect the chelate added (determined by HPLC in the chelate control solution) is presented in Figure 3. For the commercial product the chelated Fe has been considered the addition of both o,oEDDHA/Fe^{3+} and o,pEDDHA/Fe^{3+}. Results are similar to those of total Fe in solution obtained by Atomic Absorption Spectrophotometry indicating that all the iron in solution came from the chelates. Very similar behavior was observed for EDTA/Fe^{3+} and o,pEDDHA/Fe^{3+}, except for the peat. For this material, despite the use of HEPES, the pH remained in the pH 6-7 range, and o,p-EDDHA/Fe^{3+} is more retained than EDTA/Fe^{3+}. However EDTA/Fe^{3+} is completely retained in the iron oxide. o,pEDDHA/Fe^{3+} is less retained in the clays and soils than EDTA/Fe^{3+}. The novel chelate o,pEDDHA/Fe^{3+} is able to maintain around 50% of the added Fe in the three natural soils after three days of interaction. o,o-EDDHA/Fe^{3+} is, as expected, the iron chelate that is more stable. Only some significant reduction is observed when it interacts with the Fe oxide. In the commercial chelate, both o,oEDDHA/Fe^{3+} and o,pEDDHA/Fe^{3+} are present, but there is also an extra amount of iron in unknown forms. For this commercial chelate Fe oxides and the more reactive soil, the SCS, produce a higher reduction on the availability of the iron.

In figure 4 the molar concentration of Cu in solution after the interaction (determined by Atomic Absorption Spectrophotometry) respect the total Fe chelate added is presented. While the decrease of chelate from the EDTA/Fe^{3+} can be due to the displacement of the Fe by other metals (principally Ca at the working pH) for o,pEDDHA/Fe^{3+} the main competitor is Cu, as was shown in the modeling section. In fact the displacement of Fe by copper is more important for soil 3, with larger Cu availability (see figure 4). However the Cu displacement does not account for the total Fe loss from o,pEDDHA/Fe^{3+}. Sorption processes may also occur as has already been described for o,oEDDHA/Fe^{3+} (16). The ability of the plants to use the Fe from the adsorbed chelates is unknown.

Thus, while interaction experiments are useful to determine the global reaction of the chelates in the soil, they do not differentiate the processes involved. When used in conjunction with the model, where only complexation and dissolution reactions are studied, a better explanation is obtained.

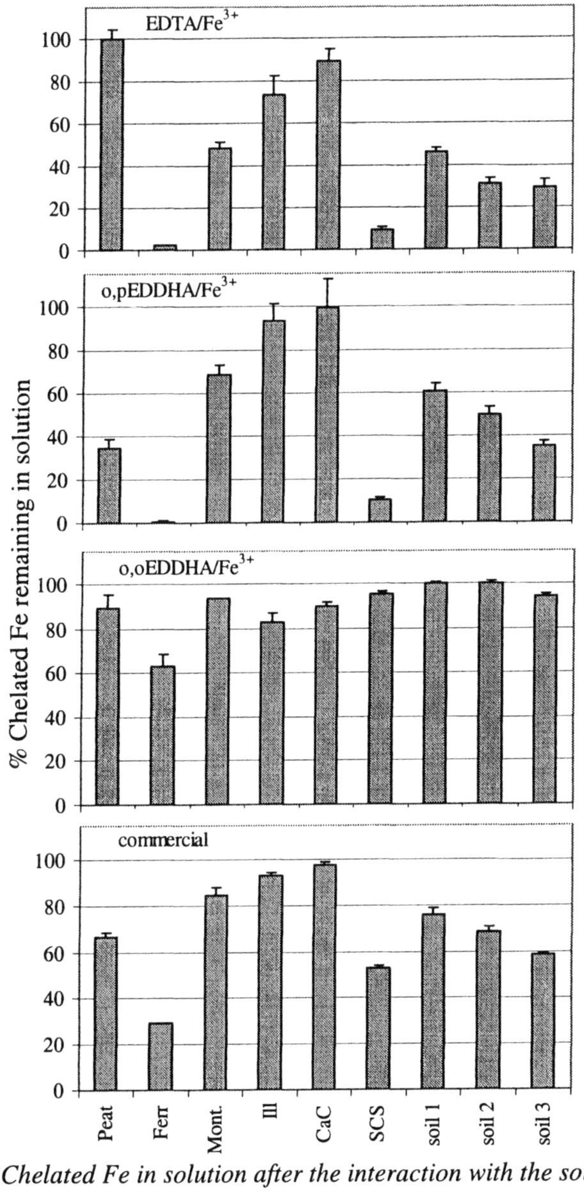

Figure 3. Chelated Fe in solution after the interaction with the soil materials.

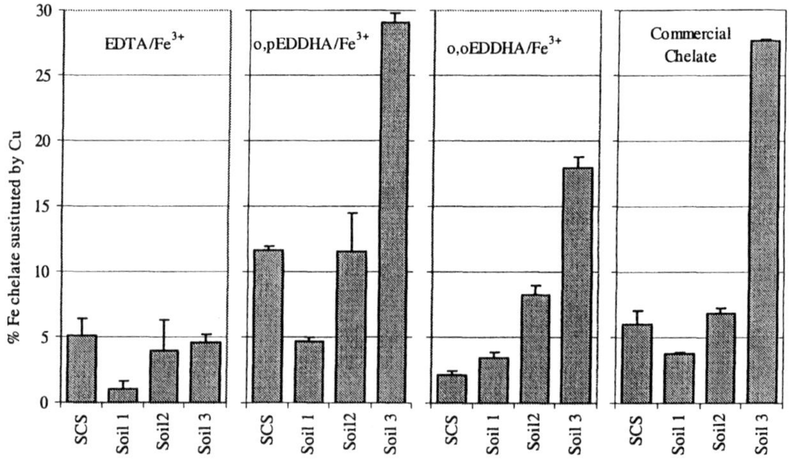

Figure 4. Percentage of Fe replaced by Cu for each chelates in the soils studiedResults of the leaching Column test

Results of the leaching Column test

The aim of this experiment was to study the mechanisms that govern the mobility of different iron fertilizers (o,oEDDHA/Fe^{3+}, o,oFe-EDDHMA/Fe^{3+} and EDDHSA/Fe^{3+}) through a calcareous soil (soil 2 in table 2).

In figure 5 the percentage of chelated Fe in the percolates with respect to the amount added is presented. It is shown that EDDHSA/Fe^{3+} is the chelate that moves fastest in the soil, and a greater amount is obtained at the end of the experiment. No significant difference was observed between EDDHA/Fe^{3+} and EDDH4MA/Fe^{3+}. In the interaction test no retention was observed for o,oEDDHA/Fe^{3+} for soil 2 (Figure 3), but in this case only the 70% of the added chelate percolates from the column. This implies that in leaching column tests a more realistic situation is considered, since the assay conditions are more similar to that occurring in agronomic conditions. From the theoretical model we know that this chelate does not suffer from significant metal competition, so the reduction observed should be due to chemical degradation or sorption onto soil material.

From the environmental point of view, EDDHSA/Fe^{3+}, with three negative charges, is the chelate that is least retained in the soil, so its application should be done avoiding leaching, such in fertirrigation where fertilization with high

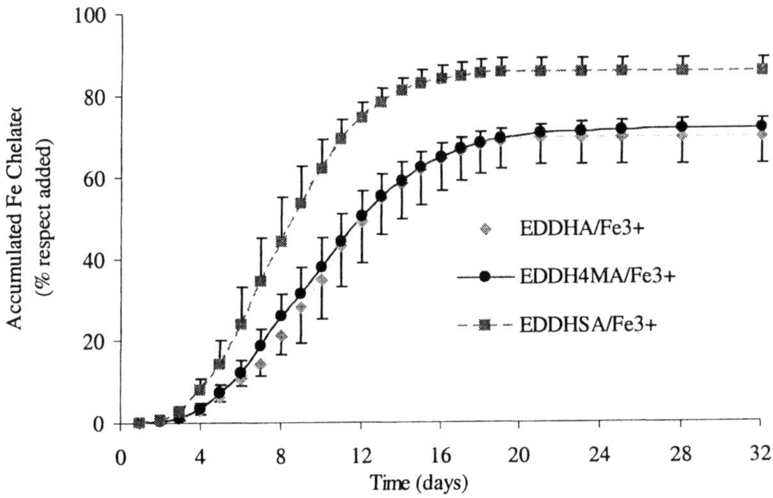

Figure 5. Percentage of accumulated chelate leached by the columns

frequency and low dosage may improve the chelate efficiency and diminish the environmental impact.

Leaching column experiments permit knowledge to be gained on the mobility of the chelates on the soil profile, so it is also a good tool in the evaluation of the Fe-chelates used as fertilizers, with the advantage that it allows consideration of different application patterns to the soils and to study the environmental impact. However it does not provide information about the plant response.

Discussion

In this paper, three types of methods have been used to test the behavior of iron chelates used as fertilizers to correct iron deficiencies in crops. Theoretical modeling is a powerful tool that allows comparing different chelates in a wide range of conditions. In fact it permits to detect the main competitors of an iron chelate. If it is an abundant element in the media, such as calcium (a macronutrient for plants), the chelate will not be stable in the conditions where the competition occurs. This is the case of EDTA/Fe^{3+}, which efficiency in soil applications is known to be quite low. However, when the main competitor is a scarce element, such as Cu^{2+} for o,pEDDHA/Fe^{3+}, the competition will be important only when the element is highly available in the soil.

Modeling is based in equilibria conditions that are not normally attained in soils, so it will help to detect ineffective chelates, but it will not distinguish among more stable ones. Then interaction tests, such as the one we have shown here, can be a better tool to compare chelates ability to maintain iron in solution. The chelates EDTA/Fe^{3+}, o,pEDDHA/Fe^{3+} and o,oEDDHA/Fe^{3+} have been used in the theoretical modeling and interaction tests and in both cases similar results are obtained. o,oEDDHA/Fe^{3+} maintains more Fe in solution than o,pEDDHA/Fe^{3+}, and o,pEDDHA/Fe^{3+} more that EDTA/Fe^{3+}. Also Cu availability in soils clearly affect o,pEDDHA/Fe^{3+} behavior. However interaction tests also consider sorption processes, which are more difficult to include in the theoretical models. It is clear from our results that different soil materials affect differently to each chelate (figure 3).

Interaction tests have been done for three days, but even when longer periods of time are considered *(11)* they are not made in realistic conditions: chelate concentration and solution/solid ratio are far from the agronomic conditions. A more realistic situation is the one presented in the leaching column tests, since they reproduce the movement of the chelates in the soil with a longer and stronger interaction. o,oEDDHA/Fe^{3+}, that was used in both laboratory experiments and in the theoretical modeling, is more retained when the column test is used, indicating that we can obtain more information of the availability of the chelate to plants and to groundwater. EDDHSA/Fe^{3+} with similar stability than EDDHA/Fe^{3+}, according to the theoretical modeling, is less retained in the soil, and then more available to the plants if no leaching occurs.

Conclusions

The theoretical speciation and the two tests proposed in this work give complementary information about the efficacy of the Fe chelates for plants and pollution capability prior their usage as fertilizers. While, the theoretical modeling gives information about the possible competition processes, interaction experiments allow consideration of the sorption processes. Leaching column experiments presents more realistic situation with respect to element availability to plants and the environment.

Acknowledgements

This study has been partially supported by Ministerio de Ciencia y Tecnología project AGL 2001/1142, Comunidad de Madrid project 07M/0042/2002 and by Syngenta Crop Protection AG.

The authors thank Tomás Navarro for his technical work.

References

1. Miller, G.; Pushnik, J.; Welkie, G. Iron chlorosis, a world wide problem. The relation of chlorophyll biosynthesis to iron. *J. Plant Nutr.* **1984.** *7*, 1-20
2. Chaney, R. Diagnostic practices to identify iron deficiency in higher plants. **1984** *J. Plant Nutr.* *7*, 47-67
3. Marschner, H.; Römheld, V. Strategies of plants for acquisition of Fe. In: *Iron nutrition in soils and plants.* Abadía J (Ed.); Kluwer Academic Publishers. The Neetherland 1995; pp.375-378
4. Lucena, J.J. Fe chelates for remediation of Fe chlorosis in strategy I plants. **2003** *J. Plant Nutr.* *26*, 1969-1984.
5. Lucena, J.J. Effects of bicarbonate, nitrate and other environmental factors on iron deficiency chlorosis. A review *J. Plant Nutr.* **2000.** *23*, 1591-1606
6. Lindsay, W.L. *Chemical Equilibria in Soils.* J. Wiley and sons, 1979
7. Mengel, K.; Kirkby, E.A.; Kosegarten, H.; Appel, T. *Iron.* In: Principles of Plant Nutrition. Kluwer Academic Pubs. Dordrecht, 2001; pp553-571
8. Álvarez-Fernández, A.; García-Marco, S.; Lucena, J.J, Evaluation of synthetic Iron(III)-chelates (EDDHA/Fe^{3+}, EDDHMA/Fe^{3+} and EDDHSA/Fe^{3+}) to correct iron chlorosis. **2003** *Europ. J. Agron.* (In press)
9. Orphanos, P.; Hadjiloucas, C. Laboratory test for screening iron chelates for use in alkaline soils. *Plant Soil* **1984**, *77*, 401-404
10. Lucena J.J.; Manzanares M.; Gárate A. A test to evaluate the efficacy of commercial Fe-chelates. *J. Plant Nutr.* **1992**, *15*, 1553-1566.
11. Cantera R.G.; Zamarreño, A.M.; García-Mina, J.M. Characterization of commercial iron chelates and their behaviour in an alkaline and calcareous soil. *J. Agr. Food. Chem.* **2002**, *50*, 7609-7615
12. Hill-Cottingham, D. G. Photosensitivity of iron chelates. Nature **1955**, *175*, 347-348.
13. Lahav, N.; Hochberg, M. Kinetics of fixation of iron and zinc applied as FeEDTA, FeEDDHA and ZnEDTA in the soil. *Soil Sci. Soc. Amer. Proc.* **1975,** *39*, 55-58
14. Yunta, F.; García-Marco, S.; Lucena, J.J.; Gómez-Gallego, M.; Alcázar, R. and Sierra, M.A. Chelating agents related to ethylenediamine bis(2-hydroxyphenyl)acetic acid (EDDHA): synthesis, characterization and equilibrium studies of the free ligands and their Mg^{2+}, Ca^{2+}, Cu^{2+} and Fe^{3+} Chelates. *Inorg. Chem.,* **2003,** *42*, 5412-5421
15. Harter, R.D. *Micronutrients adsorption-desorption reactions in soils.* In: Micronutrients in agriculture. Mortvedt, J.J.; F.R. Cox; L.M. Shuman

and R.M. Welch (Eds.). 2nd ed. SSSA Book Series No.4. Soil Sci. Soc. Am., Madison, WI. 1991. pp. 59-87.

16. Hernández-Apaolaza, L.; Lucena, J.J. Fe(III)- EDDHA and - EDDHMA sorption on Ca-Montmorillonite, Ferrihydrite and Peat. *J. Agric. Food Chem.* **2001**, *49*, 5258-5264
17. Yongming, L.; Xianliang, O.; Jing, S.; Peter C.; Minghung W. Use of a multi-layer column device for study on leachability of nitrate in sludge-amended soils. *Chemosphere* **2003**, *52*, 1483-1488
18. Lucena J.J.; Ibarreta R.; Gárate A. Chelates mobilization through different layers of the "enarenado" *Agrochimica* **1992,** *36*, 396-405.
19. Allison, J.D.; Brown, D.S.; Novo-Gradak, K.J. *MINTEQA2/ PRODEFA2. A Geochemical Assessment model for environmental systems. Ver. 3.0 User's Manual.* Environmental Research Laboratory. United States Environmental Protection Agency. Washington D.C.1990
20. Halvorson, A.D.; Lindsay, W.L. Equilibrium relationships of metal chelates in hydroponic solutions. *Soil Sci. Soc. Am. Proc.*, **1972**, *36*, 755-761
21. Pérez-Sanz, A.; Lucena, J.J. *Synthetic iron oxides as sources of Fe in a hydroponic culture of sunflower. In: Iron nutrition in soils and plants.* A. Abadía (ed.). Kluwer Academic Publishers, The Netherlands. 1995. pp.241-246
22. Sims, J.; Bingham, F. Retention of boron by layer silicates, sesquioxides and soil materials. II. Sesquioxides. *Soil Sci. Soc. Amer. Proc.* **1968**, *32*, 364-369.
23. Drits, V.A.; Sakharov, B.A.; Salyn, A.L.; Manceau, A. Structural model for ferrihydrite. *Clay Minerals* **1993,** *28,* 185-207.
24. Lindsay, W.L.; Norvell, W.A. Development of a DTPA soil test for zinc, iron, manganese and copper. *Soil Sci. Soc. Am. J.* **1978,** *42,* 421-428.
25. van Olphen, H. *Data handbook for clay minerals and other non-metallic minerals.* Pergamon Press. London. 1979
26. Soltanpour, P.N; Schwab, A.P. A new Soil Test for simultaneous extraction of macro- and micronutrients in alkaline soils. *Commun. Soil Sci.Plant Anal.* **1977**, *8*, 195-207
27. Sierra, M.A.; Gómez-Gallego, M.; Alcázar-Romero, R.; Lucena, J.J.; Álvarez-Fernández, A.; Yunta-Mezquita. F. Novel method for preparing bis(2-hydroxyaryl)aminoacetic acids using cyanide transfer agents. Patent WO 02/00604 A1
28. Álvarez-Fernández, A.; Gárate, A.; Lucena, J.J. Interaction of iron chelates with several soil materials and with a soil standard. *J. Plant Nutr.* **1997,** *20*, 559-572.

29. Lucena J.J.; Barak, P.; Hernández-Apaolaza, L. Isocratic ion-pair high-performance liquid chromatographic method for the determination of various iron(III) chelates. *J. Chromatogr. A.*, **1996,** *727,* 253-264.
30. Elgala, M. And Maier, R.H. Eddect of ethylene-diamine-di (ortho-hydroxyphenylacetic acid) application to soil columns on the distribution of certain nutrientes elements in the water-soluble acid-soluble and exchangeable forms. *Plant Soil* **1971,** *34*, 607-617

Remediation Applications

Chapter 22

Chelate-Enhanced Phytoremediation of Heavy Metal Contaminated Soil

Jing Song, Yong M. Luo, and Long H. Wu

Soil and Environment Bioremediation Research Centre, State Key Laboratory of Soil and Sustainable Agriculture, Institute of Soil Science, Chinese Academy of Sciences, Nanjing, 210008, Peoples Republic of China

Chelate enhanced phytoextraction was proposed to overcome limitations due to low metal solubility, bioavailability and low metal translocation from shoot to root. A variety of chelates have been tested for their effectiveness on enhanced phytoextraction of soil metals. Chelates have been shown to significantly increase metal concentrations in soil solution. However, the enhancement of plant uptake varies greatly, depending on the specific metal, chelate, plant combination and on soil conditions. Uncustomized addition of chelates will result in unsuccessful phytoextraction meanwhile causing negative effects on the eco-environment. Further research is needed to screen environment-friendly and cost-effective chelates and optimize processes involved in chelate induced phytoextraction to site specific conditions.

Phytoremediation is a broad term for the use of plants to treat inorganic and organic contaminated matrices including soil, sludge, sediment, surface water, groundwater and air (1-4). Based on the physical and biological processes involved, phytoremediation can be categorized into several subtypes: phytohydraulics, phytodegradation, phytoextraction, phytostabilization, phytovolatilization, rhizodegradation and rhizofiltration (5-6). Being cost-effective and environment-friendly, these plant-based technologies have attracted research attention world-wide and are being developed as alternative remediation methods suitable for use at sites with low to moderate soil contamination over large areas, and at sites where large volumes of low-level contaminated groundwater has to be cleaned to meet stringent standards.

Phytoextraction, removal of inorganic contaminants by plants, has been proposed as an alternative method for decontamination of soil with low to moderate metal pollution (7-8). A subtype of phytoextraction is phytomining. With a similar approach but different aim, phytomining involves the use of plants to mine commercially valuable metals (such as Ni and Au) from metal-rich soils or low-grade ore bodies (9-10). For phytoextraction or phytomining to occur, metals in soil have to be in the bioavailable form. It is generally established that metals in soil solution as well as metals adsorbed at ion exchange sites, which are in equilibrium with metals in soil solution, are available for plant uptake. Whereas, metals existing in non-silicate forms, such as metal bound to soil organic matter, oxides, hydroxides and carbonates etc., are less bioavailable. However, these less labile forms could partially, if not all, be rendered soluble by application of metal mobilizing agents such as organic acids, synthetic chelates, sulfur and humic substances (11-13).

Accordingly, two phytoextraction strategies exist: continuous or natural phytoextraction and chemically induced phytoextraction. The former makes use of wild metal hyperaccumulator plants or genetically improved hyperaccumulator crops to enrich metal(s) of concern in harvestable plant tissue (usually shoots). The second approach relies on metal mobilizing agents to enhance metal availability and uptake by high biomass plants.

Continuous Phytoextraction Using Metal Hyperaccumulators

An ideal plant for phytoextraction at a specific site should have the following features: tolerant to metal(s) in the soil, adapted to soil and climate conditions, spatial fitting of roots to pollution distribution, high biomass and high metal-accumulating capacity. Advantages of metal hyperaccumulator plants include high bioaccumulation factor (usually greater than 1, and in some cases reaching 50-100); high translocation coefficient (the shoot to root ratio of metal concentration is greater than 1), and hypertolerance to metals in the medium and inside plant cells (14). Hypertolerance will be of stronger consideration when selecting species for phytoextraction at sites where metals are accumulated to a level that will cause phytotoxicity to non-hyperaccumulators. However, some features of most known wild hyperaccumulator plants may limit their use in field application. For instance, with the exception of a few Ni hyperaccumulators (e.g. *Berkheya coddii*), most wild hyperaccumulator plants are slow growing, low biomass species and some tend to form small rosettes. Although low biomass can usually be compensated by exceptional high metal concentration in the shoots, low biomass plants are difficult to harvest mechanically. More importantly, they are generally endemic to specific soil and/or climate conditions and may have difficulties competing with native species or performing well under soil or climate conditions other than their natural habitat. So far, the majority of natural phytoextraction studies are focused on Ni, Zn and Cd. Laboratory and field experiments have shown that under optimized conditions, *Berkheya coddii*, *Sedum alfredii* and the Ganges ecotype of *Thlaspi caerulescens* have great potential to extract soil Ni, Zn and Cd, respectively (15-17).

Due to the complexity of hyperaccumulation phenomenon, the potential for using genetic engineering technologies to cultivate transgenic plants with desirable characteristics for phytoextraction (e.g. fast-growing, high biomass, high tolerance, and high metal-accumulating capacity etc.) has not been realized (18). However, the ever-increasing public concern over the release of genetically modified organisms could force regulators to veto their use. This is certainly the case in much of the developed world at this time (19).

As an alternative, conventional plant breeding can use the available genetic diversity within a species to combine the traits needed for successful phytoextraction. Two super performance Ni hyperaccumulators (*Alyssum murale* and *Alyssum corsicum*) have been screened by Dr. Chaney's group. Further work will focus on combining desirable characteristics using conventional plant breeding techniques to develop a genetically improved crop for commercial phytoextraction/phytomining of Ni (20-21).

Chelate Induced Phytoextraction of Heavy Metals in Soil

Although hyperaccumulators have much greater metal-accumulating capacity than non-hyperaccumulator plants, there has been little direct evidence showing that hyperaccumulators are able to acquire soil metals present in the non-labile pool. Robinson et al. (1998) attributed decreasing uptake of Ni by *Berkheya coddii* (Ni hyperaccumulator) over successive croppings to reduced soluble Ni in the soil (22). Recent studies imply that there may be significant restrictions to metal bioavailability, even to hyperaccumulator species, in contaminated soils where a large proportion of the metal is present in 'non-labile' forms (23-25).

In situations where the mobility and phytoavailability of soil metals are extremely low, the application of synthetic chelates or organic acids has been proposed to overcome the limitation due to low metal solubility, diffusion to root surface and root to shoot translocation. So far, a wide range of synthetic chelates (e.g. EDTA, CDTA, DTPA, EGTA, EDDHA, HEDTA, NTA etc.) and organic acids (e.g. citric acid, oxalic acid, malic acid) have been tested for their effectiveness for chelate-induced phytoextration of metals (e.g. Pb, Cu, Zn, Cd, Cr). With the possible exception of citric acid, which proved to be very effective in inducing plant hyperaccumulation of U (26-28), organic acids, in general, are much less effective than synthetic chelates in mobilizing or inducing phytoextraction of heavy metals such as Pb, Cu, Zn or Cd (29-30). Due to limited space, in this chapter, we will focus on synthetic chelate-induced phytoextraction of heavy metals. Phytomining of rare metals will not be discussed here.

Four major processes are involved in metal uptake by plants: metal desorption from soil particles; transport of soluble metal to roots surface via diffusion or mass flow; metal uptake by roots and metal translocation from root to shoot. All these processes will be affected by addition of chelates.

Effects of Chelates on Solubilization of Soil Metals

When added to soil, chelates will form soluble complexes with metals in the soil solution and mobilize metals from the solid phase. Excess chelate may exist in free form. The mechanisms of chelate induced metal solubilization include dissolution of soil minerals via ligand exchange reactions and remobilization of metals adsorbed onto the solid phase (31). Different sequential extraction schemes have been widely employed to investigate the effect of chelates on the solubility of metals associated with different soil fractions. A recent review of different sequential extraction procedures for fractionation of heavy metals in contaminated soil and sediment was presented by Gleyzes et al. (32). Elliot and Shastri (1999) proposed that metals that can be mobilized by EDTA were mainly from the non-detrital soil components (exchangeable fraction, organic matter and carbonate bound fractions), and EDTA is ineffective in solubilizing metals from the detrital fractions (metals in oxides and residual bound fractions) (33). However, recent studies by Sun et al. (2001) and Barona et al. (2001) seemed to show a complex picture of metal release from different fractions after EDTA addition (34-35). For instance, Barna et al. (2001) found that after EDTA extraction acetic acid-extractable soil Pb (exchangeable + carbonatic fractions) significantly increased, coupled with a marked decrease in oxides and organic bound fractions. In addition, EDTA seems to be able to release certain amounts of silicate bound Pb. However, they also pointed out that the effects of EDTA were different for different metals and soils (35).

Although it is difficult to generalize which fraction is more mobile than the others, the metal mobilizing effect of a chelate can be indicated by changes in metal concentration in the soil pore water (29, 36, 37) or in labile metal fractions determined by soil extraction with water (38-39), 1 M NH_4OAC (22), 0.1 M $NaNO_3$ (11, 40), 0.1 M $CaCl_2$ (41), 0.01 M $CaCl_2$ (42) or 1 M NH_4NO_3 (39,43). For example, Lombi et al. (2001) observed markedly increased metals in the soil pore water within 24 hours after application of 2.7 mmol/kg EDTA (added as salt), with soluble Zn having increased from 2.4 to 104 mg/L in the UK soil, soluble Pb having increased from 0.1 to 36 mg/L in the French soil (36). Wenzel et al. (2003) reported a 1500-fold increase of 1 M NH_4NO_3 extractable Pb in 5.4 mmol/kg EDTA treated soil (added as solid EDTA) relative to the control in a field lysimeter study (43).

Factors that influence the effectiveness of a given chelate to solubilize soil metals include metal species and distribution among soil fractions, metal:ligand ratio, formation constant of metal-ligand complexes, presence of competing cations, soil pH, adsorption of free and complex metals onto soil particles etc. A study by Epstein et al. (1999) showed that when sufficient EDTA was added, all $PbCO_3$ spiked in the soil could be solubilized (see Fig. 1, data from Epstein, et al. 1999). The authors also indicated that to obtain the same level of soluble Pb from Pb in more recalcitrant forms, more EDTA is needed (38). Kim et al. (2003) suggested that occlusion of Pb in Fe oxides may reduce EDTA extraction efficiency of soil Pb (44). Although the formation constant for 1:1 metal-EDTA complexes follows the order: Fe^{3+}>Cu>Pb>Zn>Fe^{2+}>Ca, major cations such as Fe and Ca present in the soil may compete for active sites of EDTA. A soil washing study by Kim et al. (2003) showed that Fe most probably competed strongly with Pb for EDTA ligand sites at pH values less than 6. In a multi-metal contaminated soil (pH<6), Cu and Zn may potentially compete with Pb for EDTA ligand sites (44). The abundant Ca^{2+} in calcareous soil may have displaced heavy metals from their EDTA complexes, leading to the formation of CaH_2EDTA and insoluble metal carbonates (41). Due to the extreme variability of metal species and soil properties, it is difficult to accurately predict all the competing reactions and conditions (45). Much of the work, therefore, relies on empirical data to determine the dosage used in chelate induced phytoextraction. Despite of large differences in metal species, concentrations, soil properties, type of chelates etc., application rates reported in the literature were up to 10 mmol/kg soil. Only in a few cases did the authors clarify how the dosage was determined. Uncustomized application rates could lead to low phytoextraction efficiency and high eco-environmental risks.

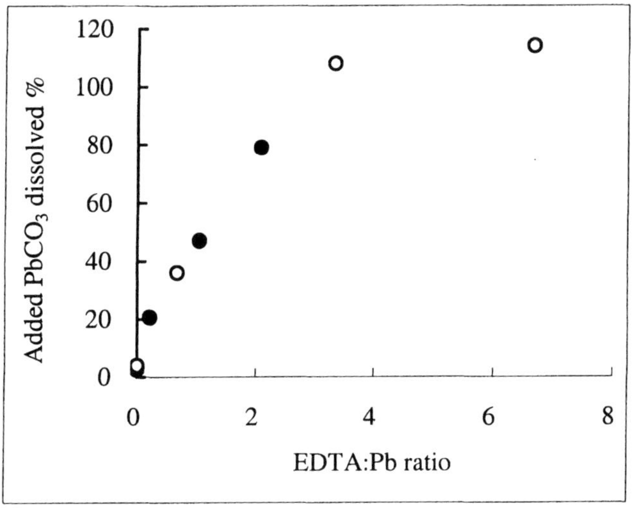

Figure 1 Effect of EDTA:Pb ratio on the percentage of added Pb solubilized (data from literature 38, solid circle: soil spiked with 4.8 mmol/kg $PbCO_3$, hollow circle: soil spiked with 1.5 mmol/kg $PbCO_3$)

Mechanism of Chelated Metal Uptake by Plants

The success of chelate-enhanced remediation requires a better understanding of the biological mechanisms involved. The predominant theory for chelate induced metal uptake is the split-uptake mechanism, where only free metals dissociated from metal-ligand complexes will be absorbed by plant roots, leaving the ligand in the soil solution. Another theory suggested that metal-ligand complexes are absorbed by plant roots, transported through the plant via the xylem and accumulated in the shoots. Recent studies using isotope tracer ([^{14}C] EDTA) (46) and analysis with IC-MS (47) seemed to support the latter theory. In addition, Wenzel et al. (2003) hypothesized that it is free protonated EDTA that enters the roots, subsequently forming metal complexes that enhance metal transport to shoots. For this to happen, the kinetics of metal–EDTA complex formation in real soil systems need to be slow enough to allow the diffusion of free protonated EDTA to the root surfaces (43).

Regarding the pathways through which metal-ligand complexes enter the xylem vessel, recent ultrastructural studies implied that both apoplastic and symplastic transport pathways may exist, depending on the plant species and chelate used. Using TEM (Transmission Electron Microscopy), Jarvis and Leung (2001) revealed that in the shoots of *Chamaecytisus palmensis* (a fast growing evergreen leguminous tree), Pb chelated with H-EDTA appears to follow an apoplastic route, as indicated by Pb deposition around the intercellular space and within the cell walls. On the other hand, EDTA chelated Pb follows a symplastic path, as indicated by occurrence of Pb within structures in the cytoplasm (48). In the needles of *Pinus radiate* (a fast growing conifer), Pb-ligand complexes were transported exactly in the opposite manner. Pb chelated with H-EDTA appears to follow a symplastic route while EDTA chelated Pb follows an apoplastic path (49).

Conventional theory suggests that once in the xylem vessel, long-distance transportation of metal-ligand complexes from root to shoot will be driven by the transpiration stream (46, 50-51). However, a study by Epstein et al. (1999) implied that transpiration appeared not to be a critical factor for the uptake of Pb and EDTA. The majority of Pb and EDTA uptake may have occurred rapidly (within hours) after the EDTA application, before the measured transpiration decreased (38).

Factors Influencing Chelates Induced Metal Accumulation by Plants

Earlier studies have reported EDTA induced hyperaccumulation of Pb by *Brassica juncea* (51), maize (52), pea and corn (53). Although many authors have recently reported enhanced metal uptake after addition of chelates (12,29,43,54-57), metal hyperaccumulation was only achieved in a few cases.

Recently we studied the effect of EDTA and biodegradable chelate EDDS on metal uptake by *Brassica juncea* grown on a multi-metal contaminated soil collected near smelters. In treatments with 3 or 6 mmol/kg EDDS, regardless of the mode of application (single or split), the average Cu concentration in leaves was well above 1000 mg/kg. Bioconcentration factors (ratio of total metal concentration in plant tissues to that in soils) were up to 7.7 for leaves and 2.5 for stems (Figure 2, unpublished results). The effectiveness of EDDS decreased in the order Cu>Zn>Pb>Cd, which follows the stability constant of EDDS-metal complexes reported in the literature (58). Compared to Cu, the bioconcentration factor of Pb was also unsatisfactory in our study. The relatively low formation constant of Pb-EDDS may partly be the reason why Pb hyperaccumulation has not been achieved by using EDDS, even at high dosage (10 mmol/kg).

Vassil et al. (1998) indicated that a threshold concentration of EDTA is required to obtain accumulation of Pb in plant shoot. They suggested that at high concentrations chelates may physiologically damage the root membranes that would normally impede the uptake of intact metal-ligand complexes. Synthetic chelates could disrupt the normal function of cell membranes by removing Zn and Ca ions that are involved in the stabilization of plasma membranes (46). Using a data set from the literature, McGrath et al. (2002) plotted extractable Pb against Pb uptake in shoots of different plant species induced by different chelates. The wide-spanned (3 orders of magnitude) data seemed to fit a linear relationship (59). However, it should be noted that there may be huge differences in plant species regarding their response to elevated metal-ligand concentration in the growth media. For instance, a study by Epstein et al. (1999) clearly showed a parabolic relationship between water extractable Pb and Pb concentration in the shoot of *Brassica juncea* (38). Robinson et al. (1999) found that the addition of chelates (NTA, DTPA and EDTA) increased 1 M NH_4OAC extractable Ni in an artificial serpentine substrate, but Ni uptake by *Berkheya coddii* decreased by about 50% relative to the control. The authors attributed decreased Ni concentration in shoots to competition with the plant's own nickel-binding agents, thereby causing the nickel to diffuse downwards to the plant's root system (60). In addition, free protonated EDTA may lead to phytotoxicity (46) and pose potential risks to eco-environment (details in next part). Therefore, it is necessary to maintain a proper concentration range of metals in soluble form to achieve high bioconcentration factors meanwhile minimize potential risks associated with chelate addition.

Potential Risks Associated with Chelate Induced Phytoextraction

As exogenous substances, chelates have risks of negative effects on the eco-environment when applied to soils. Here we will try to discuss the major risks

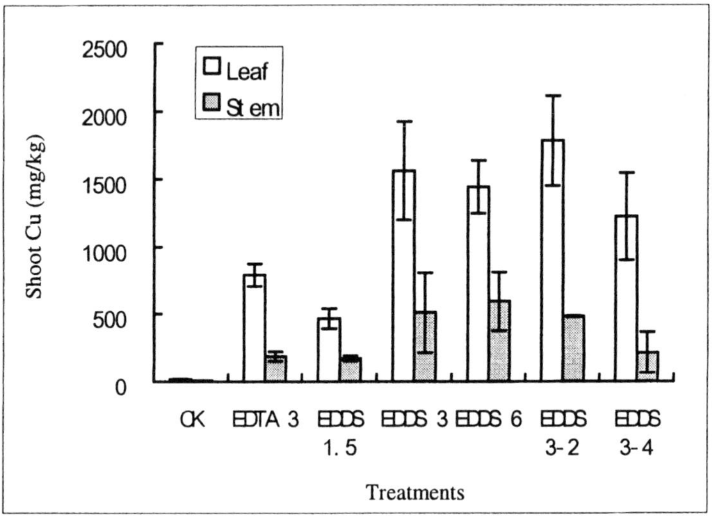

Fig. 2 Effects of EDTA and EDDS on Cu uptake by Brassica juncea

Figures after chelates represent application rate (mmol/kg), figures after dash represent number of spittings.

reported in the literature (e.g. adverse effect on plant and soil biota, metal leaching to groundwater etc.) as well as possible countermeasures.

Reported negative effects of chelates on plant growth include foliar necrosis, leaf wilt and abscission, shoot desiccation, reduced transpiration and reduced biomass (29, 36, 38, 46, 54, 57).

Soils contaminated with phytotoxic levels of Cu, Zn or Cd present a challenge for phytoremediation. In a lysimeter study by Wenzel et al. (2003), it was suggested that toxicity of Cu to Canola may be alleviated due to complexation with EDTA (43). A hydroponic study by Vassil et al. (1998) suggested that EDTA-induced foliar necrosis may be attributable to the presence of free protonated EDTA in leaves, as no phytotoxicity was observed in a treatment with equal molar of Pb and EDTA (46). HEDTA always proved to be more phytotoxic than EDTA at same concentrations (48). To avoid toxic effects at high concentrations of EDTA, it is suggested that EDTA should be applied at rates that minimize the availability of free chelates (46).

The persistence of soluble EDTA-metal complexes in soil can cause prolonged negative effects upon soil microfauna and plant growth. To reduce the long-term negative effects of recalcitrant chelates, recently, a biodegradable EDTA structural isomer EDDS was tested (55-57, 61). In a pot trial using multi-metal polluted soils, we compared the effect of EDTA and EDDS on plant growth and metal uptake by *Brassica juncea*. The third day after application, leaves of all four replicates in the 6 and 3 mmol/kg EDDS treatment started to wilt. While in soil treated with 3 mmol/kg EDTA and 1.5 mmol/kg EDDS, symptom of leaves wilt occurred a few days later and not on all four replicates (unpublished data). A study by Vassil et al. (1998) showed that increasing concentrations of EDTA caused a significant reduction in shoot water content (46). Rapid senescence of cabbage shoots occurred in treatments receiving single and weekly additions of 10 mmol/kg EDTA (57).

As pointed out by Vassil et al. (1998), the transpiration rate is not a critical factor for translocation of metal-ligand complexes from root to shoot (46), therefore, decreased transpiration rates will not be a matter of concern in the context of chelate induced phytoextraction. Whereas, wilt leaf may present a problem as leaf abscission may occur a few days after wilt. To avoid loss of metal-rich plant material, harvest should be done before leaf abscission.

Chelate induced effects on plant biomass may be largely explained by the mode of application, dosage as well as plant tolerance. The same amount of chelates can be added (i) in a single dose after plants have accumulated enough biomass or (ii) in a single dose before transplanting or (iii) gradually added at several lower dosage during the growth period.

In the first case, plant biomass will less likely be affected by chelate addition as plants will normally be harvested several days after application. Plants may die of phytotoxicity due to high concentrations of soluble metal-ligand complexes and/or free chelates. In the context of phytoextraction, dead plants are not a big problem. However, excessive soluble metals may inhibit plant growth in the follow-up croppings (36). The inhibition effect will be more pronounced for recalcitrant chelates (e.g. EDTA) than for biodegradable chelates (e.g. EDDS). A bioassay with red clover was performed to evaluate the post-treatment toxicity of EDTA and EDDS treated soil. The results showed that the biomass of red clover shoots was significantly reduced in soil that have received 5 and 10 mmol/kg weekly EDTA additions as compared to control and corresponding EDDS treatments (57). More importantly, when recalcitrant chelates are used, high soluble metals will remain in this form for a long time (e.g. several months). Leaching of soluble metals is likely to occur (to be discussed below).

When chelates are added before transplanting, plant growth may be inhibited to a extent that reduction in plant biomass can not be compensated by increased metal concentration in the shoots, thereby decreasing net metal removal (54).

The third approach is aimed to provide a maximum of soluble metals available for removal meanwhile reducing the risk of metal leaching. Studies by Wenzel et al. (2003) seemed to suggest that longer exposure of Canola plants to toxic EDTA levels in the split application treatments may limit shoot biomass production (43). To determine the time interval between two applications, Gupta et al. (2000) used 0.1 M $NaNO_3$ extraction to indicate the change in the phytoavailable pool of soil metals (11).

Microbial biomass, respiration, nitrogen mineralization, microbial diversity and functional groups of soil fauna (e.g. nematodes) are well recognized indicators for evaluating soil quality (62). A recent study by Römkens et al. (2002) showed that addition of EGTA (ethylenebis [oxyethylenetrinitrilo] tetraacetic acid) resulted in an increase of microbial biomass, bacterial activity measured as ^{14}C-leucine incorporation was slightly lower in EGTA treatment, while bacterial activity measured as ^{3}H-thymidine incorporation was not affected by EGTA. The net effect of EGTA addition on bacterial growth was limited. The effect of EGTA on the number of soil nematodes was found to be dependent on type of nematodes and plant species. EGTA significantly reduced bacterivores (up to 90% reduction) and fungivores under three crops (grass, lupin and yellow mustard), but had no direct effect on herbivores. The authors also indicated that EGTA may exert its influence on soil ecosystem through its effect on plant growth, which in turn affected the number, activity and type of soil microbes and nematodes that are critical to the function of soil ecosystem (29). In a laboratory study to assess the potential toxicity of EDTA and [S,S]-EDDS to soil microbes, Kos and Leštan (2003) observed increased glucose-induced microbial respiration with increasing rates of [S,S]-EDDS presumably due to microbial use of [S,S]-EDDS as an additional carbon or energy source. In contrast, high concentrations of EDTA decreased glucose induced respiration.

However, the authors also suggested that increased Pb leaching in a 10 mmol/kg EDDS treatment was likely due to toxic effects on soil microbes which are capable of degrading EDDS (56). A study by Grčman et al. (2003) showed that mycorrhizal infection of red clover seemed to be not affected by soil pre-treatment with 5 and 10 mmol/kg EDDS or EDTA. While the biomass of red clover significantly reduced in the 5 and 10 mmol/kg EDTA treatment. Using PLFA (phospholipid fatty acid) technique, the same authors further demonstrated that EDDS addition was less toxic to soil fungi than EDTA and caused less stress to soil microbes (57). It should be noted that the observed effects of chelate addition on soil microbes and soil biota are joint effects of metal-ligand complexes, free chelates and existing plants.

As the amount of mobilized metals are normally far beyond the amount that plants can take up in the growing season, solubilized metals may leach down the soil profile to the groundwater during rain event. Unusually high metal concentration in leachates collected in column (34,61,63) and field lysimeter studies (29, 43) provided further evidence of metal leaching induced by addition of EDTA and EGTA. The metal concentrations in the leachates after EDTA addition were clearly related to the rates of EDTA applied (43,63) and soil structures (64). The order of leaching response to EDTA application is largely consistent with the corresponding formation constant of metal-EDTA complexes (Cu, 20.5; Pb, 19.8; Zn, 18.3) (34). The effect of EDTA addition on metal concentration in leachates was reported to persist several months after EDTA application (43), and large percentages of Pb, Cd and Zn leached through the soil profile with EDTA application, Grčman et al. (2003) found that when applied at the same dosage, biodegradable [S,S]-EDDS caused much less loss of Pb and Cd. 22.7% and 39.8% of initial Pb and Cd leached in 10 mmol/kg EDTA treatments, as compared to 0.8% and 1.5% in EDDS treatment. Leaching of Zn (about 6.2% of initial total concentration) was comparable with the EDTA treatment (57).

Coupled with leaching of heavy metals, chelate addition may also cause loss of essential plant nutrients such as Fe, Ca and Mg. In a laboratory leaching experiment using soil columns, Wu et al. (2003) found the amount of Fe lost increased to 163 mg/kg in the 12 mmol/kg EDTA treatment as compared to 3.37 mg/kg in the control. The effects of volume and pH of simulated rainfall were minimal. On the other hand, EDTA treatment, volume and pH of simulated rainfall seemed to have no significant effects on the loss of Ca and Mg. The authors also suggested that rainfall pH may play a role in the long term (63).

Potential field application of chelate enhanced phytoextraction

The potential of chelate enhanced phytoextraction of soil metals has been extensively explored in laboratory hydroponic culture, pot trails and lysimeter studies. In contrast, data obtained from field trials are rather limited. Extrapolating pot trial results to the field is less likely to be successful because even using the same soils and plant-chelate combination, pot trials are different from field trials in many key parameters that influence performance. These parameters include soil physical conditions (compactness, heterogeneity, existence of preferential flow etc.), plant root development and distribution as well as soil water status (soil water content, evapotranspiration etc.). It was estimated that heavy metal removal in the field was on average only 20% of what was expected from pot results (65). Therefore, more field trials are needed to gain experience in improving the performance of chelate enhanced phytoextraction under field conditions.

From the technical point of view, with the possible exception of ex situ application, metal leaching down the soil profile is perhaps the biggest concern that limits the field application of this technology. Different strategies have been proposed to facilitate metal mobilization and plant uptake while reducing metal leaching. For example, simply by placing NTA 15 cm deep, the amount of NTA required to enhance metal accumulation was lower than when NTA was mixed into the entire field (12). The reported effect of dosage splitting on plant uptake seemed to be case dependent. A study by Pushenreiter et al. (2001) showed that Pb concentration in corn grown on a soil with 5600 mg/kg total Pb increased 8-fold to 49 mg/kg when 2.69 mmol/kg EDTA was added in a single dose three weeks before harvesting. While, Pb concentration in corn increased 18-fold when the same amount of EDTA was applied in three splits over three weeks (66). In our study, Cu uptake by *Brassica juncea* was on average higher in the treatment with 3 mmol/kg EDDS added in two splits over 8 days as compared to that in the treatment with a single dose of 3 mmol/kg EDDS 8 days before harvesting. In contrast, Cu uptake by *Brassica juncea* was on average lower in the treatments with 3 mmol/kg EDDS added in four dosages (Fig. 2). Due to limited available data, no definitive conclusion can be drawn concerning the effect of dosage splitting on metal leaching (65). Coated chelates are being developed to control the release of chelates over time (personal communication with Dr. H.F. Li). Using soil column experiments, Kos and Leštan (2003) tested the effectiveness of increasing field soil water holding capacity by using acrylamide hydrogel. The idea is to retain chelate solution in the top soil. Their results showed that acrylamide hydrogel was not particularly of use as a soil conditioner (56). In another column study, the same author showed that permeable barriers (consisting of nutrient enriched vermiculite, peat or hydrogel in combination with apatite) placed underneath polluted soil cores were effective in reducing total Pb leached after addition of 10 mmol/kg biodegradable EDDS (55). Although it was environmentally safe, Pb concentration achieved (463±112 mg/kg) in the study was far from concentration required for efficient phytoextraction of Pb (1%) within a reasonable time frame. Furthermore, the high costs of EDDS (US$7800 per ton) and cost required for installation of permeable barriers, may limit the use of this technique in field application. More pot and field experiments are needed to evaluate the effectiveness of potential measures to reduce metal leaching. In the field application, monitoring of metal leaching is needed.

Apart from the potential risk of metal leaching, time constraints may also constrain field application of chelate enhanced phytoextraction. To improve the performance and applicability of the technology, chelate enhanced phytoextraction of soil metals can be integrated with compatible ex situ and in situ technologies such as particle size separation, electrokinetic processes, energy plant production (67,68).

Acknowledgement

The authors are grateful to two anonymous reviewers and Dr. Jeanne VanBriesen for their critical comments and suggestions on the manuscript. The authors are also indebted to Chinese Ministry of Science and Technology and Natural Science Foundation of China for funding their research on chelate induced phytoextraction of soil metals (National Basic Research and Development Program-2002CB410809 and NSFC projects: 40125005 and 40301046).

References

1. EPA 600-R-99-107 February, **2000**, pp3-40.
2. EPA 542-R-01-006 July, **2001**, pp4-14.
3. Meagher, R.B. Curr. Opin. Plant Biol. **2000**, 3, 153-162.
4. Alkorta, I.; Garbisu, C. Bioresource Technology. **2001**, 79, 273-276.
5. Garbisu, C.; Alkorta, I. Bioresource Technology. **2001**, 77, 229-236.
6. Susarla, S.; Medina, V.F.; McCutcheon, S.C. Ecological Engineering. **2002**, 18, 647-658.
7. Baker, A.J.M.; McGrath, S.P.; Sidoli, C.M.D. Reeves, R.D. Resour. Conserv. Recy. **1994**, 11, 41-49.
8. Raskin, I.; Smith, R.D.; Salt, D.E. Curr. Opion. Biotech. **1997**, 8, 221-226.
9. Nicks, L.; Chambers, M.F. Mining Environ. Mgmt. **1995**, 3, 15-18.
10. Anderson, C.W.N.; Brooks, R.R.; Stewart, R.B,; Simcock, R. Nature. **1998**, 395, 553-554.
11. Gupta, S.K.; Herren, T.; Wenger, K.; Krebs, R.; Hari, T. In phytoremediation of contaminated soil and water; Terry, N.; Bañuelos, G., Ed.; Lewis Publishers: Boca Raton, FL, **2000**; pp 303-321.
12. Kayser, A.; Wenger, K.; Keller, A.; Attinger, K.; Felix, H.R.; Gupta, S.K.; Schlin, R. Environ. Sci. Technol. **2000**, 34,1778-1783.

13. Halim, M.; Conte, P.; Piccolo, A. Chemosphere. **2003**, 52, 265-275.
14. McGrath, S. P.; Zhao, F.J. Curr. Opin. in Biotechnol. **2003**, 14, 277-282.
15. Robinson, B.H.; Brooks, R.R.; Howes, A.W.; Kirkman, J.H.; Gregg, P.E.H. J. Geochem. Exploration. **1997**, 60, 115-126.
16. Yang, X.E.; Long, X.X.; Ni, W.Z.; Fu, C,X. Chinese Sci. Bull. **2002**, 47,1003-1006.
17. Zhao, F.J.; Lombi, E.; McGrath, S.P. Plant Soil. **2003**, 249, 37-43.
18. Clemens, S.; Palmgren, M.G.; Krmer, U. Trends in plant Sci. **2002**, 7, 309-315.
19. Baker A.J.M.; Whiting, S.N. New Phytologist. **2002**, 155, 1–7.
20. Li, Y. M ; Chaney, R.; Brewer, E.; Angle, J.S.; Nelkin, J. Environ. Sci. Technol. **2003**, 37, 1463-1468.
21. Li, Y.M.; Chaney, R.; Brewer, E.; Roseberg, R.; Angle, J.S.; Baker, A.J.M.; Reeves, R.; Nelkin, J. Plant and Soil. **2003**, 249, 107-115.
22. Robinson, B.H.; Brooks, R.R.; Gregg, P.E.H.; Kirkman, J.H. Geoderma. **1998**, 87, 293-304.
23. Wenzel,w.w.; Jockwer, F. Environ. Pollut. **1999**, 104: 145-155.
24. Hutchinson, J.J.; Young. S.D.; McGrath, S.P.; West, H.M.; Black, C.R.; Baker, A.J.M. New Phytol. **2000**, 146, 3, 453-460.
25. Schwartz, C.; Echevarria, G.; Morel, J.L. Plant Soil. **2003**, 249, 27-35.
26. Huang, J.W.; Blaylock. M.J.; Kapulnik. Y.; Ensley. B.D. Environ. Sci. Technol. **1998**, 32, 2004–2008.
27. Shahandeh,H.; Hossner,L.R.; Soil Sci. **2002**, 167, 269-280.
28. Vandenhove, H.; Van Hees, M. J. Environ. Radioactivity. **2004**, 72, 41-45.
29. Römkens, P.; bouwman, L.; Japenga, J.; Draaisma, C. Environ. Pollut. **2002**, 116, 109-121.
30. Wu, L.H.; Luo, Y.M.; Christie, P.; Wong, M.H. Chemosphere. **2003**, 50, 819-822.
31. Nowack, B. Environ. Sci. Technol. **2002**, 36, 4009-4016.
32. Gleyzes,C; Tellier,S; Astruc, M. Trends Analyt. Chem. **2002**, 21, 451-467.
33. Elliott, H.A., Shastri, N.L. Water Air Soil Pollut. **1999**, 110, 335-346.
34. Sun, B.; Zhao, F.J.; Lombi, E.; McGrath, S.P. Environ. Pollut. **2001**, 113, 111-120.
35. Barona, A.; Aranguiz, I.; Elias, A. Environ. Pollut. **2001**, 113, 79-85.
36. Lombi, E.; Zhao, F.J.; Dunham, S.J.; McGrath; S.P. J. Environ. Qual. **2001**, 30, 1919-1926.
37. Wu, L.H.; Luo, Y.M.; Song, J.; Christie, P.; Wong, M.H. Bull. Environ. Contam. Toxicol. **2003**, 71, 706-713.
38. Epstein, A.L.; Gussman, C.D.; Blaylock, M.J.; Yermiyahu, U.; Huang, J.W.; Kapulnik, Y.; Orser, C.S. Plant Soil. **1999**, 208, 87-94.
39. Jiang, X.J.; Luo, Y.M.; Zhao, Q.G.; Baker, A.J.M.; Christie, P.; Wong, M.H. Chemosphere. **2003**, 50, 813–818.

40. Cooper, E.M.; Sims, J.T.; Cunningham, S.D.; Huang, J.W.; Berti, W.R.; J. Environ. Qual. **1999**, 28, 1709–1719.
41. Walker, D.J.; Clemente, R.; Roig, A.; Bernal, M.P. Environ. Pollut. **2003**, 122, 303-312.
42. Degryse, F.; Broos,K.; Smolders, E.; Merckx, R. Eur. J. Soil Sci. **2003**, 54, 149-157.
43. Wenzel, W.W.; Unterbrunner, R.; Sommer, P.; Sacco, P. Plant Soil. **2003**, 249, 83-96.
44. Kim, C.; Lee, Y.; Ong, S.K. Chemosphere. **2003**, 51, 845–853.
45. Blaylock, M.J., Huang, J.W., In phytoremediation of toxic metals: using plants to clean up the environment; Raskin, I.; Ensley, B.D., Ed.; John Wiley & Son, Inc. **2000**; 53-70.
46. Vassil, A.D.; Kapulnik, Y.; Raskin, I; Salt, D.E. Plant physiol. **1998**, 117,447-453.
47. Collins, R.; Onisko, B.; McLaughlin, M.; Merrington, G. Environ. Sci. Technol. **2001**, 35, 2589-2593.
48. Jarvis, M.D.; Leung D.W.M. Plant Sci. **2001**, 161, 433–441.
49. Jarvis, M.D.; Leung D.W.M. Environ. Exptl. Botany. **2002**, 48, 21–32.
50. Salt, D.E.; Prince, R.C.; Pickering, I.J.; Raskin, I. Plant Physiol. **1995**, 109, 427-433.
51. Blaylock, J. M.; Salt, D.E.; Dushenkov, S.; Zakharova, O.; Gussman, C.; Kapulnik, Y.; Ensley, B.D.; Raskin, I. Environ. Sci. Technol. **1997**, 31, 860-865.
52. Bricker, T.J. ; Pichtel, J.; Brown, H.J.; Simmons, M.J. Environ. Sci. Heal. A **2001**, 36, 1597-1610.
53. Huang, J.W.; Chen, J.; Berti, W.R.; Cunningham, S.D. Environ. Sci. Technol.**1997**,31,800-805.
54. Chen, H; Cutright, T. Chemosphere. **2001**, 45, 21-28.
55. Kos, B.; Leštan, D. Environ. Sci. Technol. **2003**, 37, 624-629.
56. Kos, B.; Leštan, D. Plant Soil. **2003**, 253, 403-411.
57. Grěman, H.; Vodnik, D.; Velikonja-Bolta, Š.; Leštan, D. J. Environ. Qual. **2003**, 32, 500-506.
58. Bucheli-Witschel, M.; Egli, T. FEMS Microbiol. Rev. **2001**, 25, 69-106.
59. McGrath, S. P.; Zhao, F.J.; Lombi, E. Advances in Agronomy. **2002**, 75, 1-56.
60. Robinson, B.H.; Brooks, R.R.; Clothier, B.E. Annals of Botany. **1999**, 84, 689-694.
61. Grěman, H.; Velikonja-Bolta, Š.; Vodnik, D.; Kos, B.; Leštan, D. Plant Soil. **2001**, 235, 105-114.
62. Schloter, M.; Dilly, O.; Munch, J.C. Agri. Ecosys. Environ. **2003**, 2076, 1-8.
63. Wu, L.H.; Luo, Y.M.; Xing, X.R.; Christie, P. Agr. Ecosyst. Environ. **2004**, 102, 307-318.

64. Thayalakumaran T.; Robinson, B.H.; Vogeler. I.; Scotter, D.R.; Clothier, B.E.; Percival, H.J. Plant Soil. **2003**, 254, 415-423.
65. Schmidt, U. J.E.Q. **2003**, 32, 1939-1954.
66. Puschenreiter, M.; Tesar, M.; Horak, O.; Wenzel, W.W. In Proc. Int. Conf. on Interactions in the root environment-An Integrated Approach; Powlsen, D.S. et al. Ed.; Rothamsted, UK. 10–12 Apr. **2001**.
67. Blaylock, M.J. In Summary of the Phytoremediation State of the Science Conference Boston, Massachusetts 1-2 May, **2000**
68. Robinson, B.; Fernández, J.E.; Madejón, P.; Marañón, T.; Murillo, J.M.; Green, S.; Clothier, B. Plant Soil **2003**, 249, 117–125.

Chapter 23

Soil Washing Using a Biodegradable Chelator

Domen Leštan and Boštjan Kos

Agronomy Department, Biotechnical Faculty, University of Ljubljana, 1000 Ljubljana, Slovenia

Remediation of heavy metal contaminated soils with soil washing methods involves addition of water with chelators, or acids in combination with chelators. This can be done *in situ* or *ex situ* in reactors with soil slurry, and on site as a heap leaching. Chelators are used to enhance heavy metal solubility in soil solution from the soil solid phases, where the major part of soil heavy metals usually resides. A novel method of soil washing using biodegradable chelator [S,S]-stereoisomer of ethylenediamine-disuccinate (EDDS) and a horizontal permeable reactive barrier was proposed, and evaluated for Pb and Cu contaminated soil. During laboratory studies of EDDS enhanced *in situ* soil washing and heap leaching, biodegradable heavy metal-EDDS complexes were microbially degraded and the released heavy metals chemically immobilized in the barrier. After remediation the barrier material enriched with heavy metals was removed from the soil.

Heavy metals make a significant contribution to soil contamination. The contamination has resulted from a combination of industrial, urban, and modern agronomical practices. The primary sources of this pollution are the burning of fossil fuels, mining and smelting activities, and the use of fertilizers, pesticides and sewage sludge. Some heavy metals: V, Cr, Mn, Fe, Cu, Zn, Mo and Ni are considered to be essential micronutrients for at least some forms of life; others have no known biological function. All heavy metals at high concentrations have strong toxic effects and are an environmental threat. Heavy metals enter the body in food (i.e. crops grown on heavy metal contaminated soil), ingestion of soil, or inhalation of dust. An increasing body of evidence suggests that soil organisms, vitally important for soil health and fertility, are sensitive to heavy metal stress (*1*) and that the biological diversity of the soil is reduced by heavy metal contamination (*2*). In future, the availability of arable land may decrease because of stricter environmental laws limiting food production on contaminated land. Already national farmers and consumer organizations in European Union and Associated Countries do not recognize organic/ecological farming on soils contaminated with heavy metals.

Soil remediation is just one, but absolutely necessary step in rehabilitation of contaminated land (Figure 1). It is often the most expensive measure and as such crucial for a final decision on soil rehabilitation.

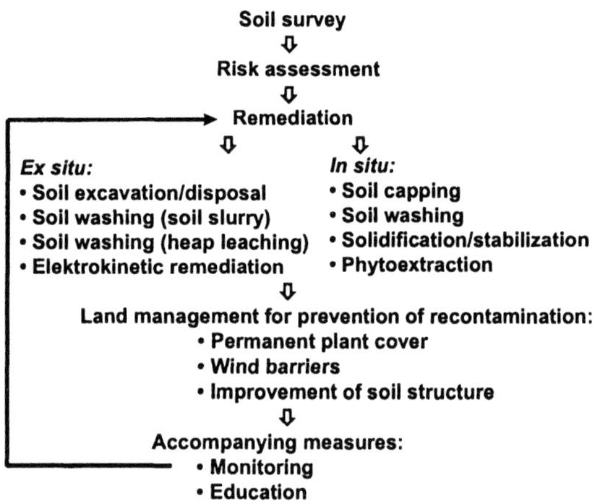

Figure 1. Steps in rehabilitation of heavy metals contaminated soil. Remediation costs are often too high and outweigh environmental impact, health hazard, legislation requirements and public sensitivity towards contaminated lands.

Remediation of Heavy Metal Contaminated Soil

Remediation of sites contaminated with toxic metals is particularly challenging. Unlike organic compounds, metals cannot be degraded, and cleanup requires their immobilization and toxicity reduction or removal. The selection of the most appropriate soil remediation method depends on the site, soil characteristics and heavy metals concentration. At present, various approaches have been suggested for remediating heavy metal contaminated soils (*3*). However, no methods are available that give satisfactory results in dealing with the problem.

Solidification/stabilization contains the contaminants in an area by mixing or injecting agents such as cement and lime, and for Pb contaminated soil various phosphate sources (i.e. apatite), which form insoluble salts with Pb (*4*). Heavy metals are less available for plants and the bioconcentration through the food chain is reduced. However, toxic metals remain in the soil and can be harmful through soil ingestion or inhalation of soil dust.

Electroremediation is proposed as an *in situ* method for remediation of blocks of heavily contaminated soil. Current designs are not appropriate for decontamination of surface soil layers. Electroremediation methods mostly involve electrokinetic movement of charged particles suspended in soil solution. It is therefore questionable whether heavy metals are removed from solid organic and inorganic fractions, where they predominantly reside (Pb in particular) in most soils (*5*).

Soil washing currently involves *ex situ* extraction of heavy metals from soils with chelators or acids, in reactors or as heap leaching. The separation of metals from waste chelator/acid solution after extraction has not yet been adequately addressed. Soil washing in reactors involves stringent physical treatments and is harsh for the soil structure and flora. Chemical agents involved can dramatically inhibit soil fertility, with subsequent negative impacts on the ecosystem.

Phytoextraction is a 'soil-friendly', publicly appealing alternative. However, for Pb, one of the most widespread contaminants, no real

Table 1. Cost of heavy metal contaminated soil remediation

Soil Treatment	*Cost ($)*	*Cost ($/ha)*
Asphalt capping		160.000
Soil capping		140.000
Excavation/ heavy metal stabilization/ disposal	600	1.500.000
Heavy metal imobilization	270	
Soil washing	450	
Soil vitrification	400-800	
Phytoextraction with hyperaccumulating plants	2-10	200-100.000
Chelator induced phytoextraction		260.000

SOURCES: References *3*, *7*, *8*.

hyperaccumulating plant, with high Pb uptake and high biomass essential for effective phytoextraction, has been reported so far. This is presumably because only a small portion of the Pb in soil is present in soil solution or exchangeable from soil colloids and thus phytoavailable (*5*). There have been reports of high Pb phytoaccumulation when Pb in soil was mobilized by a chelator addition. However, chelator mobilized heavy metals are leached through the subsoil and pose a threat of groundwater contamination (*6*).

Many of these technologies have been used full-scale but can be prohibitively expensive (Table 1).

Soil Washing Using Biodegradable Chelator and Horizontal Permeable Reactive Barrier

The literature to date reports a number of chelators that have been tested for chelator-induced soil washing. These include ethylenediamine tetraacetate (EDTA) and its structural analogues: trans-1,2-diaminocyclohexane-*N,N,N',N'*-tetraacetic acid (CDTA), dietylenetriamine pentaacetic acid (DTPA), nitrilotriacetic acid (NTA) (*9, 10*). In most cases, the EDTA treatment was superior in terms of soil washing efficiency. However, side-effects related to the soil addition of chelators cause health, safety and environmental concerns. Many synthetic chelators and their complexes with heavy metals are toxic (*11*) and poorly photo-, chemo- and biodegradable in soil environments (*12*). A combined widespread use of fertilizers and slow decomposition has led to background concentrations of EDTA in European surface waters in the range 10-50 mg L^{-1} (*13*).

Recently biodegradable chelator [S,S]-stereoisomer of ethylenediamine-disuccinate (EDDS) was proposed for remediation of heavy metal contaminated soils with soil washing (*14, 15*). The use of horizontal permeable reactive barriers to prevent leaching of heavy metals (Figure 2) was proposed in a novel *in situ* soil washing method (*15*).

EDDS was first isolated as a metabolite of the soil actinomycete *Amycolatopsis orientalis* (*16*). It is therefore naturally present in soil, where it is easily decomposed into benign degradation products (*17*). The [S,S]-isomere of EDDS is readily biodegradable using the criteria stipulated by the organization for economic co-operation and development (OECD). The OECD criteria state that 60% of the compound must biodegrade within 28 days. For EDDS the final CO_2 yield exceeded 80% after 20 days, assessed by the modified Sturm test (*17-19*). Mineralization of EDDS in sludge-amended soil was rapid and complete in 28 days. The reported calculated half-life was 2.5 days. Jaworska et al. (*20*) assessed environmental risks for use of EDDS in detergent applications. The toxicity to fish and daphnia was low (EC_{50} > 1000 mg/L). Presently, it is the only commercially available chelator that is naturally present in soil, where it is readily decomposed into benign degradation products.

Theoretically, chelator extraction efficiency depends on the stability constant (logK) of metal-chelator complex formation. Extensive database compiled by Martel et al. (*21*) indicate that logK of EDDS complexes with most heavy metals is comparable to logK of benchmark chelators such is EDTA. However, for Pb (one of the most ubiquitous soil contaminants) reported logK of EDTA (18.0 at 25 °C and ionic strength (μ) = 0.1) is substantially higher than of EDDS complex (12.7 at 20 °C, and μ = 0.1). Nevertheless, the comparison of logK data needs to be considered with caution. Speciation in a metal-chelator system is controlled by the concentration of all metals and chelators, the logK of all complexes and by the kinetic of coordination reactions. In addition, other chelator reactions, adsorption in the soil solid phases, mineral dissolution and chelator degradation, are substantially affected by the chelated metal ion (*22*). For example: Vandevivere et al. (*23*) investigated EDDS for its applicability for the *ex situ* washing extraction of Pb, Zn, and Cu from soil, sewage sludge, and harbor sediments. They reported that extraction efficiency of EDDS for Zn, Cu and also for Pb was equal or superior to those obtained with EDTA and NTA.

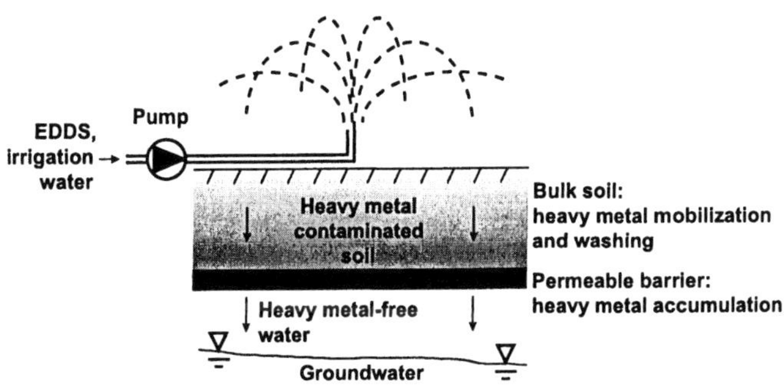

Figure 2. Flowsheet for in situ soil washing using biodegradable chelator and horizontal permeable reactive barrier.

In the proposed method a horizontal permeable reactive barrier is placed below the layer of contaminated soil. It is composed of substrates for enhanced microbial activity and sorbents for heavy metal immobilization. Excess water flows through the barrier as essentially heavy metals-clean soil water. When the targeted level of soil cleansing is reached, after one or several cycles of chelator addition, the barrier material can be excavated and deposited, and the contamination thus removed from the soil (Figure 2).

We borrowed the concept of permeable reactive barrier from groundwater remediation, where the barrier is constructed below ground as a vertical underground wall, filled with reactive materials. The barrier is built by digging a long, narrow trench in the path of the polluted groundwater. Clean groundwater flows out of the other side of the wall. Various reactive materials have been investigated and include zeolite, elemental iron and limestone. Reactive materials in the barrier trap harmful chemicals or change the chemicals into harmless ones. For example, zero-valent iron can be used for reduction of toxic Cr^{6+} to harmless Cr^{3+}, and limestone for Pb precipitation.

Chelator Induced Soil Washing of Pb

Lead contamination of soils is one of the world's most prevalent public health problems, especially through Pb intake in concentrations regarded as non-toxic over extended periods. Once released from industrial and agricultural sources of contamination into the soil matrix, Pb forms strong bonds with the solid soil fractions and is generally retained in the surface soil layer (*24*). The adverse impact of Pb on environmental quality and on human health thus persists for long periods. Pb is linked with failures of human reproduction (*4*). It also causes metabolic disorders and neurophysiological defects in children, and affects the haematological and renal systems.

The feasibility of *in situ* soil washing using EDDS and horizontal permeable reactive barriers was examined by simulating the process in 15 cm diameter soil columns, equipped with trapping devices for leachate collection. Barriers were positioned 20 cm deep in the soil (except for control columns with no barrier). Soil was collected from the 0-30 cm surface layer at an industrial site of a former Pb and Zn smelter in the Mezica Valley in Slovenia. The following soil properties were determined: pH ($CaCl_2$) 6.8, organic matter 5.2%, total N 0.25%, sand 55.4%, coarse silt 12.0%, fine silt 18.9%, clay 13.7%, total available P 81.4 mg kg^{-1}, total available K 38.2 mg kg^{-1}, CO_3^{2-} 125.6 g kg^{-1}, Pb 1400 mg kg^{-1}, Zn 800 mg kg^{-1}. The soil texture was sandy loam.

Functioning of Horizontal Permeable Reactive Barrier

Bio-degradability of the heavy metal-chelator complex and immobilization of released heavy metals is essential for horizontal permeable reactive barrier to function. The functioning of the barrier was examined by using biodegradable chelator EDDS and non-biodegradable EDTA. EDDS and EDTA were used in concentration of 10 mmol kg^{-1} of soil and applied in 200 mL of deionized water with pH 9.35 and 4.25, respectively. Soils were irrigated 2 times a week with 50 mL kg^{-1} soil tap water for 9 weeks after chelators application. Barriers were composed of 1.5 cm wide substrate layer of nutrient enriched peat, acrylamide hydrogel or vermiculite, followed by a sorption layer of apatite/soil mixture (EDDS treatment only). The mass balance of Pb leached is shown in Table 2. In columns with no chelator addition and no barrier installed, the concentrations of Pb in leachates were under the detection limit of the instrument (<0.3 mg L^{-1}). In EDTA treatments, approx. one fourth of initial soil Pb was leached from columns with barrier and more than one third from columns with no barrier. EDTA is non-biodegradable therefore none of the reactive materials in the barriers was effective. A large portion of Pb was also leached in EDDS treatments from columns with no barrier, although leaching was substantially lower than in the comparable EDTA treatment (Table 2).

Table 2. Percent of total initial Pb leached from soil after 10 mmol kg^{-1} chelator addition, during nine weeks of soil irrigation with water. Results are presented as a means of four replicates.

Barrier	*Pb leached (%)*	
	EDDS	*EDTA*
Peat	2.10 ± 1.73	23.96 ± 5.98
Peat + apatite	0.62 ± 0.87	/
Hydrogel	1.57 ± 0.76	25.26 ± 2.07
Hydrogel + apatite	0.35 ± 0.31	/
Vermiculite	2.59 ± 0.59	27.12 ± 1.53
Vermiculite + apatite	0.34 ± 0.40	/
No barrier	21.52 ± 6.13	36.17 ± 1.32

In columns with barriers, EDDS induced leaching of Pb was reduced by at least 10 times. Vermiculite, hydrogel and peat substrate layers in barriers were statistically equally effective in leaching reduction. Leaching was further reduced in treatments with an apatite layer. In columns with the barrier containing vermiculite and apatite, 11.6% of total initial Pb was washed from the soil above the barrier, and almost all leached Pb accumulated in the barrier.

These data indicate *in situ* soil washing as a two-step process (Figure 3). First, the heavy metals are mobilized by application of chelator solution into the soil. Water-soluble, biodegradable heavy metal-chelator complexes are then washed from the layer of contaminated soil into the barrier by irrigation. In the second step, heavy metal-chelator complexes are microbially degraded in the barrier. Released ions of heavy metals chemically react with the sorbents in the barrier to form insoluble products and, consequently, accumulate in the barrier.

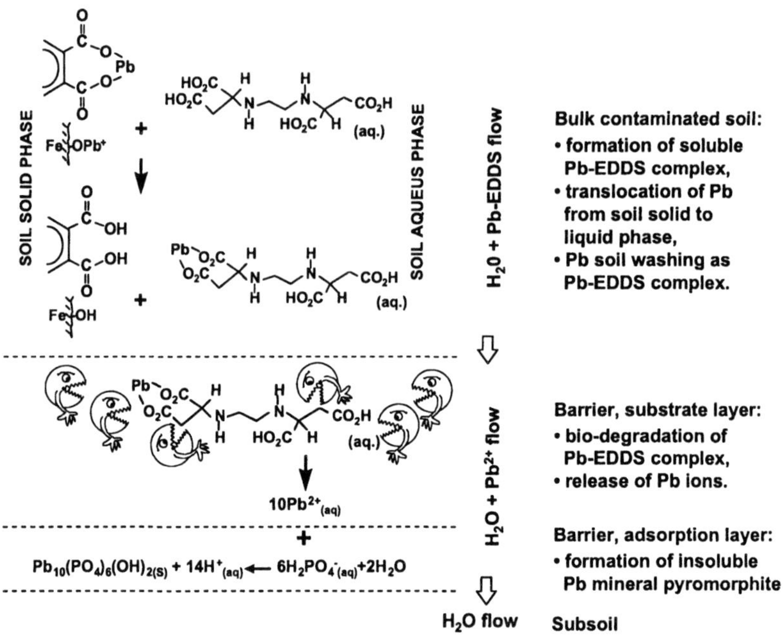

Figure 3. Pb contaminated soil washing with biodegradable chelator EDDS is a two step process. Chelator induced mobilization and washing of Pb from soil is followed by Pb adsorption in the permeable barrier.

Soil Washing with Several Cycles of EDDS Application

The novel technique was evaluated by comparing the efficiency of *in situ* soil washing with several cycles of EDDS addition followed by irrigation, to results obtained by extraction of Pb from the soil slurry in a bench-scale simulation of an *ex situ* soil washing technique.

After 4-cycles of 10 mmol kg^{-1} soil EDDS applications in 200 mL deionized water (pH 9.35) and irrigation with 50 mL kg^{-1} soil of tap water twice a week for 4 weeks after EDDS addition, 24.7% of total initial Pb was washed from the contaminated soil and accumulated into the barrier (Table 3). Most of the Pb, 14.5%, was removed during the first cycle of soil washing. In each subsequent cycle, only approx. 3.5% of Pb was removed. The probable reason for the lower efficiency of later washing cycles was mixing and repacking the soil back into the columns after each washing cycle. Repacking reduced the soil volume and thus also the soil structure and porosity, which were probably vitally important for the efficiency of Pb extraction.

Barriers reduced leaching of Pb in the first cycle of EDDS addition by more than 500-times compared to columns with no barrier. After four cycles of chelator addition, a total of 0.24% of the initial Pb was leached from the columns with barriers. The step increase in Pb leaching from the soil column after the 4th washing cycle (Table 3) suggests that material other than vermiculite, with greater capacity to immobilize Pb, preferably phosphorous compounds should be used in the barrier (*25*).

Bench-scale soil washing tests, simulation of an *ex situ* soil washing technique (Figure 4), were conducted in 50 mL plastic tubes filled with soil slurry consisting of 15 g of dry contaminated soil and 15 mL of chelator solution. The soil slurry was mixed vigorously and the soil-washing extractans obtained after centrifugation.

Table 3. Percent of total initial Pb removed from the soil and leached from the soil columns with horizontal permeable barriers in each of consecutive four cycles of *in situ* soil washing with 10 mmol kg^{-1} EDDS. Results are presented as a means of four replicates.

Soil Treatment Cycles	*Pb Removed (%)*	*Pb Leached (%)*
1	14.5 ± 2.7	0.01 ± 0.00
2	3.6 ± 3.3	0.01 ± 0.00
3	3.5 ± 3.0	0.02 ± 0.01
4	3.1 ± 2.4	0.08 ± 0.06
Total:	24.7 ± 2.4	0.24 ± 0.03

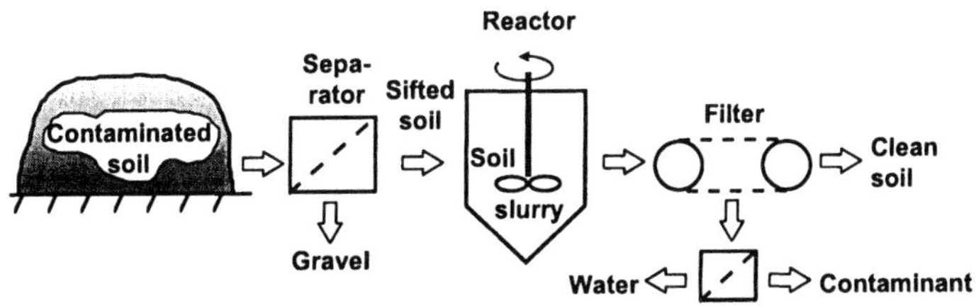

Figure 4. Flowsheet for ex situ soil washing in reactor.

The comparison of *in situ* soil washing with *ex situ* soil extraction of soil slurry showed that for the same total amount of chaletor used, the latter method was more efficient. Treatment of soil slurry with 40 mmol kg^{-1} EDDS (the total concentration of chelator used in 4-cycles of *in situ* soil washing) removed 51% of total initial Pb after 48 hour extraction (Figure 5). However, when 10 mmol kg^{-1} EDDS was used (the concentration used in each cycle of *in situ* soil washing) the efficiency of soil slurry extraction decreased with time. After 6 hours of extraction, a maximum 15.5 % of Pb was removed, comparable to the efficiency achieved in the 1st cycle of *in situ* soil washing. This presumably indicated rapid biodegradation of Pb-EDDS complexes in extracted soil, and the binding of released Pb ions back into the solid soil phase. The same process probably occurred during *in situ* soil washing. Higher concentrations of EDDS seemed to prevent biodegradation of the Pb-chelator complex, since no decline in Pb extraction efficiency with time was observed in treatments with 40 mmol kg^{-1} EDDS (Figure 5).

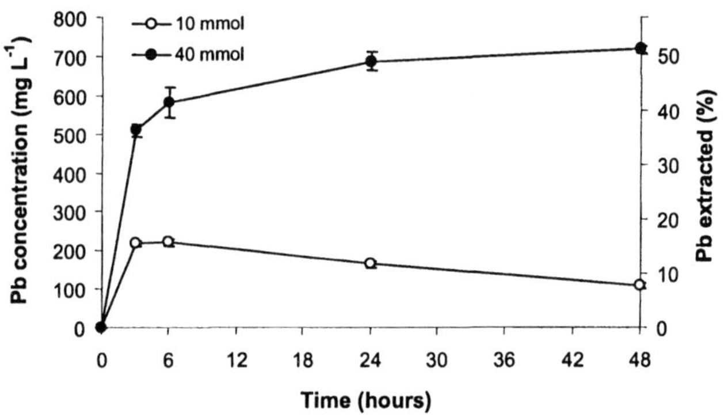

Figure 5. Pb concentration in extract and percentages of total initial Pb extracted from the soil slurry with 10 and 40 mmol kg^{-1} EDDS, after 0, 3, 6, 24 and 48 hours of intensive mixing. Means of four replicates are presented.

Chelator Induced Soil Washing of Cu

Cu enters the soil by deposition from local foundries and smelters, through manuring with contaminated sludges, and from application of fungicides. Cu is an essential element: it forms organic complexes and metalloproteins, especially haemoglobin. With its known antifungal and algicidal properties, elevated levels of Cu in soil adversely affect microbially mediated soil processes (*26*). In EU countries the warning and critical limits of Cu in soil are set at 50 and 140 mg kg^{-1}, respectively (*27*).

Cu contaminated soil (162 mg kg^{-1}) was collected from the 0-20 cm surface layer of a vineyard in the southwestern part of Slovenia, with a four-decade history of soil contamination with Cu containing pesticides. The following soil properties were determined: pH ($CaCl_2$) 7.4, organic matter 4.3%, total N 0.25%, sand 18.9%, coarse silt 16.0%, fine silt 36.4%, clay 28.7%, P 108.7 mg kg^{-1}, K 178.9 mg kg^{-1}, CO_3^{2-} 58.2 g kg^{-1}, Cu 162.6 mg kg^{-1}. The soil texture was silty loam. The feasibility of *in situ* soil washing of Cu by single treatment with 5 mmol kg^{-1} EDDS was tested in 15 cm diameter soil columns, with horizontal permeable barriers installed 18 cm below the soil surface (except for control columns). EDDS was applied in 200 mL of deionized water (pH 9.25). Soils were irrigated 2 times a week with with 50 mL kg^{-1} soil tap water for 6 weeks after chelator application. Barriers were composed of a 3 cm wide layer consisting of nutrient enriched sawdust and vermiculite for microbially enhanced degradation of Cu-EDDS complex, and 3 cm layer of soil enriched with apatite and vermiculite for adsorption of released Cu.

After EDDS addition and irrigation, 36.7 % of total initial Cu was washed from contaminated soil above the permeable barrier (Table 4). The EDDS treatment uniformly reduced the Cu concentration in the soil above the barrier, and barriers effectively reduce the concentration of Cu in leachates in EDDS treatments. Only 0.53 % of the initial total Cu was leached from the columns, 41 times less than with the treatment with no barrier (Table 4).

Table 4. Cu concentration through the soil profile in soil columns with barriers after soil washing with EDDS, and percentages of Cu removed from the soil and leached from the soil columns. Means of four replicates are presented.

Soil Profile	*Cu*
0-6 cm	103.1 ± 6.7 mg kg^{-1}
6-12 cm	96.0 ± 13.3 mg kg^{-1}
12-18 cm	114.4 ± 14.3 mg kg^{-1}
Cu removed	35.4 ± 7.1 %
Cu leached	0.528 ± 0.323 %
Cu leached (control columns)	21.6 ± 2.7 %

Heap Leaching of Pb Contaminated Soil

The main drawbacks of the proposed *in situ* soil washing method are vast consumption of water for soil irrigation after chelator treatment, and the possibility that inadequate functioning of the horizontal permeable barrier (which is difficult to control) could lead to emission of heavy metals through the barrier into the subsoil or groundwater. This issue was addressed by introducing the soil heaps with horizontal permeable barrier constructed in the heap bottom for on site soil washing with EDDS (Figure 6). In heap leaching solution exiting horizontal permeable barrier is prevented to flow into the subsoil but rather it is collected and reused for heap irrigation in a closed loop.

Soil collected from the 0-10 cm surface layer at a site of a former Pb and Zn smelter in the Mezica Valley in Slovenia had the following properties: pH ($CaCl_2$) 7.1, organic matter 10.1%, total N 0.35%, sand 53.0%, coarse silt 26.7%, fine silt 14.8%, clay 5.5%, total available P 246.2 mg kg^{-1}, total available K 22.4 mg kg^{-1}, CO_3^- 153.6 g kg^{-1}, Pb 1440 mg kg^{-1}, Zn 750 mg kg^{-1}. The soil texture was sandy loam.

On site heap leaching was simulated in 15 cm diameter soil columns, equipped with trapping devices for leachate collection, and peristaltic pumps for circling the irrigation solution. Barriers were positioned 20 cm deep in the soil (except for control columns with no barrier) and were composed of 6 cm wide layer of nutrient enriched sawdust, vermiculite and apatite for immobilization of released Pb ions after Pb-EDDS microbial degradation in the barrier. Enhanced microbial activity in the barrier of soil columns was followed as the generated metabolic heat. Temperature probes were inserted into the barrier and into the soil of the control column without barrier (Figure 7).

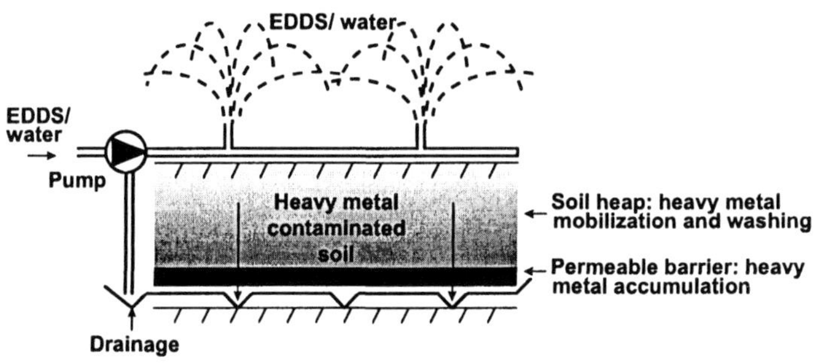

Figure 6. Flowsheet for heap soil washing of heavy metals with biodegradable chelator using horizontal permeable reactive barrier.

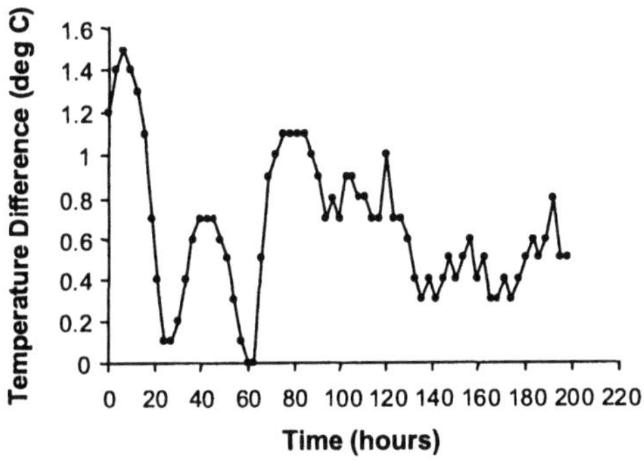

Figure 7. Difference in temperature in soil of control column and temperature in soya meal enriched sawdust of columns with barrier.

Soil was treated with the single addition of 10 mmol kg^{-1} EDDS. The amount of water used for heap irrigation was approx. 135% of the field soil water retention capacity. The dynamics of Pb removal from soil was followed by analyzing solution exiting the soil columns. After 6 days of irrigation no Pb was detected, and we stopped the heap leaching experiment. 17.8% of the initial Pb was removed from the soil and all leached Pb was accumulated in the barrier. The distribution of Pb through the soil profile is shown in Table 5.

The efficiency of soil heap leaching with single addition of EDDS was approximately the same as in simulations of *in situ* soil washing. The amount of water needed was 0.7 L kg^{-1} soil. All emissions of Pb were under control.

Table 5. Pb concentration through the profile in soil columns with horizontal permeable barrier after single cycle of EDDS induced heap leaching. Means of four replicates are presented.

Soil layer	*Pb concentration (mg kg^{-1})*	*Pb washed / accumulated (%)*
0-8 cm	1134.27 ± 58.05	7.13 ± 1.35
9-16 cm	1169.38 ± 16.19	6.32 ± 0.38
17-24 cm	1243.82 ± 77.20	4.59 ± 1.79
barrier	4905.17 ± 261.88	17.86 ± 0.95

Conclusions

Results of our studies in the laboratory scale indicated that the use of biodegradable chelator EDDS and horizontal permeable reactive barriers may lead to environmentally safe *in situ* soil washing and on site heap leaching of Pb and Cu contaminated soils. Up to now both methods were (at least for Pb) less effective than bench-scale *ex situ* soil washing in reactor, where on the other hand, stringent physicals soil treatment completely eliminated soil structure and probably inhibit soil fertility.

Soil contamination is seldom monometalic, and several heavy metals are usually simultaneously present in elevated concentrations in soil. For horizontal permeable reactive barriers to function it is essential that EDDS complexes of all present heavy metals are biodegradable. The literature on biodegradability of EDDS complexes is scarce. Vandevivere et al. (*23*) reported that activated sludge fed with EDDS as the sole C and N source readily biodegraded 1 mM concentration of Ca-, Cr(III)-, Fe(III)-, Pb-, Al-, Cd-, Mg-, Na-, or Zn-EDDS, while Ni-, Co-, Hg- and also Cu-EDDS complexes remained essentially undegraded. This is in contradiction with good barrier performance that we observed during treatment of Cu contaminated soil and which indicated that Cu-EDDS complex is readily biodegradable.

A limitation of EDDS based soil washing is high price of the chelator. The current price for 1 ton of EDDS is approx. 5000 GBP. As EDDS has been substituted for traditional chelators in a number of commercial products, e.g., industrial detergents, the price is expected to decrease. In future the biosynthesis of EDDS by *A. orientalis*, which produces chelator exclusively in the biodegradable S,S-configuration, instead of the current chemical synthesis, could significantly reduce the production costs (*28*).

In situ soil washing and on site heap leaching with EDDS and permeable reactive barriers are new methods and more work with different soils, heavy metals, barrier materials and operational conditions is needed to fully evaluate their feasibility as a soil-friendly remediation technologies.

References

1. Dahlin, S.; Witter, E.; Martensson, A.; Turner, A.; Baath, A. *Soil. Biol. Biochem.* **1997**, *29*, 1405.
2. Giller, K. E.; Witter, E.; McGrath, S. P. *Soil. Biol. Biochem.* **1998**, *30*, 1389.
3. Mulligan,C. N. *Eng. Geol.* **2001**, *60*, 193.
4. Ruby, M. V.; Schoof, R.; Brattin, W.; Goldade, M.; Post, G.; Harnois, M.; Mosby, D. E.; Casteel, S. W.; Berti, W.; Carpenter, M.; Edwards, D.; Cragin, D.; Chappell, W. *Environ. Sci. Technol.* **1999**, *32*, 3697.

5. Leštan, D.; Grčman, H.; Zupan, M.; Bačac, N. *Soil. Sediment. Contam.* **2003,** *12,* 507.
6. Grčman, H.; Velikonja-Bolta, Š.; Vodnik, D.; Kos, B.; Leštan, D. *Plant and Soil.* **2001,** *235* , 105.
7. Cunningham, S.D.; Berti, W.R. In *Phytoremediation of Contaminated Soil and Water;* Terry, N; Banuelos, G. S., Ed.; CRC Press, FL; 1999, pp 359-376.
8. Schwarzenbach, R. C.; Scholz, R. W. *Environ. Sci. Technol.* **1999,** 33, 2305.
9. Hanson, A. T.; Samani, Z.; Dwyer, B.; Jacquez, R. In *Transport and Remediation of Subsurface Contaminants*; Sabatini, D. A.; Knox, R. C., Ed.; ACS Symposium Series No. 491; American Chemical Society, Washington, DC; 1992, pp 108-121.
10. Peters, R. W.; Shem, L. In Environmental Remediation; Vandegrift, G. F.; Reed, D. T.; Tasker, I. R., Ed.; ACS Symposium Series No. 509; American Chemical Society, Washington, DC; 1992, pp 70-84.
11. Dirilgen, N. *Chemosphere* **1998,** *37,* 771.
12. Nörtemann, B. *Appl. Microbiol. Biotechnol.* **1999,** *51,* 751.
13. Kari, F. G.; Hilger, S.; Canonica, S. *Environ. Sci. Technol.* **1995,** *29,* 1008.
14. Tandy, S.; Bossart, K.; Mueller, R.; Ritschel, J.; Hauser, L.; Schulin, R.; Nowack, B. *Environ. Sci. Technol.* **2004,** *38,* 937.
15. Kos, B.; Leštan, D, *Environ. Sci. Technol.* **2003,** *37,* 624.
16. Nishikiori, T.; Okuyama, A.; Naganawa, T.; Takita, T.; Hamida, M.; Takeuchi, T.; Aoyagi, T.; Umezawa, H. *C. J. Antibiot.* **1984,** *37,* 426.
17. Bucheli-Witschel, M.; Egli, T. *FEMS Microbiol. Rev.* **2001,** *25,* 69.
18. Jones, P. W.; Williams, D. R. *Appl. Radiat. Isotopes* **2001,** *54,* 587.
19. Schowanek, D.; Feijtel, T. C. J.; Perkins, C. M.; Hartman, F. A.; Federlen, T. W.; Larson, R. J. *Chemosphere* **1997,** *34,* 2375.
20. Jaworska J. S.; Schowanek, D.; Feijtel, T. C. J. *Chemosphere* **1999,** *38,* 3597.
21. Martell, A. E.; Smith, R. M. *NIST critically selected stability constants of metal complexes*; Version 7.0: NIST, Gaithersburg, MD, 2003
22. Nowack, B. *Environ. Sci. Technol.* **2002,** *36,* 4009.
23. Vandevivere, P; Hammes, F; Verstraete, W; Feijtel, T; Schowanek, D. *J. Environ. Eng.-ASCE* **2001,** *127,* 802.
24. Rieuwerts, J. S.; Thornton, I.; Farago, M. E.; Ashmore, M. R. *Chem. Spec. Bioavailab.* **1998,** *10,* 61.
25. Ruby, M.V.; Davis, A.; Nicholson, A. *Plant Soil* **1994,** *113,* 257.
26. Wright, D. A.; Welbourn, P. *Environmental toxicology*; Cambridge University Press: Cambridge, 2002; pp 301-302.
27. Council Directive 86/278/EEC. EC Official Journal.**1986,** *L181.*
28. Zwicker, N.; Theobald,U.; Zähner,H.; Fiedler, H. P. *J. Ind. Microbiol. Biotechnol.* **1997,** *19,* 280.

Chapter 24

Enhancement of the Electrokinetic Remediation of Soil Contaminated with U(VI) by Chelating Agents

Konstantin I. Popov[1], Val G. Yachmenev[2], and Alexander Barinov[3]

[1]Physical and Colloid Chemistry, Moscow State University of Food Technologies, Volokolamskoye sh., 11, 125080 Moscow, Russia (ki-popov@mtu-net.ru)
[2]SRRC, Agricultural Research Services, U.S. Department of Agriculture, 1100 Robert E. Lee Boulevard, New Orleans, LA 70124
[3]MosNPO"Radon", Moscow, Russia

The various effects of chelating agents activity during electrokinetic remediation of clay soil contaminated with uranium (VI) in a pilot scale electrokinetic test are reported. The combination of the DC electric field (0.7~1.8 V/cm) and an inexpensive and environmentally benign complexant - citric acid, provided an effective decontamination of heavy clay soil from Oak Ridge K-25 site well below the targeted release level of 52 ppm (35 pCi/g). Uranium (VI) content in the remediated soil sample (19.4 kg) was reduced from 566 ppm to 41.3 ppm within 477 hours with estimated total power consumption of 1058 kWh/(ton of soil). In addition to significant advancement of the electrokinetic remediation technique, an implementation of the chelating agents to treated soil also revealed a variety of new phenomena such as significant change of zeta-potential of clay particles, change in mobility of background cations and a dramatic increase of the electroosmotic flow in the soil pores. The electrokinetic remediation tests also confirmed that the knowledge of the precise chemical metal/complexant speciation under various pH conditions was crucial in the prediction of the solubilization efficiency of the chelating agents.

Introduction

The contamination of land with hazardous organic and inorganic compounds, toxic heavy metals, and radionuclides is a serious and growing problem for industrially developed countries *(1-3)*. Typically, pollutants generated by industrial manufacturing facilities may consist either of particulates incorporated within the soil matrix or as compounds adsorbed onto the surface of individual soil particles. Traditional approaches to the clean-up of such contaminated soil included excavation and off-site disposal. However, these approaches rather transfer than minimize the "risk", as far as large volumes of hazardous material are transported to distant disposal facilities, increasing overall remediation and restoration cost. On the whole, the variety of identified soil remediation methods could be classified as follows *(1-20)*:

- Remediation methods based on the removal of the upper layer of soil by scrapers, followed by transportation and storage of the contaminated matter in specially equipped sites.
- Remediation methods focused on the fixation of the pollutant within the contaminated soil matrix (thermal soil treatment, fixation of pollutant with some chemicals, clay, humic acids, *etc*) or re-allocation of the pollutants to the deeper layers of soil (deep tillage, *in-situ* soil washing, *etc*).
- Remediation methods based on a selective extraction of the specific pollutant from contaminated soil (*in-situ* electrokinetic treatment, *ex-situ* reagent enhanced soil washing).

Historically, most remediation methods have been focused on the pollutant *in-situ* fixation or on the complete removal of the contaminated soil. The major shortcomings of these techniques are the risks associated with the possibility of the further spread of the pollutant (in case of insufficiently strong fixation) or the high cost related to removal, transportation and storage of the large volumes of contaminated soil. Lately, more efficient and more complete techniques such as an *ex-situ* soil washing and electrokinetic extraction have become attractive alternatives to the old technologies, mainly because they promise to dramatically minimize the volume of the secondary waste. Both techniques require the use of aggressive chemical reagents, *e.g.* inorganic acids and bases, to enhance the solubilization and extraction of contaminants from the soil matrix. Unfortunately, the use of such harsh chemicals could cause significant loss of agricultural fertility of the treated soil.

It appears that some chelating agents could be very attractive alternatives to these aggressive, soil destroying chemicals by offering an effective solubilization and extraction of a variety of heavy metals (radionuclide) pollutants from contaminated soil under "mild," "non-destructive" conditions.

Because of the wide diversity of particular site soil conditions, specificity of the inorganic/organic contaminants, their concentrations, and the intended use of the decontaminated site, none of the methods mentioned above could be treated as universal techniques. Electrokinetic remediation is a relatively new technology that employs a DC electric field induced by electrodes placed in the polluted soil to extract contaminants. Application of an electric field across a polluted soil mass instigates movement of the charged ionic species (electromigration and electrophoresis) and pore fluid (electroosmosis) through the soil medium. The mobilized contaminants are transported to and collected at corresponding electrodes resulting in an eventual reduction in the contaminant concentration at the site *(11-14, 16)*. Electrokinetic remediation is especially advantageous for treatment of the heavy loam type soils, which have an extremely low hydraulic permeability. Experimental studies indicate that for heavy, clay-rich soils, electrokinetic remediation can be effectively employed for extraction/removal of both toxic metals, radionuclides and hazardous organics, while for sandy soils other methods, such as "pump-and-treat" techniques could be more cost efficient *(10)*. To date, a variety of electrokinetic field tests have been done in USA *(18, 19)*, Netherlands *(17)* and Russia *(20)*, and the results are quite promising.

The tests demonstrate that an application of DC electric field for removal of the organic matter and weakly soil-bound cations yields good results. However, for strongly adsorbed (strongly soil-bounded) pollutants, the effectiveness of electro-reclamation is not very high, primarily because the efficiency of electrokinetic extraction is strongly dependent on the available concentration of the dissolved contaminant in the soil's pore solution. One of the most promising techniques intended to significantly increase the pollutants solubility, could be an application of chelating agents in pore solution. Obviously, similar dissolution enhancement of the strongly bounded contaminants could also be achieved by the use of strong acids (pH~1). But then the soil structure and biota are completely destroyed. It is clear that complexants represent a much more acceptable alternative to the strong acids because they could enhance the solubility of contaminant species in soil at much higher pH. Thus the destruction of the agricultural properties of the soil is eliminated. Indeed, powerful chelating agents like EDTA, DTPA and citric acid are known to form stable and highly soluble complexes with most of the heavy metals and radionuclides. The major disadvantage of these reagents arises from their low selectivity and some side effects. Currently, the majority of the experiments involving the introduction of chelating agents during an electrokinetic remediation of contaminated soil do not yet exceed the level of laboratory or demo tests due to a considerable number of unresolved problems, *e.g.* methods of complexant administration and reuse, the drastic change of pollutant electrokinetic behavior, complexant's interaction with main background cations present in soil, *etc.*

The present report is intended to summarize the results of the several DOE and ISTC projects *(21-23)* related to the advancement of the electrokinetic remediation of soil contaminated with uranium(VI). Most of the experiments have been run in New Orleans with uranium contaminated soil from K-25 Site (Oak Ridge), while numerous complimentary studies have been done in Moscow with similar but non-contaminated sod-podzolic soil as well as with uranium(VI) artificially contaminated soils. Some brief preliminary communications on the present research have been published earlier *(24-26)*, but this is the first comprehensive and detailed report.

K-25 Soil Characterisation and Chelating Agents Screening

Large representative samples of uranium(VI) contaminated soil were taken from the K-311 site adjacent to the Oak Ridge Gaseous Diffusion Plant (K-25). This site embodied an old contamination accident with over 50 years continuous weathering history. Uranium was present as U(VI) with a significant degree of enrichment: $^{235}U/^{238}U$ 30-36% (in Bq/kg). Two relatively large samples of soil with slightly different characteristics have been collected for these studies. The first sample (characterized in Table I) was used for preliminary laboratory uranium static leaching tests. The preparation of the soil samples included equilibration at ambient conditions, removal of large pieces of foreign matter (pebbles, roots, *etc*), mixing uniformly, drying in the oven at 100 °C, milling in a grinder, sieving through 0.5 mm sieve, and again equilibrating at ambient conditions. The calculated moisture content of the prepared/studied samples was *c.a.* 2%. The listed above procedure provided the uniformity of soil samples for comparative leaching tests with various chelating agents. This is done under an assumption that the best ligand selected in these fast preliminary batch leaching tests would be also the favorite in field conditions.

The second sample (19.4 kg) was prepared in a bit different manner (without drying, milling and sieving procedures) and used in a pilot scale electrokinetic decontamination test. It fitted well the field conditions. Therefore, the average uranium and moisture content in both samples was somewhat different, although the basic soil characteristics and properties remained almost the same. The clayey soil used in both experimental studies had very low hydraulic permeability. The coefficient of permeability was found to be within $4.3 \bullet 10^{-6}$ cm/sec. The data on the sequential leaching tests (presented in Table II) indicated that most of the uranium was associated with Fe/Mn oxides/hydroxides and oxidizable organic matter, while the water-soluble fraction was negligible. Thus, less than 10% of the total uranium content in the sample could be leached by the conventional cation exchange technique.

Table I. K-25 Soil Characteristics (Sample 1).

Soil Parameter	*Min*	*Max*	*Average*	*S.D.*
pH	5.3	5.5	5.4	0.1
Density, (air dry, g/cm^3)	1.11	1.29	1.20	0.03
Porosity, %	31	55	42	7
U, mg/kg	691	939	770	120
Ca, mg/kg	344	734	550	100
Mg, mg/kg	92	233	165	60
Fe, mg/kg	43000	48300	44700	1600
Mn, mg/kg	185	197	190	4
Cu, mg/kg	12	22	17	6

NOTE: soil color: 7.5YR4/6; texture: Sicl (Silt Clay); structure: 2fsbk; depth 0-30; The concentrations of **U**, **Ca**, **Mg**, **Fe**, **Mn** and **Cu** have been determined in accordance with U.S. EPA Method 200.8 in ISOTRON's Analytical Laboratory.

Therefore, straightforward application of the electric field alone, without solubilization enhancing agents, will remove only a small, easily available fraction of total uranium contamination in the sample. Indeed, the preliminary laboratory electrokinetic tests that were carried out with small electrokinetic cells without any complexing agents achieved a very modest, up to ~15%, level of uranium extraction. It became obvious, that only a combination of the conventional electrokinetic extraction technique with the solubility enhancing complexants would make possible to achieve the mandated level of decontamination.

Normally it is assumed, that those complexing agents that form the most thermodynamically stable complexes in the aqueous phase (higher $\log\beta_{ML}$, $\log\beta_{MHL}$, *etc.*) would be the most efficient as

Table II. Data on the uranium sequential extraction from the K-25 soil

Cation Fraction	U, %	Ca, %	Mg, %	Fe, %	Mn, %
Water Soluble	0.6	4.1	8.7	0.8	0.1
Ion Exchangeable	7.9	60.6	43.9	0	1.3
Bound to Fe/Mn oxides/hydroxides	76.6	24.1	23.1	4.2	80.7
Bound to organic matter	8.9	5.5	13.7	3.6	5.2
Associated with soil minerals	6.4	5.7	10.6	91.4	12.7

NOTE: sequential extraction is done according to *(27)*

solubilizers. Unfortunately, equilibrium constants of the most powerful chelating agents are either completely (partially) missing for uranium(VI) complexes, or present conflicting data *(28)*. Nevertheless, we were able to select about twenty perspective complexing/leaching reagents and their compositions with the best known affinity towards uranium, and to test them in the variety of batch leaching experiments, in a wide pH range. The results for some individual reagents are presented in Table III. In each leaching test, 3 gm of soil was mixed with 25 ml of 2% (mass) aqueous leaching solution. All tests were performed in two replicates. With an exception of DTPMP, all complexants had reagent grade purity. The soil/solubilizer mixture was shaken for 10 minutes twice a day, with a total leaching time of 3 days. The soil/solubilizer suspension was then centrifuged, the pH of equilibrium solution was measured (Table III), and the supernatant analyzed for the U(VI) concentration by ICP spectrophotometer.

The data in Table III show very different solubilization capability of complexing agents. It is reasonable to note, that solubilization efficiency of chelating agents varies greatly and does not always correlate with the stability constant values.

Table III. Static leaching tests of K-25 Soil with 2% aqueous solutions of various complexing agents

Leaching agent	*logβ_{MHL}*	*pH*	*U (VI) extraction, %*	*Comments*
De-ionized (DI)Water		7.2	~ 0.1	Colorless transparent liquid phase
Citric acid	9.68 *(29)*	3.2	82	Transparent yellow liquid
		6.5	96	Stable colloid solution
		8.0	100	
EDTA	17.16 *(30)*	3.6	28	Colorless liquid phase
		7.1	11	
		9.5	15	Stable colloid solution
DTPA	19.2 *(30)*	3.9	28	Pale yellow transparent liquid
		5.9	9	
		8.6	3	
HEDP	7.99 *(28)*	4.0	76	Brown liquid phase
		7.2	76	
		8.3	93	Stable colloid solution
NTMP		4.2	3	Pale yellow liquid phase
		7.1	58	
		8.2	52	Stable colloid solution
DTPMP	25.89 *(28)*	4.1	27	Slightly colloidal solution with pale yellow liquid phase
		7.1	59	
		9.1	54	

NOTE: Stability constants at 20-25°C and I=0.1-1.0 mol/L presented by different research groups; for NTMP no data is found; for HEDP logβ refers to UO_2+H+H_2L equilibrium; 2% solutions have been taken for uniformity, as far as along with individual chemical compounds some compositions with maximal 2% mass content (not indicated in Table) have been also tested within the frames of DOE Project *(21)*

Meanwhile, an account of the ligand's protonation constants indicates, that the real complexing abilities of citric acid, EDTA and DTPA with respect to uranium(VI) appear to be nearly identical within a wide pH range (Table IV, -log[UO_2^{2+}] equilibrium values). In the table, the lowest free uranium(VI) concentration corresponds to a higher pollutant masking and, therefore, indicates a better leaching ability. Thus, the citric acid demonstrates the best masking effect at pH 3-4, although it should be noted that by definition, the correlation between the thermodynamic model and observed rate of uranium removal is always a very approximate one.

An observed increase of the uranium leaching performance by citric acid even at higher pH is also in a good agreement with thermodynamic predictions. On the other hand, a drastic decrease of uranium extraction coinciding with the observed increase in stability of complexes for EDTA and DTPA while pH is shifted from 4 to 7 could be explained by the low solubility of the formed complexes, Table III. Indeed, it is reported *(30)*, that for both, EDTA and DTPA complexes, the precipitation occurs even at 0.0005 mol/l uranium content.

Table IV. Static leaching of K-25 soil with selected chelating agents

Chelating Agent, H_nL	*[L], mol/l*	*$log\beta_{MHL}$*	*pH*	*$-log[UO_2^{2+}]$*	*U(VI) Removal, %*
EDTA	0.07	17.16	3.6	6.5	28
DTPA	0.05	19.2	3.9	5.1	28
Citric acid	0.10	9.68	3.2	6.9	82
			6.5	9.7	96

NOTE: "free" $[UO_2^{2+}]$ values were calculated using the program "SPECIES", operating ligand's p*K* values and uranium(VI) stability constants (without account of uranium hydrolysis and solubility) *(28)*. The total uranium(VI) concentration in all model systems constitutes 0.0039 mol/l.

It is noteworthy that a certain discord between the predicted efficiency of uranium extraction by specific complexing agent based on the available thermodynamic data and the real contaminant removal demonstrated by these leaching tests for pH 8-9 still remains unexplained. Most probably this is due to uncertainty and incompleteness of available equilibrium data. Based on the static leaching tests the two most promising complexing agents, citric acid and DTPMP, were selected for preliminary electrokinetic remediation tests. It is essential that, most of the metal/ligand species formed by uranium and selected complexing agents are negatively charged. They could therefore be transported throughout soil to collection zone via the electromigration mechanism.

Data in Table III clearly demonstrate that efficiency of the uranium solubilization by specific ligand is strongly pH dependent.

This mainly is due to the pH dependent competition between the pollutant's metal hydrolysis reactions at high pH and ligand protonation reactions at low pH. Besides, the solubility of a chelated uranium species is itself influenced by pH. Therefore, the accurate monitoring (control) of the soil pH during the whole electrokinetic remediation process is of paramount importance.

The observed intensive yellow and brown colors of the liquid phase in a variety of leaching tests indicated an intensive dissolution of background cations of Fe(III)) and, possibly, of some organic matter. Evidently, these side reactions also have to be taken into consideration in the course of electrokinetic extraction.

Data in Table III also indicate that the most efficient complexing agents were prone to the formation of quite stable colloid solutions. This propensity to form colloids was also very pH dependent. It appears that an introduction of the chelating agents was able not only to increase the pollutant (uranium) solubility, but also somewhat affected the electric double layer (Zeta potential) on the surface of soil particles. This possible change of Zeta potential of soil particles caused by introduction of complexants should strongly affect both the electromigration and electroosmosis and, as result, strongly alter the electrokinetic remediation process itself. The preliminary laboratory tests confirmed these assumptions.

Preliminary Electrokinetic Laboratory Tests

The small scale laboratory electrokinetic extraction tests were conducted using the ISOTRON vertical columns equipped with electrode assembles for applying an electric field across the soil sample and corresponding chambers for electrolyte solution. The air-dry soil samples were saturated with leaching solutions and compacted into the cells. Both cathode and anode chambers have been filled with the investigated chelating agent solution. They had special outlets for collecting the effluent caused by electroosmosis. In all of these electrokinetic tests, a constant 20 volts D.C. potential was applied across the soil sample. The resulting electric current was recorded permanently during the test. After the completion of the electrokinetic treatment the triplicate sets of the analytical samples from the top, the bottom and the middle parts of the columns were taken, weighed, and digested in a MDS-200 Microwave Sample Preparation System in accordance with US EPA Method SW-3051 (uranium analysis in soil/sediments). Concurrently, the soil samples were analyzed for moisture content by drying them in an oven at 105

°C to a constant weight. Quantitative analysis of U(VI) content in the samples was performed using the Perkin Elmer Emission Spectrometer "Plasma 400", pH of aqueous solutions was measured by Beckman Φ44 pH meter. The results of the small-scale electrokinetic extraction tests are presented in Table V. Unexpectedly, the data revealed an uncharacteristically high electroosmotic flow of pore liquid towards the cathode. This dramatic enhancement of the electroosmotic flow in soil was especially pronounced for the soil sample treated with phosphonate. However, even for soil treated with citric acid, it was approximately an order of magnitude higher than was normally observed *(25)*. These small-scale electrokinetic tests also demonstrated that pollutant species (uranium) were driven by an electric field in opposite directions as far as uranium was detected in both cathodic and anodic chambers. This was in conflict with speciation models *(29)*, that predicted all the uranium (VI) species to be negatively charged at the soil pH of ~ 8-9. Therefore, irrespective of the high pollutant mobilization (100% for citric acid; 54% for DTPMP), the overall elctrokinetic remediation results appear to be rather poor.

Table V. Results of electrokinetic soil treatment in ISOTRON vertical cells

Experimental parameters	*8% Citric acid*	*8% DTPMP*
pH of complexant solution	8.8	7.5
Soil mass, g	187.4	200.0
Initial **U** content, mg	84.7	136.0
Treatment time, hours	43	~20
Final **U** content, mg:		The drying of soil near anode has been observed along with a drastic decrease of electric current. The test was abandoned.
- Cathode effluent	33.1	
- Cathode chamber	0.2	
- Soil	43.3	
- Anode chamber	11.0	
Total **U** found, mg	87.6	
U extraction, %	49	

NOTE: electroosmotic flow in all cases has direction from anode to cathode

This fact clearly indicates, that some supplementary study of the detailed mechanisms of the chelating agent/clay soil interactions were required.

Study of the Effects of the Chelating Agents on the Electroosmotic Flow and Soil Zeta-potential

To clarify the observed anomalies such as abnormally high electroosmotic flow and confusing transport behavior of the complexed uranium species observed in ISOTRONIC vertical cells, an additional set of experiments with K-25 soil and similar sod-podzolic soil (sampled in a Moscow country site) were carried out with a similar variety of the complexing agents (EDTA, HEDP, citric acid and *etc*). These new studies again confirmed a very significant increase (up to ~10 times) of the electroosmotic flow in all soil samples treated with variety of complexing agents *(31)*. These data strongly indicate that this effect appeared to be pH dependent. For example, a relative increase of electroosmotic flow in K-25 soil treated with citric acid at pH 6.6 was 9.4 higher relative to the same sample treated with DI water. For soil sample treated with citric acid at pH 2.2 the flow was only 2.3 times more intensive *(31)*. This study also demonstrated that an observed sharp increase of the electroosmotic flow coincided with a corresponding increase of zeta-potential and background cation mobilization. Figure 1 presents the results of the macro-electrophoreses measurements of Zeta-potential for HEDP solutions in sod-podzolic soil along with background cation content in a liquid phase. Similar results have been obtained by micro-electrophoreses measurements of Zeta-potential *(32)*.

Data show that the major increase of Zeta-potential takes place at quite low concentrations of complexing agents, and it is likely caused by the adsorption of complexants onto the surface of clay particles. The elevated concentrations of the chelating agents lead further to the dissolution of the amorphous hydroxo/oxides on the surface of clay particles and probably to a following re-adsorption of either ligand itself or its complexes with iron(III), aluminum(III), and *etc* back onto the surface of clay minerals. All these tentative phenomena will promote an increase of the electroosmotic flow that was observed in the earlier tests. Only at the relatively high concentrations of the complexing agents the Zeta-potential starts to decrease, probably because of the electric double layer compression at the surface of clay particles caused by high ionic strength of pore solution.

Now we can try to outline the tentative enhancement mechanism that takes place when the conventional electrokinetic remediation of soil is supplemented by an introduction of complexing agent. Electroreclamation of soil is mainly based on two electrokinetic phenomena: electroosmotic flow of the liquid phase in soil channels,

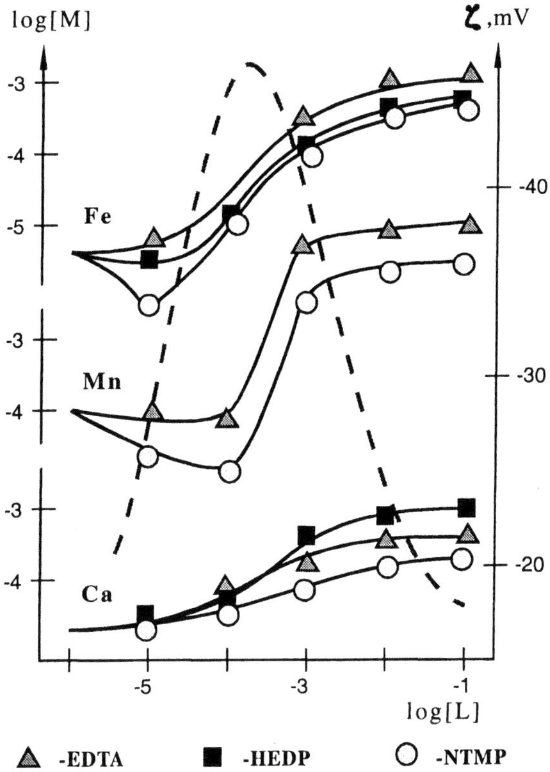

Fig.1. Effects of complexant concentration on Zeta-potential (dashed curve; sod-podzolic; HEDP; soil/water - 1:5; pH 7) and background cations content in a liquid phase (solid curves; pH 7).

and an electromigration of ions in pore solution. The surface of soil particles carries normally a negative charge within a relatively wide pH range *(16, 33)*. In the absence of complexants, the radionuclides and heavy metal ions are positively charged and are driven by the electric field gradient toward the cathode via both mechanisms, see Figure 2 (A).

Even when both transport mechanisms work together, the overall efficiency of the electrokinetic decontamination process is generally found to be quite low, mainly because of the low concentration of contaminant in pore liquid (caused by its poor solubility) and, to a certain degree, by low natural value of zeta-potential of soil particles *(16)*. On the whole, the electroosmotic transport plays secondary role, being several times lower than the electromigration velocity of the charged ionic species throughout the sample.

Introduction of strong complexing agents such as citric acid, tiron, EDTA, phosphonates into the soil matrix strongly increases solubility of the pollutant, thus significantly improving its extraction within the frames of electrokinetic technique. But in addition to the increased solubility of the targeted pollutant, the complexant will drastically change the pollutant ionic speciation, Figure 2 (B). Most of the chelated metal complexes at pH 4-10 become negatively charged and, therefore, will move towards the anode. It is noteworthy that the ionic mobility of the $[UO_2edta]^{2-}$ anion is expected to be slightly less than that of UO_2^{2+} (although the electric charge values are the same) because the chelation of the UO_2^{2+} will increase the ion size and thus impede its electromigration velocity *(34)*. For citric acid a change from UO_2^{2+} to $[UO_2cit]^-$ anion results also in direct electromigration mobility decrease due to a lower electric charge of chelated species. It should be noted, the electroosmotic flow in the system will also increase due to the increase of Zeta-potential caused by complexant interaction with soil particles (indicated by an arrow size on Figure 2, B). Therefore, the electromigration of metal complexes would take place in a counter flow with the electroosmotic stream. Obviously, the overall decontamination efficiency would be strongly dependent on the apparent velocities of electromigration and electroosmotic flow. In some peculiar cases when the apparent velocities of electromigration and electroosmotic flow could become equal, both streams will terminate each other and the pollutant will remain in the soil (this phenomenon was observed in one of our electrokinetic experiments with citric acid).

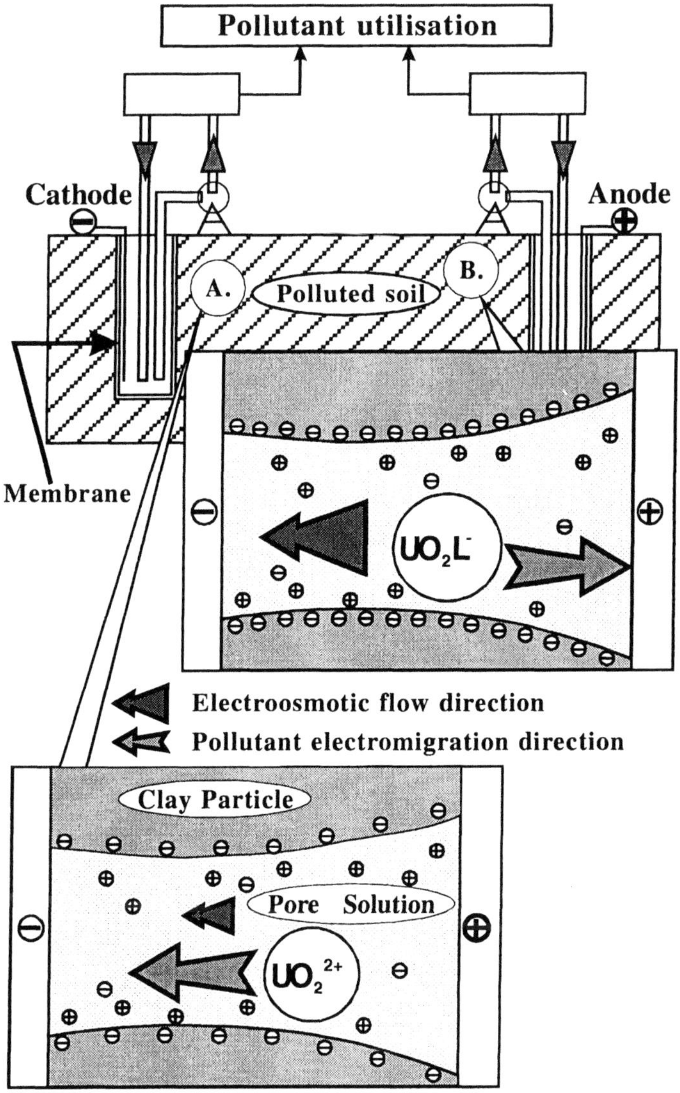

Fig. 2. Tentative scheme of the uranium(VI) electrokinetic extraction/removal from contaminated soil: without complexing agent (A), and in presence of citric acid (B).

Taking into consideration the discussed above phenomena, we have selected a solution of citric acid at pH 3-4 for the larger, pilot scale, electrokinetic remediation test of K-25 soil with the foremost objective to improve solubility of the contaminant (uranium) and also to achieve the optimum balance between counteracting transport mechanisms of the electroosmosis and electromigration. Indeed, the lower pH suppresses the elctroosmotic flow, while the chelated uranium species dominating in solution at this pH ($[UO_2cit]^-$ and $[(UO_2cit)_2]^{2-}$) *(29)* are intended to provide electromigration towards anode and to insure the most efficient decontamination rate. It is essential to note, that even small pH shift below 2.0 leads to domination of $[UO_2Hcit]^o$ and UO_2^{2+} species and therefore to an opposite side contaminant movement via electroosmosis and electromigration. Thus the pH control was of key importance in the pilot scale test.

The listed above scheme is in a general agreement with earlier attempts to combine chelating agent and electric field for pollutant removal from freshly contaminated model samples in a series of laboratory tests *(35-37)*. It was mentioned that ethylenediamine increases the electroosmotic flow, but no reasonable explanation was given to this phenomenon *(35)*. Experiments on EDTA interaction with sandy soil revealed no specific interaction of complexing agent with soil matrix *(36)*, which is not surprising for samples with a very low content of clay and ferrous oxides. A research on pure Georgia kaolinite with approximately 4.3 % of iron oxides demonstrated that EDTA is able to reverse an electroosmotic flow from "cathode to anode" (observed by authors at pH 5) to an opposite one *(37)*. All these observations can get clear when ferrous oxides dissolution by chelating agent is considered.

Pilot Scale Electrokinetic Remediation Test of K-25 Soil

The pilot scale, laboratory experiment was designed to verify the extent to which the enhancement of the dissolution/transport mechanisms offered by introduction of the chelating agent (citric acid) could advance the electrokinetic remediation of the large sample of soil heavily contaminated with uranium. The soil used in this test was taken from the area (K-311) adjacent to the Northwest wall of the Oak Ridge Gaseous Diffusion Plant (K-25) where the level of contamination was the highest. Preliminary analyses of the soil indicated that the soil consisted of two very distinctive types: top soil (black, low density, high organics content, high concentration of uranium (>1300 mg/kg) and the heavy clay soil with medium contamination of ~300 mg/kg. Accordingly, the pilot scale box test was designed to take advantage of this occurrence and to arrange the filling/compacting of the soil in a test box in such a manner that it will mimic the real case contamination scenario (rain water with dissolved uranium continuously soak the soil, resulting in gradual contamination of the soil from the top). Correspondently, the soil material was compacted into a test box with four layers of soils with increasing level of uranium contamination from the bottom to the top of the box, with the upper layer constituting the highly contaminated black soil. The soil material compacted in the test box was premixed with 200 ml of 0.15 M citric acid solution per 1.0 kg of soil. This operation was intended to imitate the delivery of complexing agent with irrigative water solution under the field conditions. Indeed, the diffusion (penetration) of a leaching composition from the top layers to the deeper ones can save electric energy consumption as an alternative to the delivery of chelating agent from cathode via electromigration mechanism.

A plastic wrap was used to cover the surface of soil in the box to maintain the soil's moisture content. The general arrangement of the pilot scale test is presented in Figure 3. Two separate circulation loops provided a continuous circulation of the anode and cathode electrolyte solutions (~0.7 liter) throughout anode and cathode compartments, which were separated from soil by porous ceramic plates. An iridium coated titanium mesh was used as both anode and cathode electrodes. Several critical factors were taken into consideration to determine the optimum voltage and current through the soil to insure maximum efficiency of the electrokinetic decontamination: temperature of the soil in a box, total energy consumption, and the rate of uranium extraction. The actual voltage and resulting electric current have been recorded during the test. The experiment was terminated at the time when the uranium extraction rate was observed to decline to nearly zero level. The results of pilot scale electrokinetic decontamination of clay soil from K-25 Site are presented in Tables VI and VII and Figures 4 and 5.

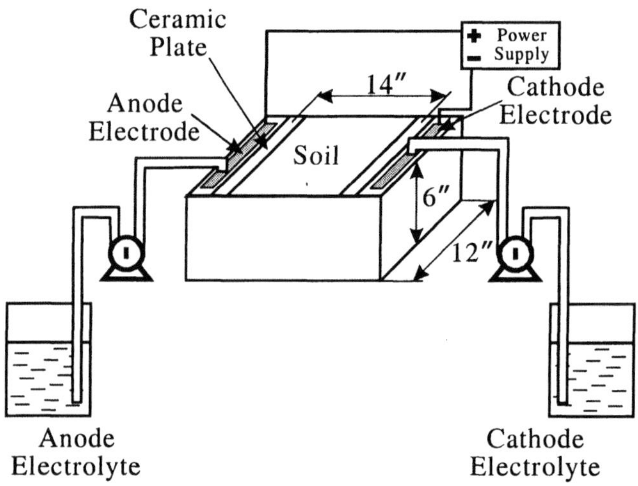

Figure 3. The general arrangement of the pilot scale test with K-25 soil

The concentration of citric acid in the pilot scale test was chosen to maintain the acidity of soil's pore solution at pH 3 with four closely bound objectives: to reduce simultaneously electroosmotic flow; to enhance solubilization of uranium, to promote ionic species electromigration to anode collection zone; and to buffer cathode chamber. The Uranium-Citrate speciation diagram indicates that most of the solubilized uranium (VI) at this pH is present as $[UO_2cit]^-$ (~60%), $[(UO_2)_2(cit)_2]^{2-}$ (~15%) and $[UO_2Hcit]^o$ (~25%), thus insuring the overwhelming transport and collection of the uranium at the anode compartment. It is reasonable to note, that the shift to pH lower than 2 leads to the accumulation of uranium(VI) as positively charged species $[UO_2H_2cit]^{2+}$ (~25%) and UO_2^{2+} (10%), which will evidently move to the cathode. And at slightly more basic conditions of pH ~ 4, almost all uranium will be present as negatively charges species, but simultaneously, the electroosmotic flow (which will counteract an electromigration) will also increase, thus causing an overall decrease of transport efficiency. Therefore an exact pH control within pH 2.9-3.1 during the electrokinetic remediation becomes a critically important requirement.

As far as uranium is preferably bound to the amorphous hydroxy/oxide films, the tentative scheme of its solubilization could be represented by the following scheme:

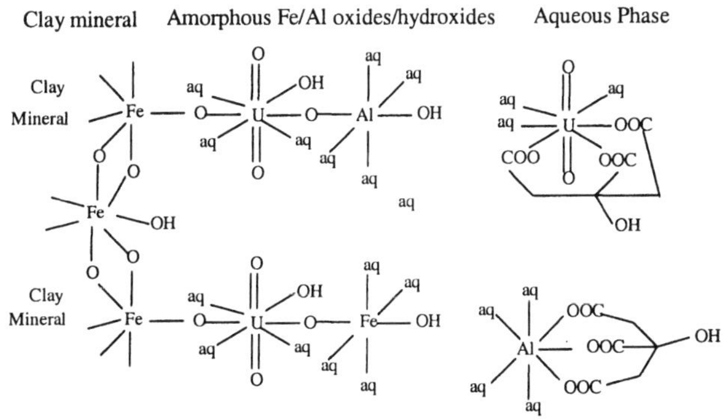

Table VI. The basic experimental parameters of the pilot scale test of electrokinetic decontamination of K-25 soil

Parameter	*Numerical value*
Dry soil mass, kg	19.4
Natural soil moisture content, %	7.7
Initial **U** content (average), mg/kg	566
Final **U** content in soil near anode, mg/kg	56.6
Final **U** content in soil near cathode, mg/kg	22.8
Final **U** content (average) in soil, mg/kg	41.3
Initial soil pH	5.5
Final soil pH near anode	2.9
Final soil pH near cathode	3.0
Treatment time, hours	477
Voltage, V/cm	0.7-1.8
Current density, amps/ft^2	2.0-3.0
Energy consumption, kWh/(ton of soil)	1058
Citric acid conc. (anode – cathode), mol/l	0.1-1.15

Table VII. Mass balance of the pilot scale electrokinetic decontamination test

Uranium	*%*
Initial Uranium Content	100
Uranium Extracted at Anode	85.2
Uranium Extracted at Cathode	2.5
Uranium in Soil after Treatment	7.3
Total Uranium found	95

NOTE: Approximately 2 liters of anolyte solution with high uranium content was lost due to a circulation tube breakage

This proposed solubilization mechanism for uranium is found to be in a good agreement with recently reported observations of solubilization of the mixed Al/U hydroxo-complexes with tentative composition: $[(H_2O)_3UO_2(OH)_2Al(OH)(H_2O)_3]^{2+}$ and $[(H_2O)_4UO_2OAl(OH)(H_2O)_4]^{2+}$ *(38)*.

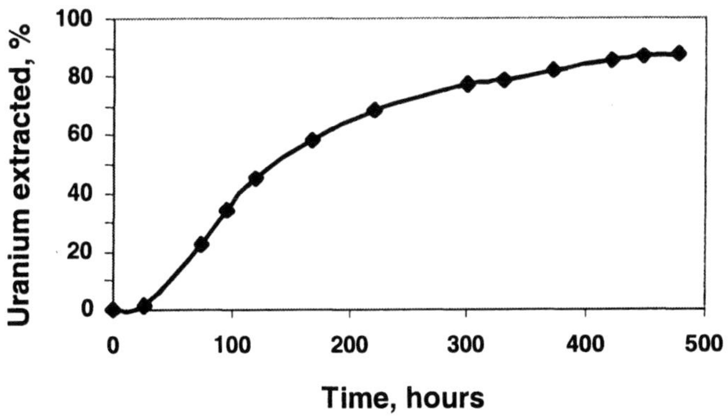

Figure 4. Uranium extraction vs. time

After the experiment was completed, nine identical core samples were taken and analyzed for post-treatment uranium content: six in ISOTRON's analytical laboratory and three at ORNL (Oak Ridge) for an independent verification. The data in Table VII indicate, that the main mechanism of uranium transportation was the electromigration of the uranium-citrate anionic species: 85% of uranium had been collected in the anode chamber. Only 2.5% of contaminant has been found in a cathode chamber, which probably was delivered by electroosmotic flow. Figure 4 and especially Figure 5 indicate that electrokinetic decontamination of K-25 soil has three distinctive phases. During the first phase (0-25 hours), the rate of uranium extraction was quite low, possibly due to incomplete chemical solubilization of uranium by citric acid. Most of the uranium (80%) was removed during the second phase (25-300 hours) with a daily rate ranging from 12% at the beginning to 6% at the end. The third phase (300-477 hours) was characterized by merely 1% per day uranium extraction with a total amount around 5%.

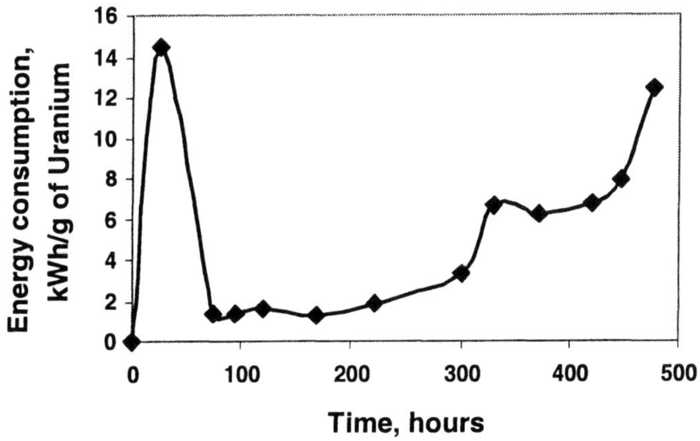

Figure 5. Energy consumption for extraction of gram of uranium vs. time

After a total of 477 hours of continuous electrokinetic extraction, the concentration of uranium in contaminated soil was successfully reduced from 566 mg/kg (average) to 57 mg/kg near anode and 23mg/kg near cathode.

Discussion and Conclusions

The pilot-scale electrokinetic remediation test demonstrated that introduction of the complexing agent (citric acid) in the course of electrokinetic extraction of uranium from highly contaminated K-25 soil greatly enhances the process. At the same time, a considerable caution is needed for a proper complexant selection and in the maintenance of the optimum treatment conditions for soils polluted with heavy metals and radionuclides. The precise metal-ligand speciation of the charged/uncharged species that can be calculated from the available thermodynamic data appears to be a very helpful tool in prediction of contaminant behavior in the soils. In addition to the significant enhancement of the electrokinetic remediation, an introduction of the chelating agents to the treated soil revealed the previously unknown phenomena of the sharp increase of electroosmotic flow in soil pores. In some electrokinetic remediation scenarios, this newly discovered phenomena could be counterproductive, mainly by counteracting the electromigration transport of ionic pollutant. On the other hand, in the case of the electrokinetic remediation of soils polluted by toxic, non-ionic, organic contaminants such as TCE, phenol, *etc*., this new phenomenon could be very beneficial *(39)*.

References

1. *Metal Speciation and Contamination of Soil*; Allen, H. E.; Huang, C. P.; Bailey, G. W.; Bowers, A. R., Eds.; Lewis Publishers: Boca Raton *e.a.*, 1995.
2. Removal of Uranium from Uranium-Contaminated Soils. PhaseI. Bench-Scale Testing. 1993. ORNL-6762. Report, DOE Contract DEAC0584OR 21400.
3. Bulatov, V.I. *Radioactive Russia*. TSERIS, Novosibirsk, 1996, (in Russian).
4. USEPA, Engineering Bulletin. *In situ soil flushing*. EPA/540/2-91/021, 1991.
5. Khodadous, A. P.; Sorial, G. A.; Wilson, G. I.; Suidan, M. T.; Griffiths, R. A. *J. Environ Engineering* **1999**, *125*, 1033.
6. Nikolaidis, N. P.; Hellerich, L. A.; Lackovic, J. A. *Environ. Sci. Technol.* **1999**, *33*, 2910.
7. Stolker, G. N. Extended Abstr. I & EC Spec. Symp. Amer. Chem. Soc. Atlanta, **1994**, 484.
8. Arevalo, E. F.; Thoming, J.; Calmano, W. *Environ. Technol.* **2002**, *23*, 571.
9. Neumaier, H. *Abwasser* **1992**, *39*, 1511.
10. Kelsh, D. J.; Parsons, M. W. *J. Hazardous Materials* **1997**, *55*, 109.
11. Alshawabkeh, A. N.; McGrath, C. J. *Environ. Sci. Pollut. Contr. Ser.* **2000**, *23*, 155.
12. Acar, Y. B.; Alshawabkeh, A. *Environ. Sci. Technol.* **1993** *27*, 2638.
13. Zelina, J. P.; Rusling, J. F. "Electrochemical Remediation of Soils" in R. A. Mayers, Ed., *Encyclopedia of Environmental Analysis and Remediation*, Vol. 3, Wiley: 1998, p.p.1567-1483.
14. Popov, K. I.; Kruglov, S. V.; Tarasova, N. P. ; Komarova, N. I. *Mendeleev Chem. J.* **1997**, *40*, 282; Engl. transl. from *Zh. Ross. Khim. Ob-va D. I. Mendeleeva* **1997**,*40*, 179.
15. Wilson, J. T.; Wilson, B. H. *Appl. Environ. Microbiol.* **1985**, *49*, 242; Strand, S. E.; Shippert, L. *Appl. Environ. Microbiol.* **1986**, *52*, 203.
16. Shapiro, A. P.; Probstein, R. F. *Environ. Sci. Technol.* **1993**, *27*, 283.

17. Lageman, R. *Environ. Sci. Technol.* **1993**, *27*, 2648.
18. Ho, S. V.; Athmer, C.; Sheridan, P. W.; Huges, B. M.; Orth, R.; McKenzie, D.; Brodsky, P. H.; Shapiro, A. M.; Sivavec, T. M.; Salvo, J.; Schultz, D.; Landis, R.; Griffith, R.; Shoemaker, S. *Environ. Sci. Technol.* **1999**, *33*, 1092.
19. Ho, S. V.; Athmer, C.; Sheridan, P. W.; Huges, B. M.; Orth, R.; McKenzie, D.; Brodsky, P. H.; Shapiro, A. M.; Thornton, R.; Salvo, J.; Schultz, D.; Landis, R.; Griffith, R.; Shoemaker, S. *Environ. Sci. Technol.* **1999**, *33*, 1086.
20. Sobolev, I.; Prozorov, L.B.; Martyanov, V. V. Proc. of V International Conf. "Management of low-level Waste and Remediation of Contaminated Sites and Facilities", Berlin, **1995**, V.2, P.1489.
21. DOE Project No 70K-EKQ-72C, Electrokinetic extraction of Uranium from K-25 Soil. 1994-1996, New Orleans, US.
22. ISTC Project 16**,** Development of Electrokinetic and Chemical methods for Rehabilitation of Soil and Groundwater Contaminated with Nuclides and Heavy Metals" 1995-1997 Moscow, RU.
23. ISTC Project 897, "Complexant Enhanced Elecrokinetic Remediation of Soil Contaminated with Organics", 1998-1999, Moscow, RU.
24. Popov, K. I.; Kruglov, S.; Yachmenev, V. G.; Lomasney, H. L. *Extended Abstr. Spec.Symp.*"Emerging Technologies in Hazardous Waste Management VIII", Birmingham, AL, **1996**, pp. 715-717.
25 Popov, K. I.; Yachmenev, V. G.; Lomasney, H. L. *Extended Abstr. Spec.Symp.* "Emerging Technologies in Hazardous Waste Management VIII", Birmingham, AL, **1996**, pp. 600-603.
26. Yachmenev, V. G.; Lomasney, H. L. *Extended Abstr. Spec.Symp.* "Emerging Technologies in Hazardous Waste Management VIII", Birmingham, AL, **1996**, p.p. 578-581.
27. Tessier, A.; Campbell, P. G. C.; Bisson, M. *Anal. Chem.* **1979**, *51*, 844.
28. *Stability Constants Database and Mini-SCDatabase.* IUPAC and Academic Software. Version 5.3. 2003. Sourby Old Farm, Timble, Otley, Yorks. UK. scdbase@acadsoft.co.uk.
29. Vanura, P.; Kuca, L. *Coll. Czech. Chem. Comm.* **1980**, *45*, 41.
30. Overvoll, P. A.; Lund, W. *Anal. Chim. Acta* **1982**, *143*, 153
31. Popov, K. I.; Yachmenev, V.; Kolosov, A.; Shabanova, N. *Colloids and Surfaces. A* **1999,** *160*, 135.
32. Popov, K.; Kolosov, A.; Ermakov, Yu.; Yachmenev, V.; Yusipovich, A.; Shabanova, N.; Kogut, B.; Frid, A. *Colloids and Surfaces. A,* *in press.*

33. Lindsay, W. L. *Chemical Equilibria in Soils.* Wiley: New York, Chichester, Brisbane, Toronto. 1979.
34. Lebedev, I. A.; Maximova, A. M.; Stepanov, A. V.; Shalinets, A. B. *Radiokhimiya (Russ. Radiochemistry)* **1967**, *9*, 707.
35. Whittle, J. K.; Pamukcu, S. Electrokinetic treatment of contaminated soils, sludges and lagoons. Report No. DOE/CH-9206, Office of Research and Development, Technology Development, Environmental Restoration and Waste Management, US Department of Energy, Washington, DC, April **1993**.
36. Wong, J. S. H.; Hicks, R. E.; Probstein, R. F. *J. Hazardous Materials* **1997**, *55*, 61.
37. Yeung, A. T.; Cheng-non Hsu, Menon, R. M. *J. Geotechnical Egineering*, **1996**, *122*, 666.
38. Usov, A. B.; Budantseva, N. V.; Fedoseev, A. M.; Astafurova, L. N. *Koord Khim.* (*Russ. J.Coord. Chem.)* **2001,** *27*, 824.
39. Popov, K. I.; Kolosov, A.; Yachmenev, V. G.; Shabanova, N.; Artemyeva, A.; Frid, A.; Kogut, B. *Sep. Sci. Technol.* **2001**, *36*, 2971.

Chapter 25

Factors in the Selection of Chelating Agents for Extraction of Lead from Contaminated Soil: Effectiveness, Selectivity, and Recoverability

P. K. Andrew Hong and Weimin Jiang

Department of Civil and Environmental Engineering, University of Utah, Salt Lake City, UT 84112

Heavy metal contamination of soil poses a serious concern for human and environmental health in industrial countries. Remediation of contaminated soil by chelating extraction is seen as a practical technique. This paper highlighted several factors relevant in selecting suitable chelating agents, which from the viewpoint of published thermodynamic constants include the chelating agent's strength and selectiveness toward the target metal as well as its viability of recovery and reuse. In these regards, six chelating agents including ethylenediaminetetraacetic acid (EDTA), trimethylenedinitrilotetraacetic acid (TMDTA), nitrilotris(methylene)triphosphonic acid (NTMP), L-5-glutamyl-L-cysteinylglycine (GCG), S-carboxymethylcysteine (SCMC), and N-(2-acetamido)iminodiacetic acid (ADA) were examined for their extraction of Pb from soil and for their subsequent recovery. All tested chelating agents proved to be amenable to recovery under appropriate conditions with added precipitants, such as $FeCl_3$, $Ca(OH)_2$, or sulfide. Demonstrated in this work was viable recovery with these precipitants for even some of the strong chelating agents available (e.g., EDTA and DTPA). In light of these methods, strong chelating agents such as EDTA and DTPA are seen as good choices for effective remediation.

Introduction

Heavy metal contamination of soil and groundwater resulting from industrial activities including mining and smelting operations is common among developed and developing countries. Lead, chromium, cadmium, zinc, copper, and mercury are among the most frequently found in contaminated sites; they are prevalent at many sites on the US National Priority List as well as military installations (*1*). Heavy metals are deleterious to human and many organisms. Unlike organics that undergo degradation, heavy metals once they are released into soil intermingle and remain in the soil matrix indefinitely with the potential to release and contaminate the natural resources of land and groundwater.

Chelating extraction of heavy metals from contaminated soils is seen as a viable remediation technique. Many chelating agents have been used for different metal contaminants. Ellis et al. (*2*) demonstrated the sequential treatment of soil contaminated with cadmium, chromium, copper, lead, and nickel using ethylenediaminetetraacetic acid (EDTA), hydroxylamine hydrochloride, and citrate buffer. Elliott et al. (*3*) studied the effects of pH, ionic strength, and Pb/EDTA ratio on extraction of lead from soil using EDTA. Davis and Singh (*4*) extracted zinc from contaminated soils in soil columns using HCl, EDTA, diethylenetriamine pentaacetic acid (DTPA), and chlorine. Burckhard et al. (*5*) mobilized zinc from mine tailings using formic, succinic, oxalic, and citric acid. Tsang et al. (*6*) mobilized bismuth, cadmium, lead, thorium, and uranium from contaminated soil with the use of water, glycine, cysteine, and thioglycollate solutions. Bulman et al. (*7*) examined the use of humic and fulvic acids in mobilizing plutonium (^{239}Pu) from saltmarsh soil. Hessling et al. (*8*) investigated soil washing techniques employing tap water, anionic surfactant, and EDTA for lead-contaminated soils at battery recycling facilities. A series of laboratory bench-scale soil washing studies using water, EDTA, or a surfactant was conducted to treat soils from metal recycling sites (*9*). Peters and Shem (*10*) conducted batch feasibility/treatability studies examining the effectiveness of pH-adjusted water, EDTA, and nitrilotriacetic acid (NTA) in removing lead from soil. Doepker (*11*) examined the mobilization of metal contaminants (cadmium, copper, and lead) from soil using acetic acid solution. Kocher (*12*) tested the feasibility of using EDTA to separate lead and chromium from spent foundry sands. In a feasibility/treatability study to investigate the leaching potential of heavy metals (particularly lead) from soil, Peters (*13*) used deionized water, EDTA, and citric acid. Sistani et al. (*14*) compared the performance of four commonly used chemical extractants, which included the use of strong and weak acids, along with acetic acid, EDTA, and DTPA, to extract various metals (Zn, Fe, Mn, Ca, Mg, K, Na, Al, Cd, Cu, and Pb) from a silt loam topsoil and from two different mine spoil materials. Brewster et al. (*15)* evaluated the remediation of a heavy metals-contaminated soil resulting from open burning and open detonation of chemical agents and munitions using ammonium acetate, EDTA, citric acid, Citranox, gluconic acid, phosphoric acid, oxalic acid, and NTA, in addition to acidified water. Neale et al. (*16*) examined the use of acids and chelating agents such as fluorosilicis acid, citric acid, EDTA, DTPA, and NTA to treat soils contaminated with lead, cadmium, and chromium from low levels up to 30,000 mg/kg. Lim et al. (*17*) recently examined familiar, strong chelating agents EDTA, NTA, and DTPA in the extraction of metals (Pb, Cd, Cr) from soil. The use of chelating agents in extracting heavy metals from contaminated soils had been tested for the past 2 decades, and varying levels of success were reported that sustained interest in remediation application.

The selection of chelating agents for soil remediation had often been intuitive, and been in want of an assessment methodology. In addition, many contaminant metals form very stable complexes with strong chelating agents such as EDTA, DTPA and NTA; this makes subsequent recovery of the metals and reuse of the chelating agents very difficult. Chelating extraction of heavy metals from soils followed by recovery and reuse of the chelating agents had not been a major focus, although several studies demonstrated such. Allen and Chen (*18*) investigated the recovery of EDTA by an electrochemical method following chelating extraction of lead from soil. Various other chelating agents pyridine-2,6-dicarboxylic acid, N-(2-acetamido)iminodiacetic acid, and S-carboxymethyl-L-cysteine were tested for extraction of lead, followed by recovery with simple adjustment of the solution pH and subsequent reuse of the chelating agents for several more cycles (*19-24*). Other spent chelating agents such as EDTA and DTPA were recovered by both pH adjustment and addition of precipitation aids (*25-27*). For the chelating agent to be used for at least several consecutive cycles, it must be relatively stable with respect to biodegradation during the remediation duration. The stability of the employed chelating agent relative to biodegradation was thus the subject of multiple studies (*26,27*). Many of the employed chelating agents are aminopolycarboxylates and possibly phosphonates, as is nitrilotris(methylene)triphosphonic acid (NTMP) presented in this work; the impact of them on metal speciation and mobility in the natural environment warrants further attention and have only begun to be appreciated (*28,29*).

We championed the recovery and reuse of spent chelating agents (*19-27*). Using different chelating agents including EDTA, TMDTA, NTMP, GCG, SCMC, and ADA (formulae listed in Table I) for the extraction of Pb from a contaminated soil, this paper attempts to highlight important issues gleaned from our previous work (*19-27*). The important issues concerning the deployment of chelating agents are: 1) extraction strength based on consideration of complexation constants, 2) extraction selectivity toward target metal based on a developed selectivity ratio (being proposed here), and 3) the potential of recovering the spent chelating agents for reuse.

Experimental

Deionized water (18 MΩ-cm) was obtained from a Milli-Q system (Millipore) and used throughout. Chelating agents EDTA (Sigma), GCG (Aldrich), NTMP (Aldrich), TMDTA (Fluka), SCMC (Fluka), and ADA (Fluka) were used as received. Soils in which lead was the major contaminant were taken from a contaminated site, air-dried for one month, then passed through a 2 mm sieve. Soil properties were characterized by Utah State University Analytical Laboratories following well-accepted methods (*30*): loamy sand (sand, 78%; silt, 14%; clay, 8%); very coarse, 7.5%; coarse, 25.5; medium, 20.1%; fine, 30.1%; very fine, 0.1%; organic matter, 2.67%; organic carbon, 1.55%; Cu, 440 mg/kg; Pb, 1900 mg/kg; Ni, 23.3 mg/kg; Zn, 1800 mg/kg; and As, 685 mg/kg. Typical experiments were conducted in 125 mL glass Erlenmeyer flasks using a batch solution volume (V) of 100 mL. All flasks were sealed with stoppers to reduce exchange with the atmosphere during experiments. All pH adjustments were performed manually by addition of either a 5 M HNO_3 or NaOH solution. pH measurements were with an Orion model SA 720 pH meter. Stock metal solution (1000 mg/L) was prepared according to ASTM Method D3559 (*31*). A gyratory shaker table (New Brunswick Scientific Co., Model G-2) provided moderate agitation and soil suspension during extraction procedures. All experiments were conducted at the room temperature of 23±1 °C. Total dissolved metal concentrations were measured in aliquots withdrawn from the reaction mixtures and after filtration through a 0.45 μm filter (Gelman Sciences sterile aerodisc), then acidified with nitric acid. Metal analyses were by atomic absorption (AA) spectrometry (Perkin Elmer Model 280). All equilibrium constants were obtained from Smith and Martell (*32*).

Extraction experiments were conducted with 5% soil slurries (i.e., 5 g of soil in 100 mL of chelator solution). To begin extraction, soil was added to the chelator solutions of selected concentration, and the pH was initially adjusted to about pH 7.5 if necessary. The mixture was continuously maintained in suspension by a shaker table. Either 4 or 24 hours were allowed for extraction. The tests of chelator recovery were performed by adjusting solution pH and/or adding various amounts of precipitant solids (cationic or anionic) into solutions of metal-ligand complexes. The mixtures were continually agitated for 1 or 2 hours to promote precipitation, then centrifuged, filtered, and the concentrations of metals in the filtrate analyzed.

Results and Discussion

Lead extraction was quantified according to various parameters, and the results were to be discussed in view of available equilibrium data. Figure 1 compares the extraction of lead by EDTA, TMDTA, NTMP, and GCG at different concentrations. The amounts of extracted lead increased with chelator concentrations as expected, but followed the order of GCG, NTMP, TMDTA, and EDTA in increasing order. The order of effectiveness in extraction agreed and increased with increasing complexation constants with lead, as shown in Table I (i.e., 10.6 < 13.7 < 18 for GCG, TMDTA, and EDTA, respectively).

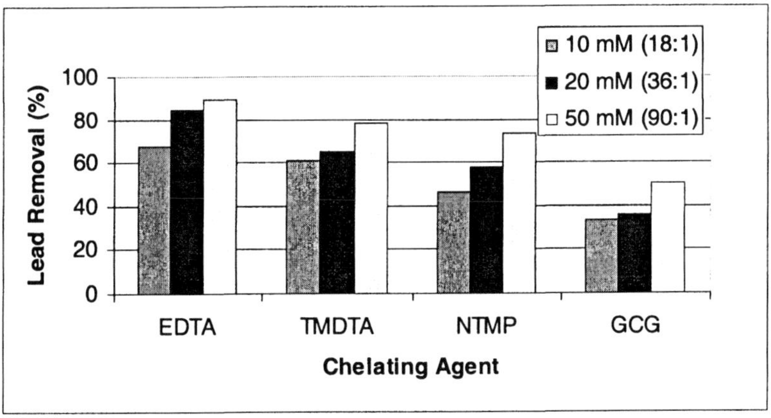

Figure 1. Removal of Pb from soil by various chelating agents at different concentrations (total ligand-to-lead mole ratio as shown). (Conditions: 100 mL of 5% soil slurry, pH 7.4-8.4, 24 hrs. Total Pb = 2320 mg/kg.)

Among the extracted metals was calcium, which is an abundant, constituent ion of soil. Figure 2 shows the mole ratio of extracted lead to calcium in the increasing order of EDTA, TMDTA, NTMP, and GCG. This means that NTMP and GCG, despite their lower complexing effectiveness, are more effective in extracting the target Pb rather than ambient Ca ion. Table I shows the chelating agents' complexation constants for Pb and Ca, as well as a computed selectivity ratio SR (taken as the ratio of complexation constants pK(Pb)/pK(Ca)). Hong et al. (*25*) developed the concept of selectivity ratio (SR) taken as the ratio of the average complexation constants (pKs) of the chelating agent for 6 target metals (Pb, Cu, Cd, Zn, Ni, and Hg) to the average complexation constants for Ca and Mg. In this work that addresses only lead contaminant, the SR is simplified to be the ratio of pK for Pb to that for Ca. The selectivity results of Figure 2 are consistent with computed SR values (i.e., 1.7 < 1.9 < 2.8 for EDTA, TMDTA, and GCG, respectively).

Table I. Complexation Constants and Selectivity Ratio of Chelating Agents for Lead.

Chelating Agent	*Formula*	*pK (Pb)*	*pK (Ca)*	*SR (Pb/Ca)*
Ethylenediaminetetraacetic acid (EDTA)	$C_{10}H_{16}O_8N_2$	18	10.6	1.7
Trimethylenedinitrilotetraacetic acid (TMDTA)	$C_{11}H_{18}O_8N_2$	13.7	7.3	1.9
Nitrilotris (methylene) triphosphonic acid (NTMP)	$C_3H_{12}O_9NP_3$	na	7.5	na
L-5-Glutamyl-L-cysteinylglycine (Glutathione) (GCG)	$C_{10}H_{17}O_6N_3S$	10.6	3.8	2.8
S-Carboxymethylcysteine (SCMC)	$C_5H_9O_4NS$	5.8	na	na
N-(2-Acetamido) iminodiacetic acid (ADA)	$C_6H_{10}N_2O_5$	8.4	3.96	2.1

SOURCE: Complexation constant data obtained from reference 13 and selectivity ratio (SR) computed as pK(Pb)/pK(Ca), adopted from reference 10. (na = not available.)

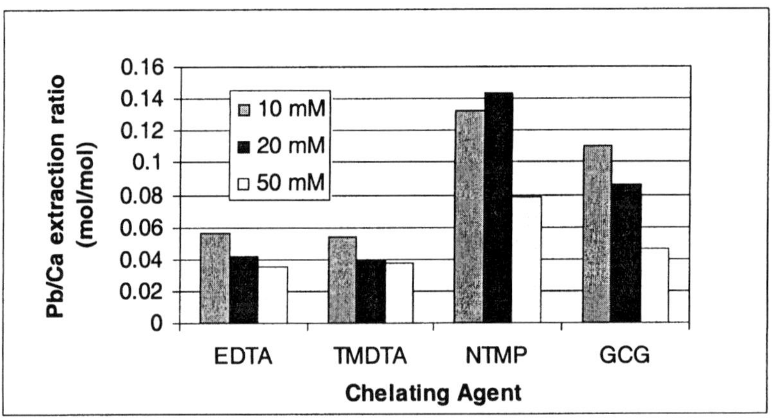

Figure 2. Extracted Pb/Ca ratio (by mol.) by various chelating agents at different concentrations. (Same conditions as in figure 1.

Figure 3 shows the extracted amounts of lead by the EDTA, TMDTA, and NTMP during consecutive extraction runs. As shown, EDTA and TMDTA at 20 mM approaches complete (100 and 97%, respectively) removal of lead from soil whereas NTMP achieves 85% removal after 3 consecutive runs.

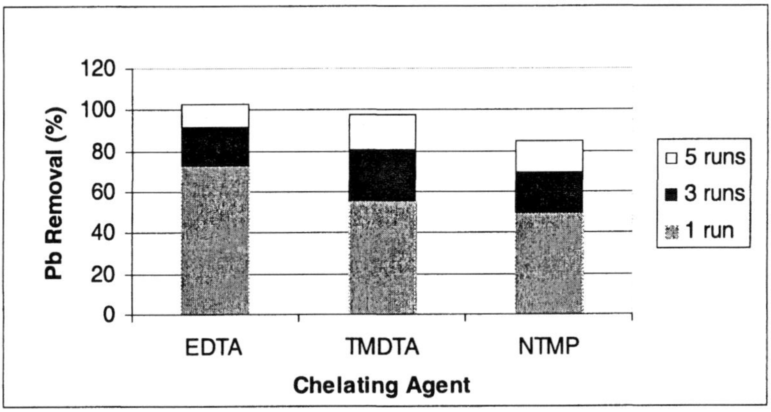

Figure 3. Removal of Pb by consecutive extraction with various chelating agents. (Conditions: 20 mM chelator, 100 mL of 5% soil slurry, pH 7.4-8.1, 4 hrs. Total Pb = 2320 mg/kg.)

SCMC and ADA (pK with Pb of 5.8 and 8.4, respectively) are weaker chelating agents. Figure 4 shows extracted lead at different chelator concentrations. As shown, ADA extracted at least 10 times more lead than SCMC, which is consistent with the higher complexation constant of ADA with Pb than that of SCMC with Pb.

The cost of remediation with chelating agents can be significantly lower if the chelating agents can be recovered and reused. This calls for disrupting the chelator-metal complex by separating the chelating agents from the metals. For weak complexes such as those mediated by SCMC and ADA, separation can be readily achieved by increasing the solution pH by base addition. This results in hydroxide ligands competing for the metal for the precipitation and formation of metal hydroxide solids. Figure 5 shows the recovery of Pb as a hydroxide solid at increasing solution pH. As shown, adding 10 mM $Ca(NO_3)_2$ and adjusting solution pH to about 10 lead to a 98% and 89% recovery of Pb from ADA and SCMC, respectively. Without Ca addition, separation of Pb from SCMC was similarly viable; however, it was only about 20% from ADA. Once precipitation occurres, the chelating agent that remains in solution can be readily separated from the precipitate and thus be available for reuse.

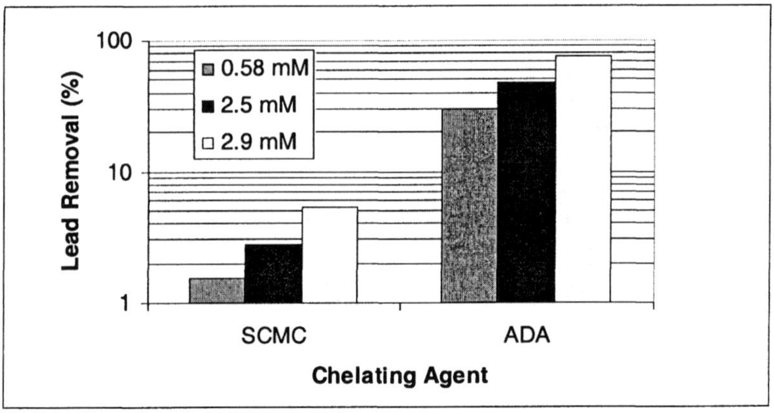

Figure 4. Removal of Pb from soil by SCMC and ADA at different concentrations. (Conditions: 200 mL of 5% soil slurry, pH 7.6-7.9, 12 hrs, ionic strength (I) = 0.1 M, total carbonate concentration (C_T) added = 1 mM.)

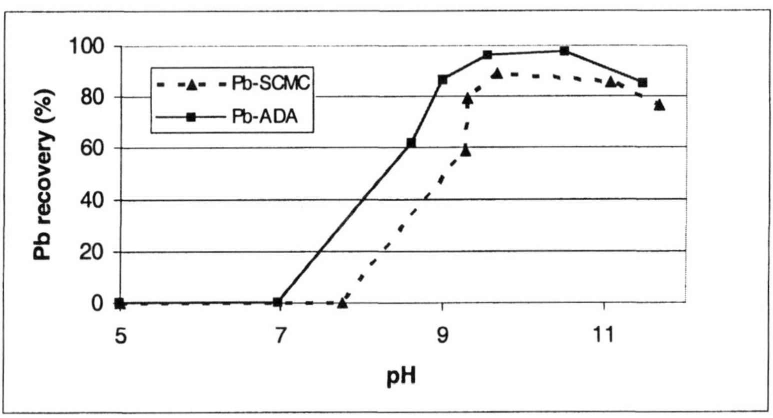

Figure 5. Separation and recovery of Pb and Chelating agents SCMC and ADA by elevation of pH. (Conditions: 600 mL of 20% soil slurry, chelator = 3 mM, $Ca(NO_3)_2$ added = 10 mM, 12 hrs, initial Pb = 7.2 and 176 mg/L; low and high curves, respectively.)

For strong complexes such as those of EDTA, it would be necessary to add base and/or competing precipitants such as sulfide, a strong precursor of metal sulfide precipitate. Figure 6 shows enhanced recovery of lead from EDTA, the strongest test chelating agent EDTA, at various sulfide doses. The results show that at a concentration of 10 mM of added sulfide (i.e., 2 times the 5 mM EDTA employed) enabled a complete recovery of lead, and hence EDTA for reuse.

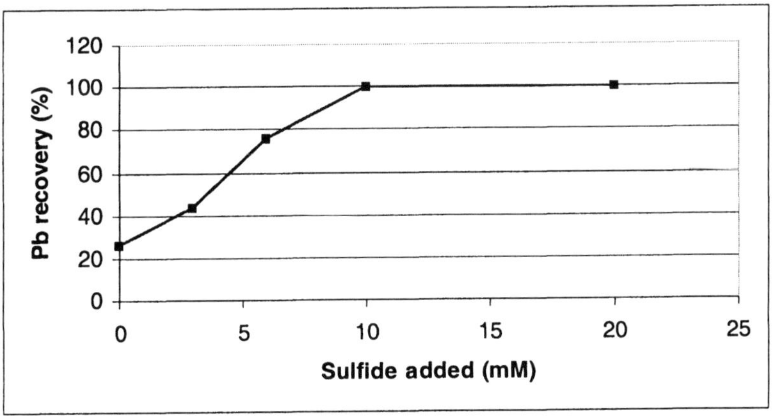

Figure 6. Separation and recovery of Pb from strong chelating agent EDTA with added anionic sulfide precipitant. (Conditions: initial Pb = 1 mM, EDTA = 5 mM. Separation pH = 10.5, 1 hr.)

Other common water treatment agents such as $Ca(OH)_2$ and $FeCl_3$ are useful for recovery of extracted metals from chelating agents. Table II shows the combined addition of $Ca(OH)_2$ and $FeCl_3$ for the separation of Pb from EDTA. The results show little recovery (7%) at elevated pH 10, or incomplete recovery (50%) when $Ca(OH)_2$ or $FeCl_3$ was used alone. However, effective recovery (>97%) resulted when both were used. It is likely that Ca competes with Pb for EDTA, while the coagulation/co-precipitation of Fe (mainly as $Fe(OH)_3(s)$) and Pb (e.g., as $Pb(OH)_2(s)$) significantly removes Pb from solution. Minimum workable doses should be determined and used in the interest of less sludge production and chemical costs.

Considered in this work are factors important to the selection of chelating agents for extraction of lead from contaminated soil. These include the chelating agent's effectiveness toward the target metal as revealed by the complexation equilibrium constants, selectivity toward the target metal as predicted by the SR ratio, and the ability to recover the chelating agent after extraction. Separation and recovery of chelating agent of weak or moderate strength such as SCMC and ADA can be readily achieved by elevation of pH through addition of base, while for strong chelating agent such as EDTA additional precipitants such as $FeCl_3$, $Ca(OH)_2$, and/or sulfide will be required. Strong chelating agents effect higher extraction effectiveness at a reduced number of washing cycles, which is likely most relevant to cost-effectiveness.

Table II. Separation and Recovery of Pb from Strong Chelating Agent EDTA with the Aid of Added Ca and Fe Precipitants.

$Ca(OH)_2$ added (mg/L)	*$FeCl_3$ added (mg/L)*	*Pb recovery (%)*
none	none	6.7
none	600	52
100	none	50
100	200	97
100	300	99

CONDITIONS: initial pH < 2.0, initial Pb = 300 mg/L (1.45 mM), EDTA = 1 mM. Separation pH = 10-10.5, 2 hrs. (Note that initial Pb is in excess of EDTA, thus recovery values (%) include initially dissolved but not chelated Pb in solution.)

In light of viable recovery through added precipitants, strong chelating agents such as EDTA and DTPA make excellent choices for remediation.

Acknowledgement

This work was made possible through funding by the Great Plain Rocky/Mountain Hazardous Substance Research Center.

References

[1] Bricka, R.; Williford, C.; Jones, L. *Heavy metal soil contamination at U.S. Army installations: Proposed research and strategy for technology development*, Technical report IRRP-94-1, US Army Engineer Waterways Experiment Station, Vicksburg, M.S., 1994.

[2] Ellis, W.D.; Fogg, T.R.; Tafuri, A.N., In *Land Disposal, Remedial Action, Incineration and Treatment of Hazardous Waste, 12th Ann. Res. Sympos. - Treatment of Soils Contaminated with Heavy Metals*, EPA 600/9-86/022, Cincinnati, OH, 1986, pp 201-207.

[3] Elliott, H.A.; Linn, J.H.; Shields, G.A. *Haz. Waste & Haz. Mater.*, **1989**, 6, 223-229.

[4] Davis, A.P.; Singh, I. *J. Environ. Engrg.*, **1995**, 121, 174-185.

[5] Burckhard, S.R.; Schwab, A.P.; Banks, M.K. *J. Haz. Mater.*, **1995**, 41, 135.

[6] Tsang, K.W.; Dugan, P.R.; Pfister, R.M. (1992) In *Emerging Technologies in Hazardous Waste Management IV, ACS Sympos. Series No. 554 - Mobilization of Bi, Cd, Pb, Th, and U Ions from Contaminated Soils and the Influence of Bacteria on the Process*, D.W. Tedder and F.G. Pohland, Eds., American Chemical Society, Washington, DC, 1992, pp 78-93.

[7] Bulman, R.A.; Wedgwood, A.J.; Sabo, G. *Sci. Total Environ.*, **1992**, 114, 215-226.

[8] Hessling, J.L.; Esposito, M.P.; Traver, R.P.; Snow, R.H. In *Metals Speciation, Separation, and Recovery, Vol. II - Results of Bench-Scale Research Efforts to Wash Contaminated Soils at Battery-Recycling Facilities*, Patterson J.W.; Passino, R., Eds.; Lewis Publishers, Inc., Chelsea, MI, 1989, pp 497-514.

[9] Royer, M.D.; Selvakumar, A.; Gaire, R. *J. Air Waste Manage. Assoc.*, **1992**, 42, 970-980.

[10] Peters, R.W.; Shem, L. In *ACS Sympos. Series 509 on Environmental Remediation: Removing Organic and Metal Ion Pollutants—Use of Chelating Agents for Remediation of Heavy Metal Contaminated Soil,* Vandegrift, G.F.; Reed, D.T; Tasker, I.R.; Eds., American Chemical Society, Washington, DC, 1992, *509*, pp 70-84.

[11] Doepker, R.D. In *Emerging Technologies in Hazardous Waste Management II, ACS Sympos. Series 468* - Enhanced Metal Mobiliation through Leachants Containing Acetate Ion, Tedder D.W.; Pohland, F.G. Eds., American Chemical Society, Washington, DC, 1991, pp 365-381.

[12] Kocher, W.M. *The Use of Soil Washing Processes for the Reclamation and Reuse of Foundry Waste Sands, Proc. 27th Mid-Atlantic Indus. Waste Conf.*, *27*:1995, 450-459.

[13] Peters, R.W. *Feasibility/Treatability Studies for Removal of Heavy Metals from Training Range Soils at the Grafenwöhr Training Area, Germany,* ANL/ESD/TM-81, Argonne National Laboratory, Argonne, IL, 1995.

[14] Sistani, K.R.; Mays, D.A.; Taylor, R.W.; Buford, C. *Commun. Soil Sci. Plant Anal.*, **1995**, 26, 2167-2180.

[15] Brewster, M. D.; Peters, R. W.; Miller, G. A.; Li, W.; Patton, T. L.; Martino, L. E. Physical/chemical treatment of metals-contaminated soils. Waste Processing and Recycling in Mineral and Metallurgical Industries II, Proceedings of the International Symposium on Waste Processing and Recycling in Mineral and Metallurgical Industries, 2nd, Vancouver, B. C., 1995, 539-565.

[16] Neale, C.N.; Bricka, R.M.; Chao, A.C. *Environmental Progress,* 1997, 16, 274-280.

[17] Lim, T.T.; Tay, J.H.; Wang, J.Y. *J. Environ. Eng.* **2004,** 130, 59-66.

[18] Allen, H.E.; Chen, P.-H. *Environ. Prog.* **1993**, 12, 284-293.

[19] Hong A.; Chen, T.C.; Okey, R. *Water Environ. Res.* **1995,** 67, 971-978.

[20] Hong, A.; Chen, T.-C.; Okey, R. In *Emerging Technologies in Hazardous Waste Management 5: Chelating Extraction of Zinc from Soil Using N-(2-Acetamido)iminodiacetic Acid,* Tedder D.W.; Pohland, F.G., Eds.; American Chemical Society, Washington, DC 1995, pp 210-223.

[21] Macauley, E.; Hong, A. *J. Hazard. Mater.* **1995,** 40, 257-270.

[22] Chen, T.C.; Hong, A. *J. Hazard. Mater.* **1995,** 41, 147-160.

[23] Chen, T.C.; Macauley, E.; Hong, A. *Can. J. Civil Eng.* **1995,** 22, 1185-1197.

[24] Hong, A.; Chen, T.C. *Water, Air, Soil Pollut.* **1996,** 86, 335-346.

[25] Hong, P.K.A.; Li, C.; Jiang, W.; Chen, T.C.; Peters, R.W. In *Emerging Technologies in Hazardous Waste Management 8: Chelating Agents for Extraction of Heavy Metals from Soil—Complexing Power, Selectivity, and Recoverability*, Tedder D.W.; Pohland, F.G., Eds.; Kluwer Academic/Plenum Publishers, New York, NY, 2000, pp 9-20.

[26] Hong, P.K.A.; Li, C.; Banerji, S.K.; Regmi, T. *J. Soil Contam.* **1999,** 8, 81-103.

[27] Hong, P.K.A.; Li, C.; Banerji, S.K.; Wang, Y. *J. Hazard. Mater.* **2002,** B94, 253-272.

[28] Nowack, B. *Env. Sci. & Technol.* **2002**, 36, 4009-4016.

[29] Nowack, B. *Water Res.* **2003**, 37, 2533-2546.

[30] *Western States Laboratory Proficiency Testing Program – Soil and Plant Analytical Methods (Ver. 4.10, 1998)*, based on *Plant, Soil, and Water Reference Methods for the Western Region*, Gavlak, R.G.; Horneck, D.A.; Miller, R.O., 1998.

[31] ASTM Annual Standards, , American Society for Testing and Materials, Philadelphia, PA, 1993.

[32] Smith, R.E.; Martell A.E. *Critical Stability Constants;* Plenum Press; New York, NY, 1974, 1976, 1982, 1989; Vol. 1, 4, 5, and 6.

Indexes

Author Index

Subject Index

E

F

H

I

M

O

R